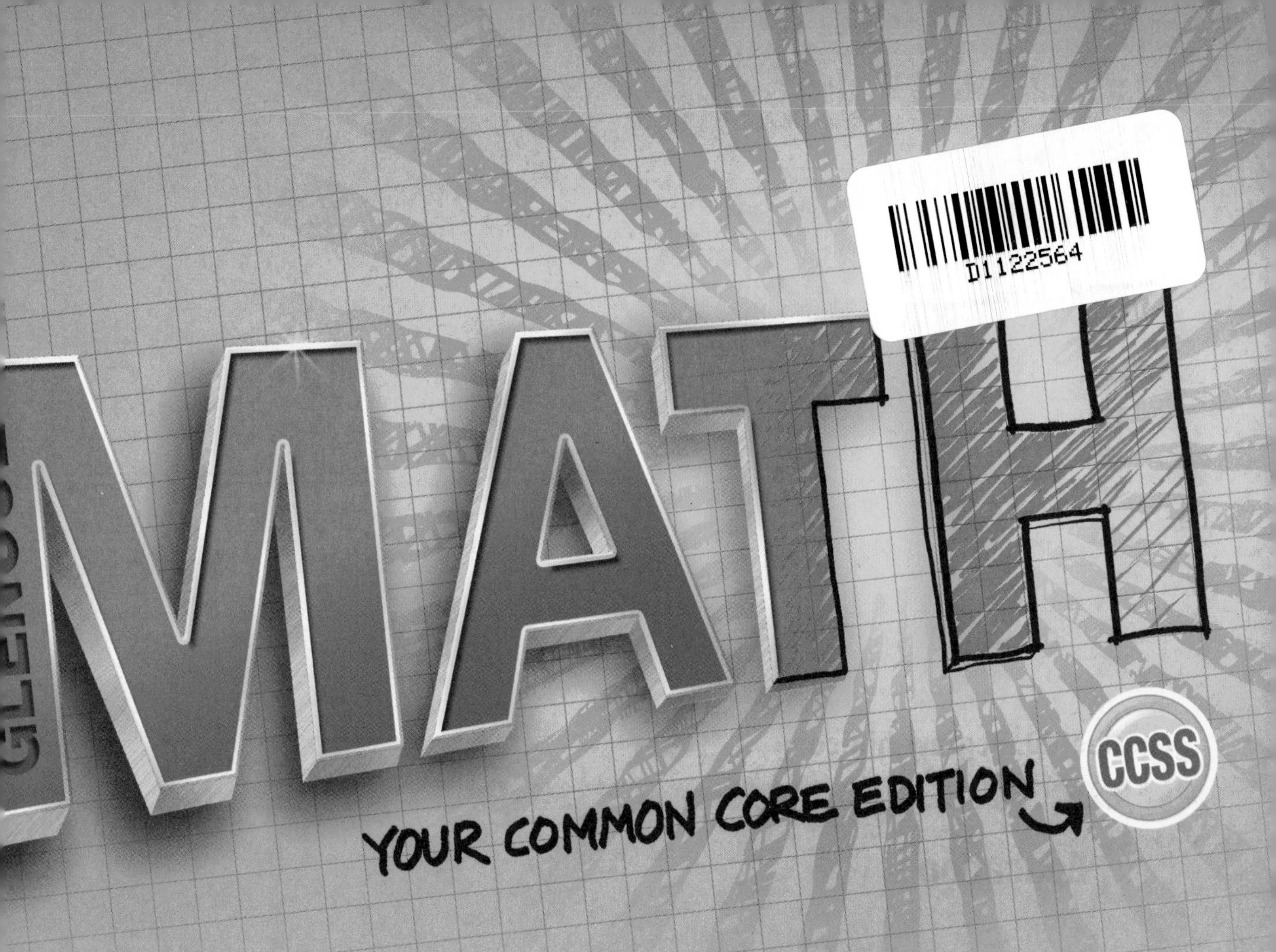

AUTHORS

Carter • Cuevas • Day • Malloy • Kersaint • Luchin • McClain
Molix-Bailey • Price • Reynosa • Silbey • Vielhaber • Willard

Bothell, WA • Chicago, IL • Columbus, OH • New York, NY

connectED.mcgraw-hill.com

The McGraw-Hill Companies

Mc Graw Hill **Education**

McGraw-Hill is committed to providing instructional materials in Science, Technology, Engineering, and Mathematics (STEM) that give all students a solid foundation, one that prepares them for college and careers in the 21st century.

Send all inquiries to:
McGraw-Hill Education
8787 Orion Place
Columbus, OH 43240

ISBN: 978-0-07-661530-8 (*Volume 1*)
MHID: 0-07-661530-8

Printed in the United States of America.

16 17 18 19 QSX 19 18 17 16

Our mission is to provide educational resources that enable students to become the problem solvers of the 21st century and inspire them to explore careers within Science, Technology, Engineering, and Mathematics (STEM) related fields.

CONTENTS IN BRIEF

Master the Common Core State Standards

Glencoe Math is organized into units based on groups of related standards called domains. This year, you will study and understand the five domains shown below.

it's all at connectED.mcgraw-hill.com

Go to the Student Center for your eBook, Resources, Homework, and Messages.

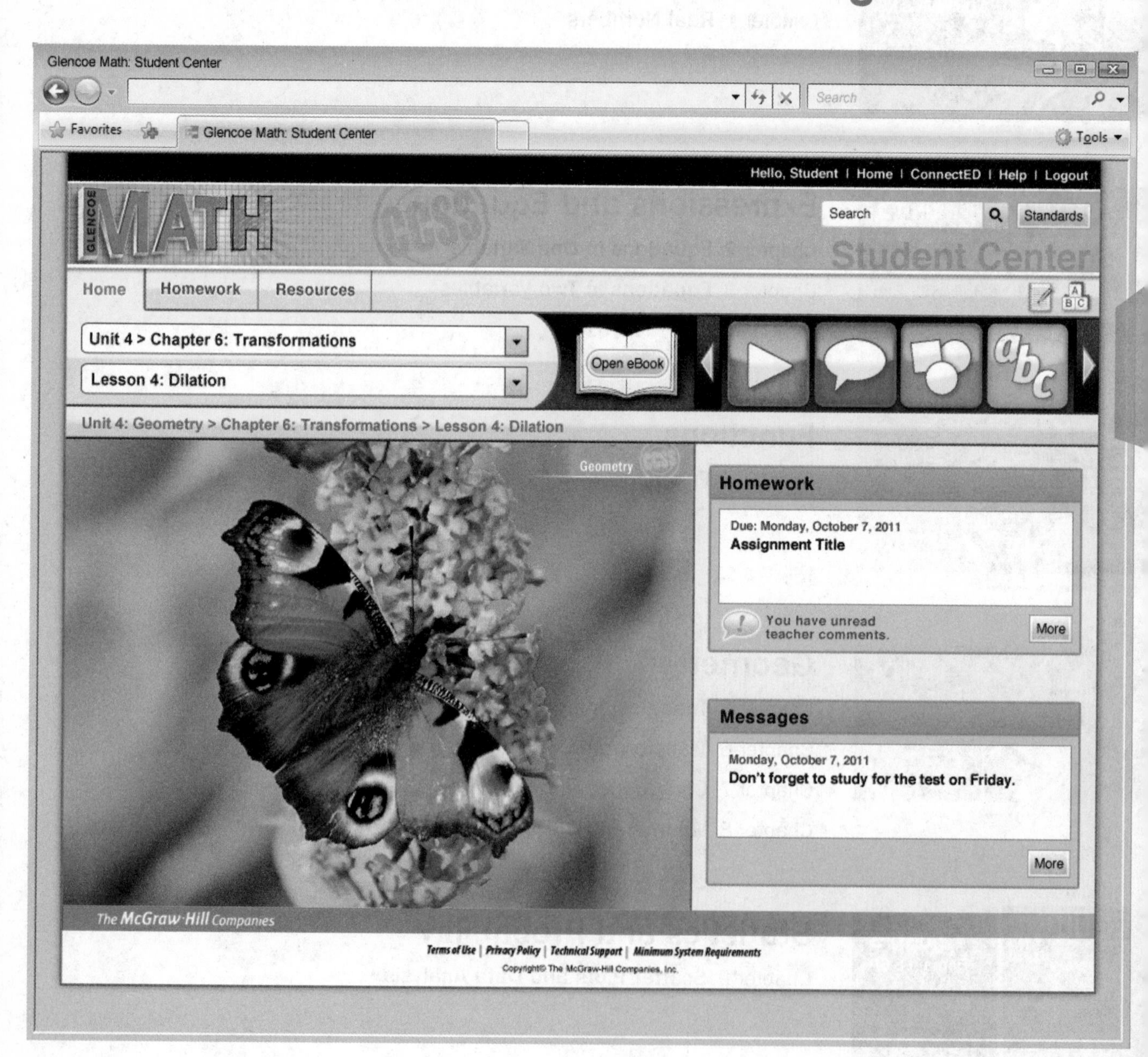

Write your Username ____________ Password ____________

Get your resources online to help you in class and at home.

Vocab

Find activities for building vocabulary.

Watch

Watch animations and videos.

Tutor

See a teacher illustrate examples and problems.

Tools

Explore concepts with virtual manipulatives.

Check

Self-assess your progress.

eHelp

Get targeted homework help.

Masters

Provides practice worksheets.

GO mobile

Scan this QR code with your smart phone* or visit mheonline.com/apps.

*May require quick response code reader app.

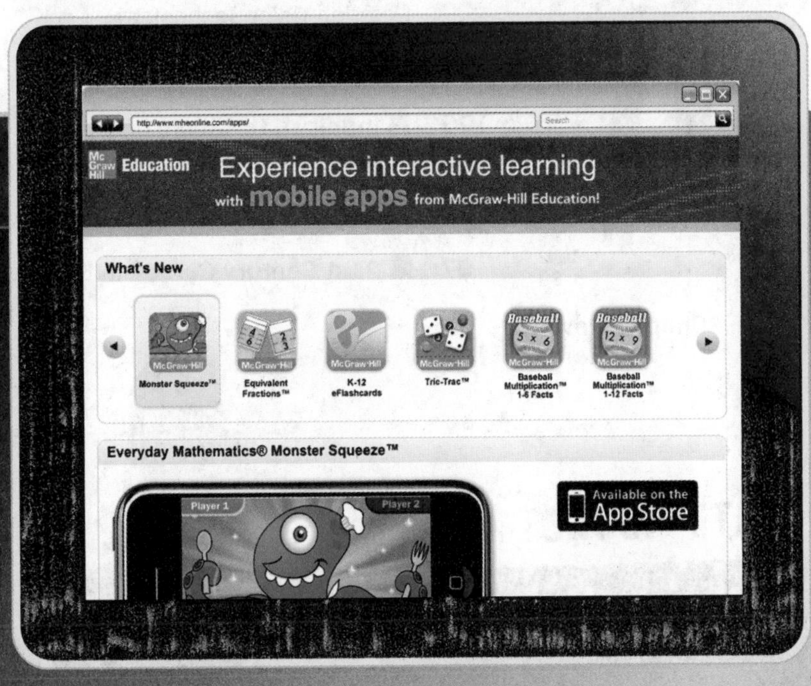

CCSS **UNIT 1** The Number System

UNIT PROJECT PREVIEW
page 2

Chapter 1
Real Numbers

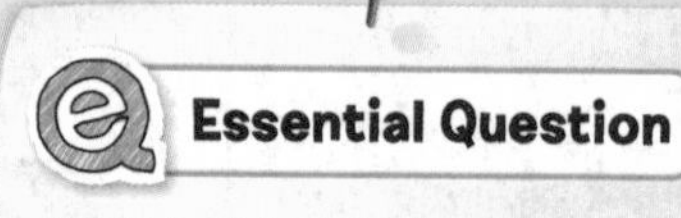

WHY is it helpful to write numbers in different ways?

Real World
p. 89

UNIT 2 Expressions and Equations

UNIT PROJECT PREVIEW page 106

Chapter 2 Equations in One Variable

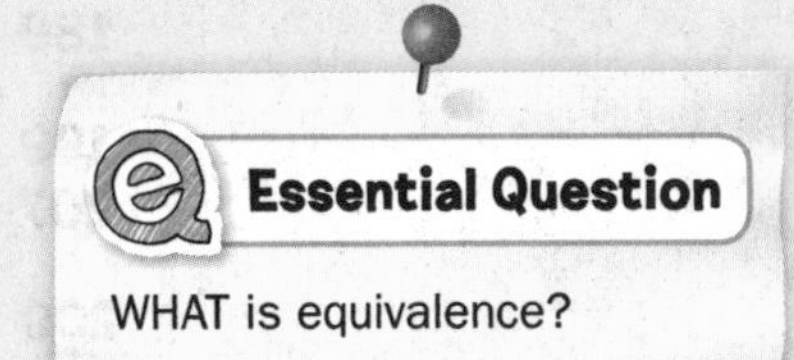

Essential Question

WHAT is equivalence?

Chapter 3
Equations in Two Variables

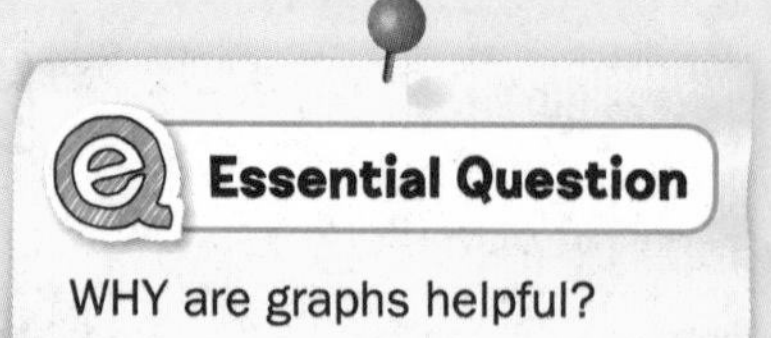

UNIT 3 Functions

UNIT PROJECT PREVIEW
page 262

Chapter 4
Functions

Essential Question

HOW can we model relationships between quantities?

Real World
p. 327

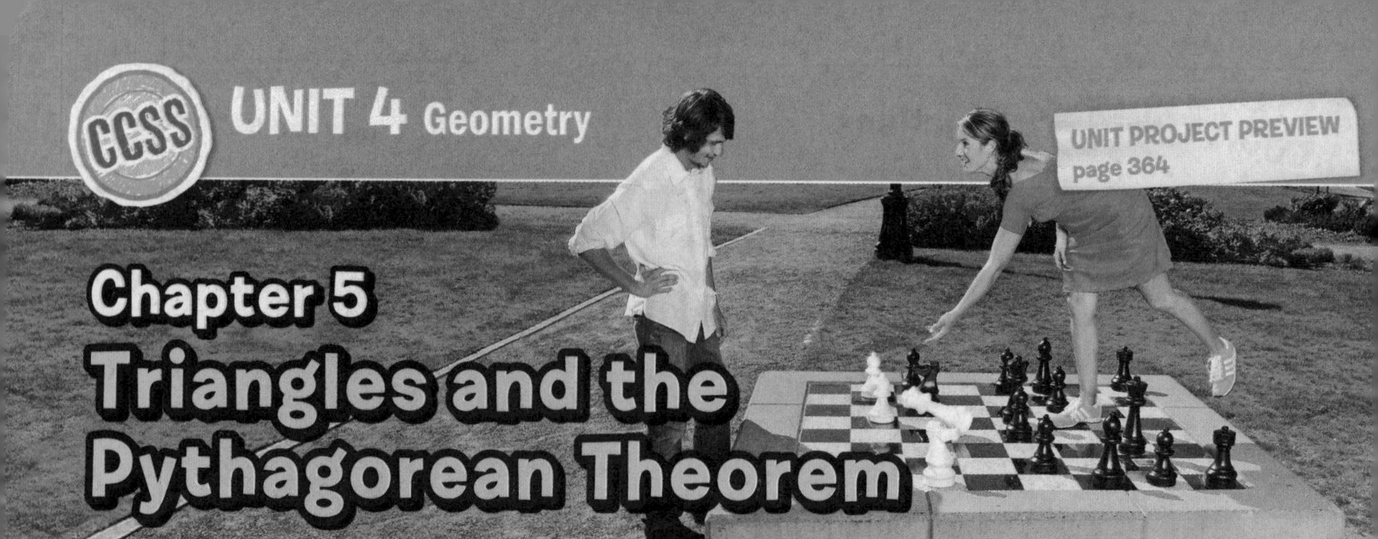

Chapter 5 Triangles and the Pythagorean Theorem

Essential Question

HOW can algebraic concepts be applied to geometry?

p. 431

Chapter 6 Transformations

Essential Question

HOW can we best show or describe the change in position of a figure?

p. 487

Chapter 7 Congruence and Similarity

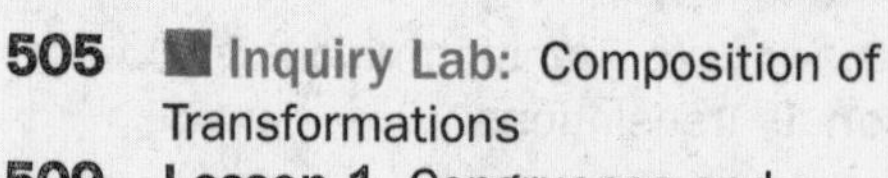

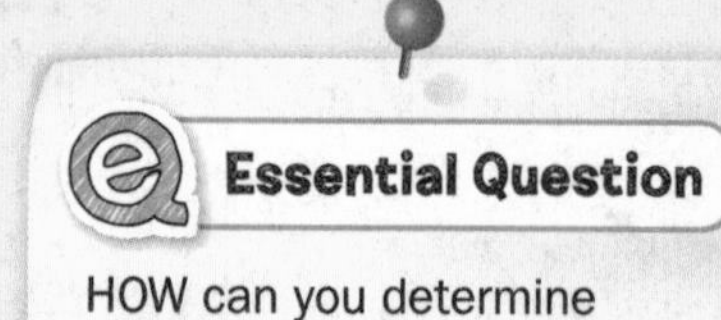

Essential Question

HOW can you determine congruence and similarity?

p. 545

Chapter 8
Volume and Surface Area

Essential Question

WHY are formulas important in math and science?

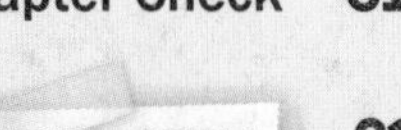

p. 597

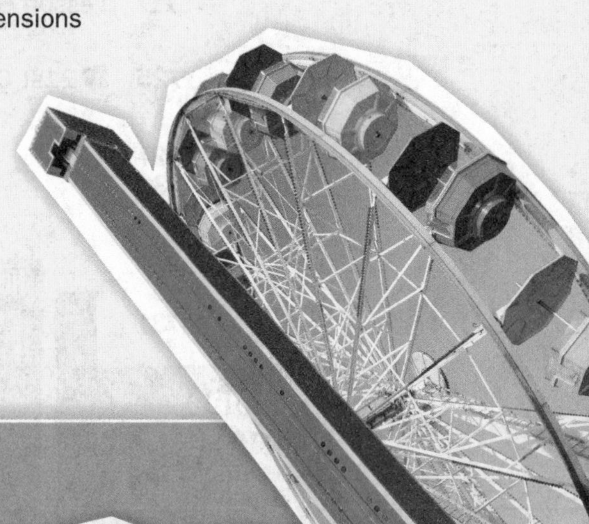

UNIT 5 Statistics and Probability

UNIT PROJECT PREVIEW page 658

Chapter 9 Scatter Plots and Data Analysis

Essential Question

HOW are patterns used when comparing two quantities?

Common Core State Standards for MATHEMATICS, Grade 8

Glencoe Math, Course 3, focuses on three critical areas: (1) applying equations in one and two variables; (2) understanding the concept of a function and using functions to describe quantitative relationships; (3) applying the Pythagorean Theorem and the concepts of similarity and congruence.

Content Standards

Domain 8.NS

The Number System

- Know that there are numbers that are not rational, and approximate them by rational numbers.

Domain 8.EE

Expressions and Equations

- Work with radicals and integer exponents.
- Understand the connections between proportional relationships, lines, and linear equations.
- Analyze and solve linear equations and pairs of simultaneous linear equations.

Domain 8.F

Functions

- Define, evaluate, and compare functions.
- Use functions to model relationships between quantities.

Domain 8.G

Geometry

- Understand congruence and similarity using physical models, transparencies, or geometry software.
- Understand and apply the Pythagorean Theorem.
- Solve real-world and mathematical problems involving volume of cylinders, cones and spheres.

Domain 8.SP

Statistics and Probability

- Investigate patterns of association in bivariate data.

Mathematical Practices

1. Make sense of problems and persevere in solving them.
2. Reason abstractly and quantitatively.
3. Construct viable arguments and critique the reasoning of others.
4. Model with mathematics.
5. Use appropriate tools strategically.
6. Attend to precision.
7. Look for and make use of structure.
8. Look for and express regularity in repeated reasoning.

Domain 8.NS

Related Unit Project: Unit 1
Music to My Ears

The Number System

Know that there are numbers that are not rational, and approximate them by rational numbers.

1. Know that numbers that are not rational are called irrational. Understand informally that every number has a decimal expansion; for rational numbers show that the decimal expansion repeats eventually, and convert a decimal expansion which repeats eventually into a rational number.

2. Use rational approximations of irrational numbers to compare the size of irrational numbers, locate them approximately on a number line diagram, and estimate the value of expressions (e.g., π^2).

Domain 8.EE

Expressions and Equations

Work with radicals and integer exponents.

1. Know and apply the properties of integer exponents to generate equivalent numerical expressions.

2. Use square root and cube root symbols to represent solutions to equations of the form $x^2 = p$ and $x^3 = p$, where p is a positive rational number. Evaluate square roots of small perfect squares and cube roots of small perfect cubes. Know that $\sqrt{2}$ is irrational.

3. Use numbers expressed in the form of a single digit times an integer power of 10 to estimate very large or very small quantities, and to express how many times as much one is than the other.

Related Unit Project: Unit 2
Web Design 101

4. Perform operations with numbers expressed in scientific notation, including problems where both decimal and scientific notation are used. Use scientific notation and choose units of appropriate size for measurements of very large or very small quantities (e.g., use millimeters per year for seafloor spreading). Interpret scientific notation that has been generated by technology.

Understand the connections between proportional relationships, lines, and linear equations.

5. Graph proportional relationships, interpreting the unit rate as the slope of the graph. Compare two different proportional relationships represented in different ways.

6. Use similar triangles to explain why the slope m is the same between any two distinct points on a non-vertical line in the coordinate plane; derive the equation $y = mx$ for a line through the origin and the equation $y = mx + b$ for a line intercepting the vertical axis at b.

Analyze and solve linear equations and pairs of simultaneous linear equations.

7. Solve linear equations in one variable.

a. Give examples of linear equations in one variable with one solution, infinitely many solutions, or no solutions. Show which of these possibilities is the case by successively transforming the given equation into simpler forms, until an equivalent equation of the form $x = a$, $a = a$, or $a = b$ results (where a and b are different numbers).

b. Solve linear equations with rational number coefficients, including equations whose solutions require expanding expressions using the distributive property and collecting like terms.

8. Analyze and solve pairs of simultaneous linear equations.

a. Understand that solutions to a system of two linear equations in two variables correspond to points of intersection of their graphs, because points of intersection satisfy both equations simultaneously.

b. Solve systems of two linear equations in two variables algebraically, and estimate solutions by graphing the equations. Solve simple cases by inspection.

c. Solve real-world and mathematical problems leading to two linear equations in two variables.

Functions

Domain 8.F

Define, evaluate, and compare functions.

1. Understand that a function is a rule that assigns to each input exactly one output. The graph of a function is the set of ordered pairs consisting of an input and the corresponding output.

2. Compare properties of two functions each represented in a different way (algebraically, graphically, numerically in tables, or by verbal descriptions).

3. Interpret the equation $y = mx + b$ as defining a linear function, whose graph is a straight line; give examples of functions that are not linear.

Use functions to model relationships between quantities.

4. Construct a function to model a linear relationship between two quantities. Determine the rate of change and initial value of the function from a description of a relationship or from two (x, y) values, including reading these from a table or from a graph. Interpret the rate of change and initial value of a linear function in terms of the situation it models, and in terms of its graph or a table of values.

5. Describe qualitatively the functional relationship between two quantities by analyzing a graph (e.g., where the function is increasing or decreasing, linear or nonlinear). Sketch a graph that exhibits the qualitative features of a function that has been described verbally.

Related Unit Project: Unit 3
Green Thumb

Geometry

Domain 8.G

Understand congruence and similarity using physical models, transparencies, or geometry software.

1. Verify experimentally the properties of rotations, reflections, and translations:

a. Lines are taken to lines, and line segments to line segments of the same length.

b. Angles are taken to angles of the same measure.

c. Parallel lines are taken to parallel lines.

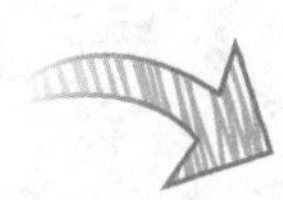

Related Unit Project: Unit 4
Design That Ride

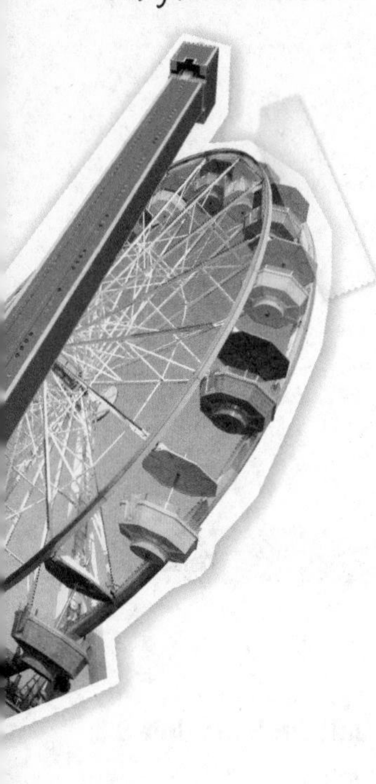

2. Understand that a two-dimensional figure is congruent to another if the second can be obtained from the first by a sequence of rotations, reflections, and translations; given two congruent figures, describe a sequence that exhibits the congruence between them.

3. Describe the effect of dilations, translations, rotations, and reflections on two-dimensional figures using coordinates.

4. Understand that a two-dimensional figure is similar to another if the second can be obtained from the first by a sequence of rotations, reflections, translations, and dilations; given two similar two-dimensional figures, describe a sequence that exhibits the similarity between them.

5. Use informal arguments to establish facts about the angle sum and exterior angle of triangles, about the angles created when parallel lines are cut by a transversal, and the angle-angle criterion for similarity of triangles.

Understand and apply the Pythagorean Theorem.

6. Explain a proof of the Pythagorean Theorem and its converse.

7. Apply the Pythagorean Theorem to determine unknown side lengths in right triangles in real-world and mathematical problems in two and three dimensions.

8. Apply the Pythagorean Theorem to find the distance between two points in a coordinate system.

Solve real-world and mathematical problems involving volume of cylinders, cones, and spheres.

9. Know the formulas for the volumes of cones, cylinders, and spheres and use them to solve real-world and mathematical problems.

Domain 8.SP

Statistics and Probability

Investigate patterns of association in bivariate data.

Related Unit Project: Unit 5
Olympic Games

1. Construct and interpret scatter plots for bivariate measurement data to investigate patterns of association between two quantities. Describe patterns such as clustering, outliers, positive or negative association, linear association, and nonlinear association.

2. Know that straight lines are widely used to model relationships between two quantitative variables. For scatter plots that suggest a linear association, informally fit a straight line, and informally assess the model fit by judging the closeness of the data points to the line.

3. Use the equation of a linear model to solve problems in the context of bivariate measurement data, interpreting the slope and intercept.

4. Understand that patterns of association can also be seen in bivariate categorical data by displaying frequencies and relative frequencies in a two-way table. Construct and interpret a two-way table summarizing data on two categorical variables collected from the same subjects. Use relative frequencies calculated for rows or columns to describe possible association between the two variables.

UNIT 1

The Number System

Essential Question

HOW can mathematical ideas be represented?

Chapter 1
Real Numbers

Rational numbers can be used to approximate the value of irrational numbers. In this chapter, you will perform operations on monomials and numbers written in scientific notation. You will then use rational approximations to estimate roots and to compare real numbers.

Unit Project Preview

Music to My Ears Listening to music can be both fun and relaxing. But did you know that many interesting relationships exist between math and music? Even the ancient Greek mathematician Pythagoras observed and wrote about many of these relationships.

At the end of Chapter 1, you'll complete a project to find how math and music are connected. But for now, write about the connections between math and music in the space provided.

Chapter 1
Real Numbers

Essential Question

WHY is it helpful to write numbers in different ways?

Common Core State Standards

Content Standards
8.NS.1, 8.NS.2, 8.EE.1, 8.EE.2, 8.EE.3, 8.EE.4

Mathematical Practices
1. 3, 4, 5, 6, 7, 8

Math in the Real World

Space The average distance from Earth to the Moon is about 384,403 kilometers. The Sun is the closest star to Earth and is about 150 million kilometers away. The next closest star is Proxima Centauri which is about 4.22 light years away from Earth.

A light year is defined as 9,461 billion kilometers. Find and label the distance in kilometers from Earth to Proxima Centauri.

FOLDABLES®
Study Organizer

Cut out the Foldable on page FL3 of this book.

Place your Foldable on page 100.

Use the Foldable throughout this chapter to help you learn about real numbers.

What Tools Do You Need?

Vocabulary

base
cube root
exponent
irrational number
monomial
perfect cube
perfect square
power
radical sign
rational number
repeating decimal
scientific notation
square root
terminating decimal

Use a Mnemonic Device

When a mathematical expression has a combination of operations, the order of operations tells you which operation to perform first. How can you remember the orders easily? A mnemonic device is a verse or phrase to help you remember something.

In this case, it is *P*lease *E*xcuse *M*y *D*ear *A*unt *S*ally. On each rung of the ladder, fill in the operation that the mnemonic device represents. Then evaluate the numerical expression step-by-step.

$3(5 - 15)^2 - 7 \cdot 3 + 24 \div 6$

Please ____________________

Excuse ____________________

My Dear ____________________

Aunt Sally ____________________

When Will You Use This?

Your Turn! You will solve this problem in the chapter.

Are You Ready?

Try the Quick Check below.
Or, take the Online Readiness Quiz.

Common Core Review 7.NS.2

Example 1

Find 5 • 4 • 5 • 4 • 5.

$$
\begin{aligned}
5 \cdot 4 \cdot 5 \cdot 4 \cdot 5 &= 4 \cdot 4 \cdot 5 \cdot 5 \cdot 5 \\
&= (4 \cdot 4) \cdot (5 \cdot 5 \cdot 5) \\
&= 16 \cdot 125 \\
&= 2{,}000
\end{aligned}
$$

Example 2

Find the prime factorization of 60.

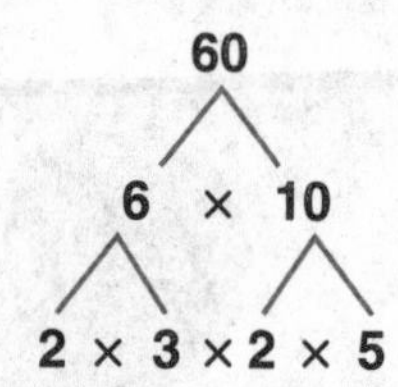

The prime factorization of 60 is $2 \times 2 \times 3 \times 5$.

Quick Check

Simplify Expressions **Find each product.**

1. $2 \cdot 2 \cdot 4 \cdot 4 \cdot 4 =$ ______

2. $(-8)(-8)(5)(5)(-8) =$ ______

3. The students at Hampton Middle School raised $8 \cdot 8 \cdot 2 \cdot 8 \cdot 2$ dollars to help build a new community center. How much money did they raise?

______

Prime Factorization **Find the prime factorization of each number.**

4. 36 ______

5. 24 ______

6. 18 ______

7. 100 ______

8. 121 ______

9. −42 ______

Which problems did you answer correctly in the Quick Check? Shade those exercise numbers below.

1 2 3     8 9

Lesson 1

Rational Numbers

What You'll Learn

Scan the lesson. Write the definitions of terminating decimal and repeating decimal.

-
-

Essential Question

WHY is it helpful to write numbers in different ways?

Vocabulary

rational number
repeating decimal
terminating decimal

Common Core State Standards

Content Standards
8.NS.1

Mathematical Practices
1, 3, 4, 6, 7, 8

Vocabulary Start-Up

Numbers that can be written as a comparison of two integers, expressed as a fraction, are called **rational numbers**.

Complete the graphic organizer.

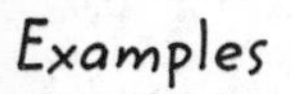

Examples: Percent; Decimal

Rational Number

Define in your own words

Examples: Fraction; Mixed Numbers

The root of the word *rational* is *ratio*. Describe the relationship between rational numbers and ratios.

Real-World Link

During a recent regular season, a Texas Ranger baseball player had 126 hits and was at bat 399 times. Write a fraction in simplest form to represent the ratio of the number of hits to the number of at bats.

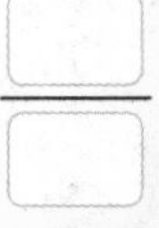

Key Concept

Rational Numbers

Words A rational number is a number that can be written as the ratio of two integers in which the denominator is not zero.

Symbols $\frac{a}{b}$, where a and b are integers and $b \neq 0$

Model

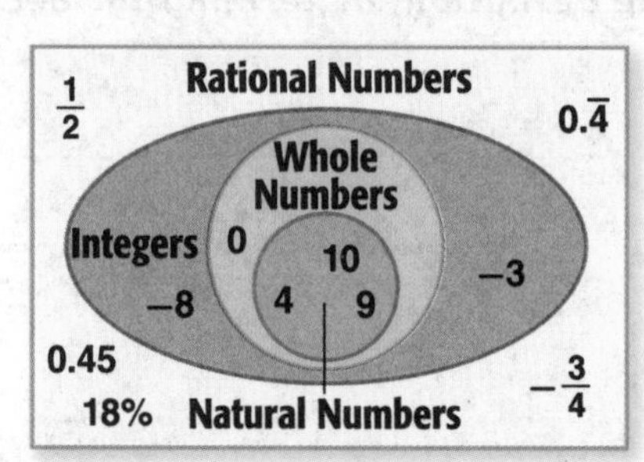

Work Zone

Bar Notation

Bar notation is often used to indicate that a digit or group of digits repeats. The bar is placed above the repeating part. To write 8.636363... in bar notation, write $8.\overline{63}$, not $8.\overline{6}$ or $8.\overline{636}$. To write 0.3444... in bar notation, write $0.3\overline{4}$, not $0.\overline{34}$.

Every rational number can be expressed as a decimal by dividing the numerator by the denominator. The decimal form of a rational number is called a **repeating decimal**. If the repeating digit is zero, then the decimal is a **terminating decimal**.

Rational Number	Repeating Decimal	Terminating Decimal
$\frac{1}{2}$	0.5000...	0.5
$\frac{2}{5}$	0.400...	0.4
$\frac{5}{6}$	0.833...	does not terminate

Examples

Write each fraction or mixed number as a decimal.

1. $\frac{5}{8}$

$\frac{5}{8}$ means $5 \div 8$.

$$\begin{array}{r} 0.625 \\ 8\overline{)5.000} \\ -48 \\ \hline 20 \\ -16 \\ \hline 40 \\ -40 \\ \hline 0 \end{array}$$

Divide 5 by 8.

2. $-1\frac{2}{3}$

$-1\frac{2}{3}$ can be rewritten as $\frac{-5}{3}$.

Divide 5 by 3 and add a negative sign.

$$\begin{array}{r} 1.6... \\ 3\overline{)5.0} \\ -3 \\ \hline 2\,0 \\ -1\,8 \\ \hline 2 \end{array}$$

The mixed number $-1\frac{2}{3}$ can be written as $-1.\overline{6}$.

Show your work.

a. ______

b. ______

c. ______

d. ______

Got It? Do these problems to find out.

a. $\frac{3}{4}$

b. $-\frac{2}{9}$

c. $4\frac{13}{25}$

d. $3\frac{1}{11}$

Example

3. In a recent season, St. Louis Cardinals first baseman Albert Pujols had 175 hits in 530 at bats. To the nearest thousandth, find his batting average.

To find his batting average, divide the number of hits, 175, by the number of at bats, 530.

175 [÷] 530 [ENTER] 0.3301886792

Look at the digit to the right of the thousandths place. Since $1 < 5$, round down.

Albert Pujols's batting average was 0.330.

Got It? **Do this problem to find out.**

e. In a recent season, NASCAR driver Jimmie Johnson won 6 of the 36 total races held. To the nearest thousandth, find the part of races he won.

e. ________

Examples

4. Write 0.45 as a fraction.

$0.45 = \frac{45}{100}$ — 0.45 is 45 hundredths.

$= \frac{9}{20}$ — Simplify.

5. Write $0.\overline{5}$ as a fraction in simplest form.

Assign a variable to the value $0.\overline{5}$. Let $N = 0.555...$. Then perform operations on N to determine its fractional value.

$N = 0.555...$

$10(N) = 10(0.555...)$ — Multiply each side by 10 because 1 digit repeats.

$10N = 5.555...$ — Multiplying by 10 moves the decimal point 1 place to the right.

$-N = 0.555...$ — Subtract $N = 0.555...$ to eliminate the repeating part.

$9N = 5$ — Simplify.

$N = \frac{5}{9}$ — Divide each side by 9.

The decimal $0.\overline{5}$ can be written as $\frac{5}{9}$.

6. Write $2.\overline{18}$ as a mixed number in simplest form.

Assign a variable to the value $2.\overline{18}$. Let $N = 2.181818\ldots$. Then perform operations on N to determine its fractional value.

$N = 2.181818\ldots$	
$100(N) = 100(2.181818\ldots)$	Multiply each side by 100 because 2 digits repeat.
$100N = 218.181818$	Multiplying by 100 moves the decimal point 2 places to the right.
$-N = \quad 2.181818\ldots$	Subtract $N = 2.181818\ldots$ to eliminate the repeating part.
$99N = 216$	Simplify.
$N = \frac{216}{99}$ or $2\frac{2}{11}$	Divide each side by 99. Simplify.

The decimal $2.\overline{18}$ can be written as $2\frac{2}{11}$.

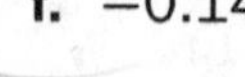

Got It? **Do these problems to find out.**

Write each decimal as a fraction or mixed number in simplest form.

f. -0.14 **g.** $0.\overline{27}$

f. ______

g. ______

Guided Practice

Write each fraction or mixed number as a decimal. (Examples 1 and 2)

1. $\frac{9}{16} =$ ______

2. $-1\frac{29}{40} =$ ______

3. $4\frac{5}{6} =$ ______

4. Monica won 7 of the 16 science competitions she entered. To the nearest thousandth, find her winning average. (Example 3) ______

Write each decimal as a fraction or mixed number in simplest form. (Examples 4–6)

5. $0.32 =$ ______

6. $-0.\overline{7} =$ ______

7. **Building on the Essential Question** How can you determine if a number is a rational number?

______

______

Rate Yourself!

I understand how to write a repeating decimal as a fraction.

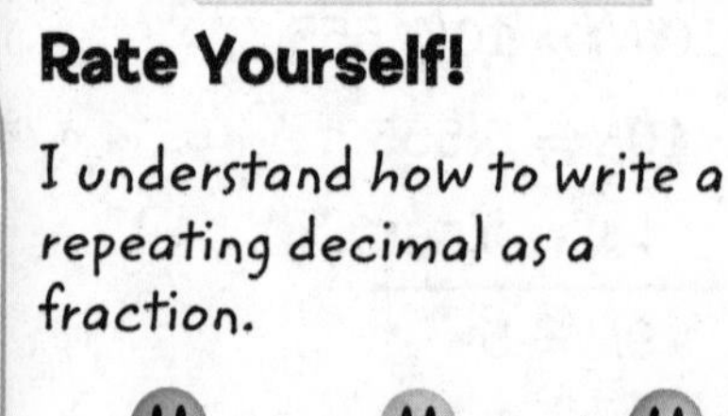

For more help, go online to access a Personal Tutor.

Name ________________ My Homework ________________

Independent Practice

Go online for Step-by-Step Solutions

Write each fraction or mixed number as a decimal. (Examples 1 and 2)

1. $\frac{2}{5}$ = ______

2. $2\frac{1}{8}$ = ______

3. $\frac{33}{40}$ = ______

4. $\frac{4}{33}$ = ______

5. $-\frac{6}{11}$ = ______

6. $-7\frac{8}{45}$ = ______

7. CCSS **Identify Repeated Reasoning** The table shows statistics about the students at Carter Junior High. (Example 3)

Number of Siblings	Fraction of Students
None	$\frac{1}{15}$
One	$\frac{1}{3}$
Two	$\frac{5}{12}$
Three	$\frac{1}{6}$
Four or more	$\frac{1}{60}$

a. Express the fraction of students with no siblings as a decimal.

______

b. Find the decimal equivalent for the fraction of students with three siblings. ______

c. Write the fraction of students with one sibling as a decimal. Round to the nearest thousandth. ______

d. Write the fraction of students with two siblings as a decimal. Round to the nearest thousandth. ______

Write each decimal as a fraction or mixed number in simplest form.
(Examples 4–6)

8. -0.4 = ______

9. -7.32 = ______

10. $0.\overline{2}$ = ______

Copy and Solve Write each decimal as a fraction or mixed number in simplest form. Show your work on a separate piece of paper. (Examples 4–6)

11. $-0.\overline{45}$

12. $2.\overline{7}$

13. 5.55

CCSS **Be Precise Write the length of each insect as a fraction or mixed number and as a decimal.**

14.

15.

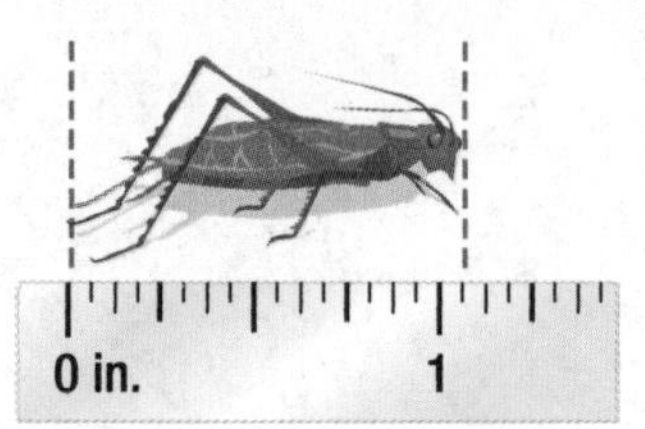

H.O.T. Problems Higher Order Thinking

16. CCSS **Identify Structure** Give an example of a repeating decimal where two digits repeat. Explain why your number is a rational number.

17. CCSS **Persevere with Problems** Explain why any rational number is either a terminating or repeating decimal.

18. CCSS **Make a Conjecture** Compare 0.1 and $0.\overline{1}$, 0.13 and $0.\overline{13}$, and 0.157 and $0.\overline{157}$ when written as fractions. Make a conjecture about expressing repeating decimals like these as fractions.

Standardized Test Practice

19. Which of the following is equivalent to the fraction below?

$$\frac{13}{5}$$

Ⓐ 2.4 Ⓑ 2.45 Ⓒ 2.55 Ⓓ 2.6

Extra Practice

20. Write $\frac{5}{9}$ as a decimal. $0.\overline{5}$

$$\begin{array}{r} 0.55 \\ 9\overline{)5.00} \\ -45 \\ \hline 50 \\ -45 \\ \hline 5... \end{array}$$

Homework Help

21. Write $7.\overline{15}$ as a mixed number in simplest form. $7\frac{5}{33}$

$N = 7.151515...$

$100(N) = 100(7.151515...)$

$100N = 715.151515...$

$-N = \quad 7.151515...$

$99N = 708$

$N = \frac{708}{99}$ or $7\frac{5}{33}$

CCSS Identify Repeated Reasoning **Write each fraction or mixed number as a decimal.**

22. $\frac{4}{5} =$ ______

23. $5\frac{5}{16} =$ ______

24. $-6\frac{13}{15} =$ ______

Write each decimal as a fraction or mixed number in simplest form.

25. $-1.55 =$ ______

26. $3.\overline{8} =$ ______

27. $-0.\overline{09} =$ ______

Write the rainfall amount for each day as a fraction or mixed number.

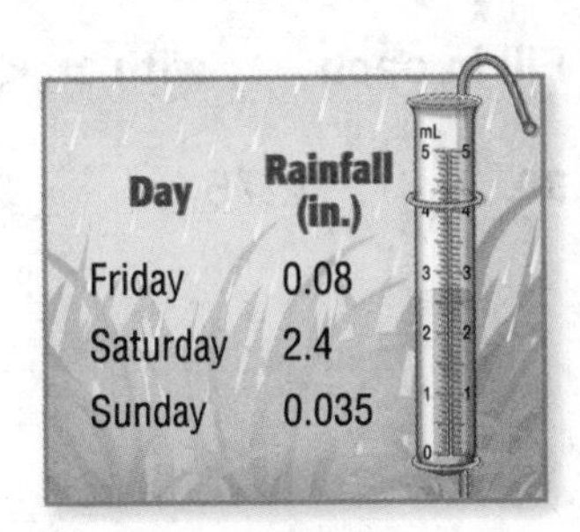

Day	Rainfall (in.)
Friday	0.08
Saturday	2.4
Sunday	0.035

28. Friday ______

29. Saturday ______

30. Sunday ______

31. The table shows three popular flavors according to the results of a survey. What is the decimal value of those who liked vanilla, chocolate, or strawberry? Round to the nearest hundredth. ______

Flavor	Fraction
Vanilla	$\frac{3}{10}$
Chocolate	$\frac{1}{11}$
Strawberry	$\frac{1}{18}$

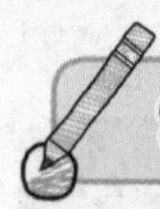

Standardized Test Practice

32. Short Response The table shows the number of free throws each player made during the last basketball season.

Player	Free Throws Made	Free Throws Attempted
Felisa	18	20
Morgan	13	24
Yasmine	15	22
Gail	10	14

Write the fraction of free throws made in simplest form for each player.

33. Short Response Write each fraction from Exercise 32 as a decimal. Round to the nearest thousandth if necessary.

34. While shopping for a new pair of jeans, Janet notices the sign below.

Which of the following expressions can be used to estimate the total discount on a pair of jeans?

Ⓐ 0.033 × $30

Ⓑ 0.33 × $30

Ⓒ 1.3 × $30

Ⓓ 33.3 × $30

35. Which of the following is *not* an example of a rational number?

Ⓕ $\frac{-6}{11}$

Ⓖ 15

Ⓗ 18%

Ⓘ 4.23242526. . .

Common Core Review

Fill in each ◯ with >, <, or = to make a true statement. 6.NS.7

36. $2\frac{7}{8}$ ◯ 2.75

37. $\frac{-1}{3}$ ◯ $\frac{-7}{3}$

38. $\frac{5}{7}$ ◯ $\frac{4}{5}$

39. $3\frac{6}{11}$ ◯ $3.\overline{54}$

40. At the grocery store, Karen was comparing the unit price for two different packages of laundry detergent. One package was $0.0733 per ounce. The other package was $3.64 for 52 ounces. Which package had the lower unit price? Explain. 6.RP.2

Lesson 2

Powers and Exponents

What You'll Learn

Scan the lesson. Write the definitions of power, base, and exponent.

- ____
  ____
- ____
  ____

Essential Question

WHY is it helpful to write numbers in different ways?

Vocabulary

power
base
exponent

Common Core State Standards

Content Standards
8.EE.1

Mathematical Practices
1, 3, 4, 8

Real-World Link

Savings Yogi decided to start saving money by putting a penny in his piggy bank, then doubling the amount he saves each week. Use the questions below to find how much money Yogi will save in 8 weeks.

1. Complete the table below to find the amount Yogi saved each week and the total amount in his piggy bank.

Week	0	1	2	3	4	5	6
Weekly Savings	1¢	2¢					
Total Savings	1¢	3¢					

2. How many 2s are multiplied to find his savings in Week 4? ☐ Week 5? ☐

3. How much money will Yogi save in Week 8? ____________

4. Continue the table to find when he will have enough to buy a pair of shoes for $80. ____________

Week	7	8	9	10	11	12
Weekly Savings						
Total Savings						

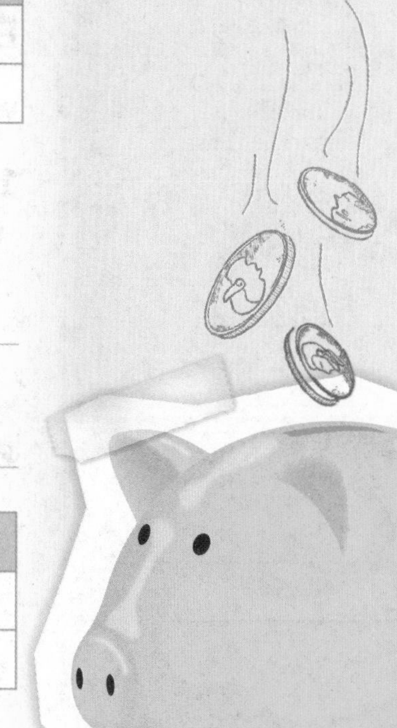

Work Zone

Write and Evaluate Powers

A product of repeated factors can be expressed as a **power**, that is, using an exponent and a base.

$$\underbrace{2 \cdot 2 \cdot 2 \cdot 2}_{\text{4 factors}} = 2^4$$

The **base** is the common factor.

The **exponent** tells how many times the base is used as a factor.

Powers are read in a certain way.

Read and Write Powers

Power	Words	Factors
3^1	3 to the first power	3
3^2	3 to the second power or 3 squared	$3 \cdot 3$
3^3	3 to the third power or 3 cubed	$3 \cdot 3 \cdot 3$
3^4	3 to the fourth power or 3 to the fourth	$3 \cdot 3 \cdot 3 \cdot 3$
⋮	⋮	⋮
3^n	3 to the *n*th power or 3 to the *n*th	$\underbrace{3 \cdot 3 \cdot 3 \cdot \ldots \cdot 3}_{n \text{ factors}}$

Examples

Write each expression using exponents.

1. $(-2) \cdot (-2) \cdot (-2) \cdot 3 \cdot 3 \cdot 3 \cdot 3$

The base −2 is a factor 3 times, and the base 3 is a factor 4 times.

$(-2) \cdot (-2) \cdot (-2) \cdot 3 \cdot 3 \cdot 3 \cdot 3 = (-2)^3 \cdot 3^4$

2. $a \cdot b \cdot b \cdot a \cdot b$

Use the properties of operations to rewrite and group like bases together. The base *a* is a factor 2 times, and the base *b* is a factor 3 times.

$a \cdot b \cdot b \cdot a \cdot b = a \cdot a \cdot b \cdot b \cdot b$

$= a^2 \cdot b^3$

Got It? Do these problems to find out.

Show your work.

a. $\frac{1}{2} \cdot \frac{1}{2} \cdot \frac{1}{2} \cdot \frac{1}{2}$ **b.** $4 \cdot 4 \cdot 4 \cdot 5 \cdot 5$ **c.** $m \cdot m \cdot n \cdot n \cdot m$

a. ________

b. ________

c. ________

Example

3. **Evaluate $\left(-\frac{2}{3}\right)^4$.**

$$\left(-\frac{2}{3}\right)^4 = \left(-\frac{2}{3}\right) \cdot \left(-\frac{2}{3}\right) \cdot \left(-\frac{2}{3}\right) \cdot \left(-\frac{2}{3}\right)$$ Write the power as a product.

$$= \frac{16}{81}$$ Multiply.

Evaluate

Remember that to evaluate an expression means to find its value.

Got It? Do these problems to find out.

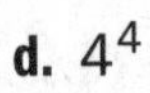

d. 4^4 **e.** $(-2)^6$ **f.** $\left(\frac{1}{5}\right)^3$

d. ______

e. ______

f. ______

Example

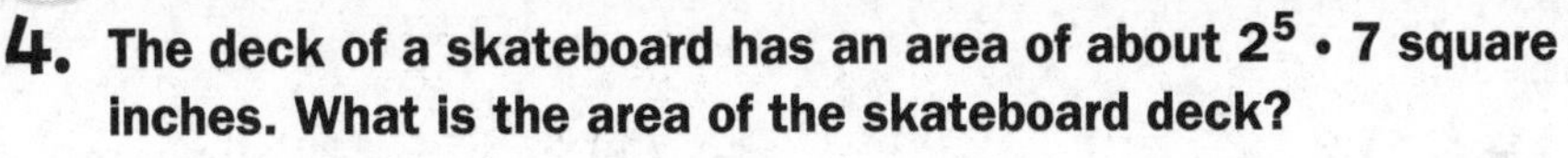

4. **The deck of a skateboard has an area of about $2^5 \cdot 7$ square inches. What is the area of the skateboard deck?**

$2^5 \cdot 7 = 2 \cdot 2 \cdot 2 \cdot 2 \cdot 2 \cdot 7$ Write the power as a product.

$= (2 \cdot 2 \cdot 2 \cdot 2 \cdot 2) \cdot 7$ Associative Property

$= 32 \cdot 7$ or 224 Multiply.

The area of the skateboard deck is about 224 square inches.

Got It? Do this problem to find out.

g. A school basketball court has an area of $2^3 \cdot 3 \cdot 5^2 \cdot 7$ square feet. What is the area of a school basketball court?

g. ______

Examples

Evaluate each expression if $a = 3$ and $b = 5$.

5. $a^2 + b^4$

$a^2 + b^4 = 3^2 + 5^4$ Replace a with 3 and b with 5.

$= (3 \cdot 3) + (5 \cdot 5 \cdot 5 \cdot 5)$ Write the powers as products.

$= 9 + 625$ or 634 Add.

6. $(a - b)^2$

$(a - b)^2 = (3 - 5)^2$ Replace a with 3 and b with 5.

$= (-2)^2$ Perform operations in the parentheses first.

$= (-2) \cdot (-2)$ or 4 Write the powers as products. Then simplify.

h. ______

Show your work.

i. ______

j. ______

Got It? Do these problems to find out.

Evaluate each expression if $c = -4$ and $d = 9$.

h. $c^3 + d^2$

i. $(c + d)^3$

j. $d^3 - (c^2 - 2)$

Guided Practice

Write each expression using exponents. (Examples 1 and 2)

1. $(-11)(-11)(-11) =$ ______

2. $2 \cdot 2 \cdot 2 \cdot 3 \cdot 3 \cdot 3 =$ ______

3. $r \cdot s \cdot r \cdot r \cdot s \cdot s \cdot r \cdot r =$ ______

Evaluate each expression. (Example 3)

4. $2^6 =$ ______

5. $(-4)^4 =$ ______

6. $\left(\frac{1}{7}\right)^3 =$ ______

7. The table shows the average weights of some endangered mammals. What is the weight of each animal? (Example 4)

______

Animal	Weight (lb)
Black bear	$2 \cdot 5^2 \cdot 7$
Key deer	$3 \cdot 5^2$
Panther	$2^3 \cdot 3 \cdot 5$

Evaluate each expression if $x = 2$ and $y = 10$. (Examples 5 and 6)

8. $x^2 + y^4 =$ ______

9. $(x^2 + y)^3 =$ ______

10. **Building on the Essential Question** How can I write repeated multiplication using powers? ______

______

______

______

Rate Yourself!

Are you ready to move on? Shade the section that applies.

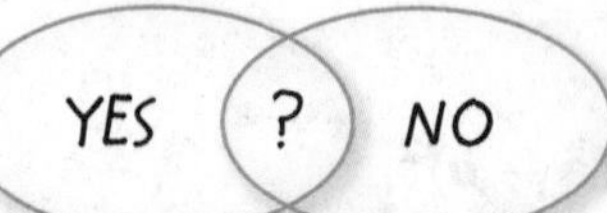

For more help, go online to access a Personal Tutor.

Name ____________________ My Homework ____________________

Independent Practice

Go online for Step-by-Step Solutions 

Write each expression using exponents. (Examples 1 and 2)

1. $(-5)(-5)(-5)(-5) =$ ______

2. $3 \cdot 3 \cdot 5 \cdot q \cdot q \cdot q =$ ______

3. $m \cdot m \cdot m \cdot m \cdot m =$ ______

Evaluate each expression. (Example 3)

4. $(-9)^4 =$ ______

5. $\left(\frac{1}{3}\right)^4 =$ ______

6. $\left(\frac{5}{7}\right)^3 =$ ______

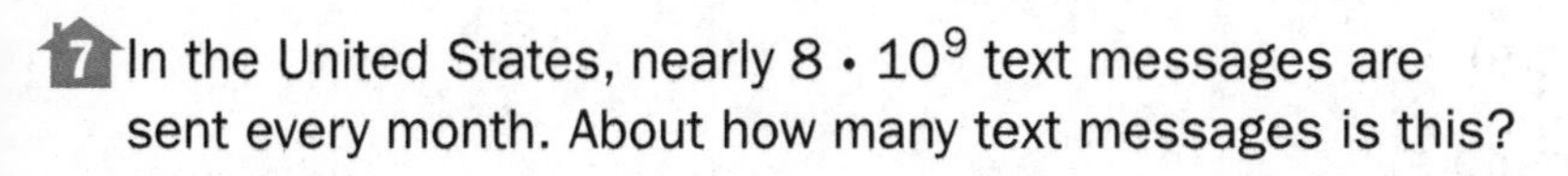

7 In the United States, nearly $8 \cdot 10^9$ text messages are sent every month. About how many text messages is this?

(Example 4) ______

8. Interstate 70 stretches almost $2^3 \cdot 5^2 \cdot 11$ miles across the United States. About how many miles long is Interstate 70?

(Example 4) ______

Evaluate each expression. (Examples 5 and 6)

9 $g^5 - h^3$ if $g = 2$ and $h = 7$ ______

10. $c^2 + d^3$, if $c = 8$ and $d = -3$ ______

11. $a^2 \cdot b^6$ if $a = \frac{1}{2}$ and $b = 2$ ______

12. $(r - s)^3 + r^2$ if $r = -3$ and $s = -4$ ______

13. **Model with Mathematics** Refer to the graphic novel frame below for Exercises a–c.

The metric system is based on powers of 10. For example, one kilometer is equal to 1,000 meters or 10^3 meters. Write each measurement in meters as a power of 10.

a. megameter (1,000,000 meters) ______

b. gigameter (1,000,000,000 meters) ______

c. petameter (1,000,000,000,000,000 meters) ______

H.O.T. Problems Higher Order Thinking

14. **Identify Structure** Write an expression with an exponent that has a value between 0 and 1. ______

15. **Identify Repeated Reasoning** Describe the following pattern: $3^4 = 81, 3^3 = 27, 3^2 = 9, 3^1 = 3$. Then use a similar pattern to predict the value of 2^{-1}. ______

Standardized Test Practice

16. Which expression is equivalent to the expression below?

$$2^3 \cdot 3^4$$

Ⓐ $3 \cdot 3 \cdot 4 \cdot 4 \cdot 4$

Ⓑ $2 \cdot 2 \cdot 2 \cdot 3 \cdot 3 \cdot 3$

Ⓒ $2 \cdot 2 \cdot 2 \cdot 3 \cdot 3 \cdot 3 \cdot 3$

Ⓓ $6 \cdot 12$

Extra Practice

17. Write $3 \cdot p \cdot p \cdot p \cdot 3 \cdot 3$ using exponents.

$3^3 \cdot p^3$

$3 \cdot p \cdot p \cdot p \cdot 3 \cdot 3 = 3 \cdot 3 \cdot 3 \cdot p \cdot p \cdot p$

$= 3^3 \cdot p^3$

18. Evaluate $x^3 + y^4$ if $x = -3$ and $y = 4$.

229

$x^3 + y^4 = (-3)^3 + 4^4$

$= (-3) \cdot (-3) \cdot (-3) + 4 \cdot 4 \cdot 4 \cdot 4$

$= (-27) + 256$

$= 229$

Write each expression using exponents.

19. $\left(-\frac{5}{6}\right)\left(-\frac{5}{6}\right)\left(-\frac{5}{6}\right) =$ ______

20. $s \cdot (7) \cdot s \cdot (7) \cdot (7) =$ ______

21. $4 \cdot b \cdot b \cdot 4 \cdot b \cdot b =$ ______

Evaluate each expression.

22. $k^4 \cdot m$, if $k = 3$ and $m = \frac{5}{6}$

23. $(c^3 + d^4)^2 - (c + d)^3$, if $c = -1$ and $d = 2$

Fill in each ◯ with <, >, or = to make a true statement.

24. $(6 - 2)^2 + 3 \cdot 4$ ◯ 5^2

25. $5 + 7^2 + 3^3$ ◯ 3^4

26. $\left(\frac{1}{2}\right)^4$ ◯ $\left(\frac{1}{4}\right)^2$

27. CCSS **Multiple Representations** A square has a side length of *s* inches.

a. Tables Copy and complete the table showing the side length, perimeter, and area of the square on a separate piece of paper.

b. Graphs On a separate piece of grid paper, graph the ordered pairs (side length, perimeter) and (side length, area) on the same coordinate plane. Then connect the points for each set.

c. Words On a separate sheet of paper, compare and contrast the graphs of the perimeter and area of the square. Which graph is a line?

Side Length (in.)	Perimeter (in.)	Area (in^2)
1	4	1
2		
3		
4		
5		
⋮		
10		

Standardized Test Practice

28. To find the volume of a cube, multiply its base, its height, and its width.

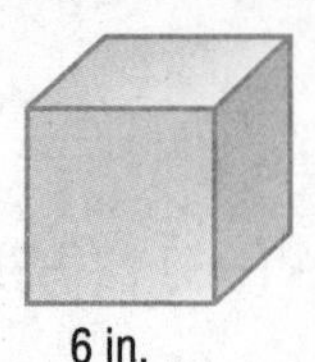

What is the volume of the cube expressed as a power?

Ⓐ 6^2 in^3 Ⓒ 6^4 in^3

Ⓑ 6^3 in^3 Ⓓ 6^6 in^3

29. Short Response The volume of an ice cube in cubic millimeters is represented by the term 11^3. What is 11^3 in standard form? ________

30. What is the value of $x^2 - y^4$ if $x = -3$ and $y = -2$?

Ⓕ −7 Ⓗ 2

Ⓖ −2 Ⓘ 7

Common Core Review

31. The table below shows the number of ants in an ant farm on different days. The number of ants doubles every ten days. 7.EE.3

Day	51	61	71
Number of Ants	320	640	1,280

a. How many ants were in the farm on Day 1? ________

b. How many ants will be in the farm on Day 91? ________

32. Nieves and her three friends are playing a video game. The table shows their scores at the end of the first round. 7.NS.1

Player	Score
Nieves	−189
Polly	−142
Saul	230
Harry	−48

a. What is the difference between the highest and lowest scores?

____

b. By how many points is Nieves losing to Polly? ________

Add. 7.NS.1

33. $-12 + (-19) =$ ________

34. $-8 + (-11) =$ ________

35. $-5 + 6 =$ ________

Lesson 3

Multiply and Divide Monomials

What You'll Learn

Scan the lesson. List two headings you would use to make an outline of the lesson.

- ______________________________
- ______________________________

Essential Question

WHY is it helpful to write numbers in different ways?

Vocabulary

monomial

Common Core State Standards

Content Standards
8.EE.1

Mathematical Practices
1, 3, 4, 7

Real-World Link

Arachnids Spiders in North America can range in size from 1 millimeter in length to 7.6 centimeters in length. Use the table to see how other metric measurements of length are related to the millimeter.

Unit of Length	Times Longer than a Millimeter	Written Using Powers
Millimeter	1	10^0
Centimeter	$1 \times 10 =$ ☐	10^1
Decimeter	$10 \times 10 =$ ☐	$10^1 \times 10^1 = 10^2$
Meter	$100 \times 10 = 1{,}000$	$10^2 \times 10^1 = 10^{☐}$
Dekameter	$1{,}000 \times 10 = 10{,}000$	$10^3 \times 10^1 = 10^{☐}$
Hectometer	$10{,}000 \times 10 =$ ☐	$10^4 \times 10^1 = 10^5$
Kilometer	$100{,}000 \times 10 =$ ☐	$10^5 \times 10^1 = 10^{☐}$

1. Look at the entries in the last column. What do you observe about the exponents of the factors and the exponent of the product for each entry? ______________________________

2. A *megameter* is $100{,}000{,}000 \times 10$ or 1,000,000,000 times longer than a millimeter. Extend the pattern to write this number using powers. ______________________________

Key Concept — Product of Powers

Words To multiply powers with the same base, add their exponents.

Examples

Numbers: $2^4 \cdot 2^3 = 2^{4+3}$ or 2^7

Algebra: $a^m \cdot a^n = a^{m+n}$

Work Zone

A **monomial** is a number, a variable, or a product of a number and one or more variables. You can use the Laws of Exponents to simplify monomials.

$$3^2 \cdot 3^4 = \underbrace{\overbrace{(3 \cdot 3)}^{\text{2 factors}} \cdot \overbrace{(3 \cdot 3 \cdot 3 \cdot 3 \cdot)}^{\text{4 factors}}}_{\text{6 factors}} \text{ or } 3^6$$

Notice that the sum of the original exponents is the exponent in the final product.

Examples

Tutor 

Simplify using the Laws of Exponents.

1. $5^2 \cdot 5$

$5^2 \cdot 5 = 5^2 \cdot 5^1$ — $5 = 5^1$

$= 5^{2+1}$ — The common base is 5.

$= 5^3$ or 125 — Add the exponents. Simplify.

Check $5^2 \cdot 5 = (5 \cdot 5) \cdot 5$

$= 5 \cdot 5 \cdot 5$

$= 5^3$ ✓

2. $c^3 \cdot c^5$

$c^3 \cdot c^5 = c^{3+5}$ — The common base is c.

$= c^8$ — Add the exponents.

3. $-3x^2 \cdot 4x^5$

$-3x^2 \cdot 4x^5 = (-3 \cdot 4)(x^2 \cdot x^5)$ — Commutative and Associative Properties

$= (-12)(x^{2+5})$ — The common base is x.

$= -12x^7$ — Add the exponents.

Show your work.

a. ________

b. ________

c. ________

Got It? **Do these problems to find out.**

a. $9^3 \cdot 9^2$ **b.** $a^3 \cdot a^2$ **c.** $-2m(-8m^5)$

Quotient of Powers

Key Concept

Words To divide powers with the same base, subtract their exponents.

Examples

Numbers: $\frac{3^7}{3^3} = 3^{7-3}$ or 3^4

Algebra: $\frac{a^m}{a^n} = a^{m-n}$, where $a \neq 0$

and Reflect

Explain below why the Quotient of Powers rule cannot be used to simplify the expression $\frac{x^5}{y^3}$.

There is also a Law of Exponents for dividing powers with the same base.

$$\frac{5^7}{5^4} = \frac{\overbrace{5 \cdot 5 \cdot 5 \cdot \cancel{5} \cdot \cancel{5} \cdot \cancel{5} \cdot \cancel{5}}^{\text{7 factors}}}{\underbrace{\cancel{5} \cdot \cancel{5} \cdot \cancel{5} \cdot \cancel{5}}_{\text{4 factors}}} \text{ or } 5^3$$

Notice that the difference of the original exponents is the exponent in the final quotient.

Examples

Tutor

Simplify using the Laws of Exponents.

4. $\frac{4^8}{4^2}$

$\frac{4^8}{4^2} = 4^{8-2}$ The common base is 4.

$= 4^6$ or 4,096 Simplify.

5. $\frac{n^9}{n^4}$

$\frac{n^9}{n^4} = n^{9-4}$ The common base is n.

$= n^5$ Simplify.

6. $\frac{2^5 \cdot 3^5 \cdot 5^2}{2^2 \cdot 3^4 \cdot 5}$

$\frac{2^5 \cdot 3^5 \cdot 5^2}{2^2 \cdot 3^4 \cdot 5} = \left(\frac{2^5}{2^2}\right)\left(\frac{3^5}{3^4}\right)\left(\frac{5^2}{5}\right)$ Group by common base.

$= 2^3 \cdot 3^1 \cdot 5^1$ Subtract the exponents.

$= 8 \cdot 3 \cdot 5$ $2^3 = 8$

$= 120$ Simplify.

Got It? Do these problems to find out.

d. $\frac{5^7}{5^4}$ **e.** $\frac{x^{10}}{x^3}$ **f.** $\frac{12w^5}{2w}$

g. $\frac{3^4 \cdot 5^2 \cdot 7^5}{3^2 \cdot 5 \cdot 7^3}$ **h.** $\frac{5^6 \cdot 7^4 \cdot 8^3}{5^4 \cdot 7^2 \cdot 8^2}$ **i.** $\frac{(-2)^5 \cdot 3^4 \cdot 5^7}{(-2)^2 \cdot 3 \cdot 5^4}$

d. ______

e. ______

f. ______

g. ______

h. ______

i. ______

Example

7. Hawaii's total shoreline is about 2^{10} miles long. New Hampshire's shoreline is about 2^7 miles long. About how many times longer is Hawaii's shoreline than New Hampshire's?

To find how many times longer, divide 2^{10} by 2^7.

$\frac{2^{10}}{2^7} = 2^{10-7}$ or 2^3 Quotient of Powers

Hawaii's shoreline is about 2^3 or 8 times longer.

Guided Practice

Simplify using the Laws of Exponents. (Examples 1–6)

1. $4^5 \cdot 4^3 =$ ______

2. $-2a(3a^4) =$ ______

3. $\frac{y^8}{y^5} =$ ______

4. $\frac{24k^9}{6k^6} =$ ______

5. $\frac{2^2 \cdot 3^3 \cdot 4^5}{2 \cdot 3 \cdot 4^4} =$ ______

6. $\frac{(-3)^4 \cdot (-4)^3 \cdot 5^2}{(-3)^2 \cdot (-4) \cdot 5} =$ ______

7. The table shows the number of people worldwide that speak certain languages. How many times as many people speak French than Sicilian? (Example 7) ______

Language	Total (millions)
French	2^6
Sicilian	2^2

8. **Building on the Essential Question** How can I use the properties of integer exponents to simplify algebraic and numeric expressions? ______

Rate Yourself!

Are you ready to move on? Shade the section that applies.

YES ? NO

For more help, go online to access a Personal Tutor.

FOLDABLES *Time to update your Foldable!*

Name ______________________ My Homework ______________________

Independent Practice

Go online for Step-by-Step Solutions

Simplify using the Laws of Exponents. (Examples 1–6)

1. $(-6)^2 \cdot (-6)^5 =$ ____________

2. $-4a^5(6a^5) =$ ____________

3. $(-7a^4bc^3)(5ab^4c^2) =$ ____________

4. $\frac{8^{15}}{8^{13}} =$ ____________

5. $\frac{16t^4}{8t} =$ ____________

6. $\frac{x^6y^{14}}{x^4y^9} =$ ____________

7. $\frac{3^4x^4}{3x^2} =$ ____________

8. $\frac{4^5 \cdot 5^3 \cdot 6^2}{4^4 \cdot 5^2 \cdot 6} =$ ____________

9. $\frac{6^3 \cdot 6^6 \cdot 6^4}{6^2 \cdot 6^3 \cdot 6^3} =$ ____________

10. $\frac{(-2)^5 \cdot (-3)^4 \cdot (-5)^3}{(-2)^3 \cdot (-3) \cdot (-5)^2} =$ ____________

11. The processing speed of a certain computer is 10^{11} instructions per second. Another computer has a processing speed that is 10^3 times as fast. How many instructions per second can the faster computer process? (Example 7)

____

12. The table shows the seating capacity of two different facilities. About how many times as great is the capacity of Madison Square Garden in New York than a typical movie theater? (Example 7)

____

Place	Seating Capacity
Movie theater	3^5
Madison Square Garden	3^9

13. Refer to the information in the table.

Power of Ten	U.S. Name
10^3	one thousand
10^6	one million
10^9	one billion
10^{12}	one trillion
10^{15}	one quadrillion
10^{18}	one quintillion

a. How many times as great is one quadrillion than one million?

b. One quintillion is one trillion times as great as what number?

CCSS **Persevere with Problems** **Find each missing exponent.**

14. $(6^{\blacksquare})(6^3) = 6^5$

15. $3x^{\blacksquare} \cdot 4x^3 = 12x^{12}$

16. $p^3 \cdot p^{\blacksquare} \cdot p^2 = p^9$

17. $\frac{3^{\blacksquare}}{3^2} = 3^4$

18. $\frac{5^9}{5^{\blacksquare}} = 5^4$

19. $2x^{\blacksquare} \cdot \frac{3x^2}{x^6} = 6x^3$

H.O.T. Problems Higher Order Thinking

20. CCSS **Identify Structure** Write a multiplication expression with a product of 5^{13}.

21. CCSS **Justify Conclusions** Is $\frac{3^{100}}{3^{99}}$ *greater than, less than,* or *equal to* 3? Explain your reasoning to a classmate.

22. CCSS **Persevere with Problems** What is twice 2^{30}? Write using exponents. Explain your reasoning.

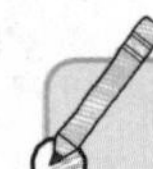

Standardized Test Practice

23. Which expression is equivalent to $8x^2y \cdot 8yz^2$?

Ⓐ $64x^2y^2z^2$

Ⓑ $64x^2yz^2$

Ⓒ $16x^2y^2z^2$

Ⓓ $384x^2y^2z^2$

Extra Practice

Simplify using the Laws of Exponents.

24. $(3x^8)(5x) =$ $15x^9$

$(3x^8)(5x) = 3 \cdot 5 \cdot x^8 \cdot x$

$= 15 \cdot x^{8+1}$

$= 15x^9$

Homework Help

25. $\frac{h^7}{h^6} =$ h^1 or h

$\frac{h^7}{h^6} = h^{7-6}$

$= h^1$ or h

26. $2g^2 \cdot 7g^6 =$ ______

27. $(8w^4)(-w^7) =$ ______

28. $(-p)(-9p^2) =$ ______

29. $\frac{2^9}{2} =$ ______

30. $\frac{36d^{10}}{6d^5} =$ ______

31. $\frac{5^3 \cdot 7^4 \cdot 10}{5 \cdot 7^4} =$ ______

32. $\frac{(-3)^2 \cdot 4^3 \cdot (-1)^8}{4 \cdot (-1)^5} =$ ______

33. CCSS **Persevere with Problems** The figure at the right is composed of a circle and a square. The circle touches the square at the midpoints of the four sides.

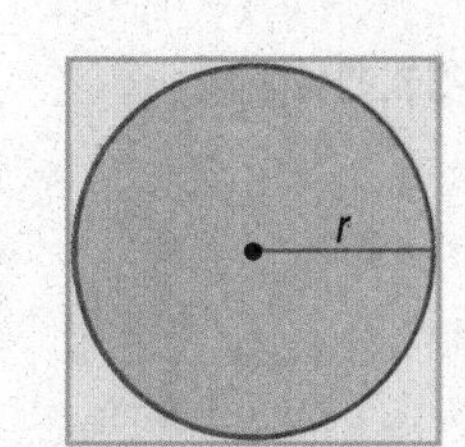

a. What is the length of one side of the square? ______

b. The formula $A = \pi r^2$ is used to find the area of a circle. The formula $A = 4r^2$ can be used to find the area of the square. Write the ratio of the area of the circle to the area of the square in simplest form.

______

c. Complete the table.

Radius (units)	2	3	4	2r
Area of Circle (units²)	$\pi(2)^2$ or 4π			
Length of 1 Side of the Square	4			
Area of Square (units²)	4^2 or 16			
Ratio $\frac{\text{(Area of circle}}{\text{Area of square)}}$				

d. What can you conclude about the relationship between the areas of the circle and the square? ______

Standardized Test Practice

34. One meter is 10^3 times longer than one millimeter. One kilometer is 10^6 times longer than one millimeter. How many times longer is one kilometer than one meter?

Ⓐ 10^9
Ⓑ 10^6
Ⓒ 10^3
Ⓓ 10

35. Which of the following is equivalent to $\left(-\frac{2}{3}\right)^3$?

Ⓕ $-\frac{6}{9}$
Ⓖ $-\frac{8}{27}$
Ⓗ $\frac{8}{27}$
Ⓘ $\frac{6}{9}$

36. Short Response What is the area of the rectangle below?

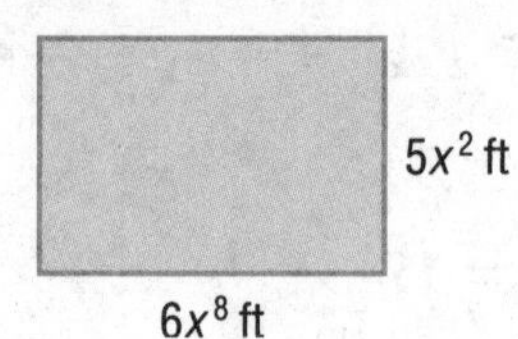

Common Core Review

Multiply or divide. 7.NS.2

37. $14(-2) =$ ______

38. $-20(-3) =$ ______

39. $-5(7) =$ ______

40. $-12 \div (-4) =$ ______

41. $63 \div (-7) =$ ______

42. $250 \div (-50) =$ ______

43. Three-fourths of a pan of lasagna is to be divided equally among 6 people. What part of the lasagna will each person receive? 6.NS.1

44. The tallest mountain in the United States is Mount McKinley in Alaska. The elevation is about $2^2 \cdot 5 \cdot 10^3$ feet above sea level. What is the height of Mount McKinley? 6.EE.1

Lesson 4

Powers of Monomials

What You'll Learn

Scan the lesson. Predict two things you will learn about the Laws of Exponents.

- ____
- ____

Essential Question

WHY is it helpful to write numbers in different ways?

Common Core State Standards

Content Standards
8.EE.1

Mathematical Practices
1, 3, 4, 7

Real-World Link

Aquariums The Marine Club at Westview Middle School purchased an aquarium. The aquarium is in the shape of a cube with a side length of 2^4 inches. Use the questions to find the amount of water the aquarium willl hold.

1. Write a multiplication expression to represent the volume of the aquarium. ____________

2. Simplify the expression. Write as a single power of 2. ☐

3. Using 2^4 as the base, write the multiplication expression $2^4 \cdot 2^4 \cdot 2^4$ using an exponent. ☐

4. Explain why $(2^4)^3 = 2^{12}$. ____________

5. Use a calculator to find the volume of the tank. ☐ cubic inches

6. One gallon of water is equal to 231 cubic inches. Write an expression to find how many gallons of water the tank will hold if it is filled to the top. $\frac{\square}{\square}$

7. How many gallons of water will the aquarium hold? Round your answer to the nearest gallon. ☐ gallons

Key Concept

Power of a Power

Words To find the power of a power, multiply the exponents.

Examples Numbers $(5^2)^3 = 5^{2 \cdot 3}$ or 5^6 Algebra $(a^m)^n = a^{m \cdot n}$

Work Zone

You can use the rule for finding the *product* of powers to discover another Law of Exponents for finding the *power* of a power.

$$(6^4)^5 = \overbrace{(6^4)(6^4)(6^4)(6^4)(6^4)}^{\text{5 factors}}$$

$$= 6^{4+4+4+4+4}$$ Apply the rule for the product of powers.

$$= 6^{20}$$

Notice that the product of the original exponents, 4 and 5, is the final power 20.

Examples

Simplify using the Laws of Exponents.

1. $(8^4)^3$

$(8^4)^3 = 8^{4 \cdot 3}$ Power of a Power

$= 8^{12}$ Simplify.

2. $(k^7)^5$

$(k^7)^5 = k^{7 \cdot 5}$ Power of a Power

$= k^{35}$ Simplify.

Show your work.

Got It? Do these problems to find out.

a. $(2^5)^2$ **b.** $(w^4)^6$ **c.** $[(3^2)^3]^2$

a. ______

b. ______

c. ______

Power of a Product

Key Concept

Words To find the power of a product, find the power of each factor and multiply.

Examples

Numbers: $(6x^2)^3 = (6)^3 \cdot (x^2)^3$ or $216x^6$

Algebra: $(ab)^m = a^m b^m$

Extend the power of a *power* rule to find the Laws of Exponents for the power of a *product*.

$$(3a^2)^5 = \overbrace{(3a^2)(3a^2)(3a^2)(3a^2)(3a^2)}^{\text{5 factors}}$$

$$= 3 \cdot 3 \cdot 3 \cdot 3 \cdot 3 \cdot a^2 \cdot a^2 \cdot a^2 \cdot a^2 \cdot a^2$$

$$= 3^5 \cdot (a^2)^5 \quad \text{Write using powers.}$$

$$= 243 \cdot a^{10} \text{ or } 243a^{10} \quad \text{Power of a Power}$$

Common Error

When finding the power of a power, do not add the exponents. $(8^4)^3 = 8^{12}$, not 8^7.

Examples

Simplify using the Laws of Exponents.

3. $(4p^3)^4$

$(4p^3)^4 = 4^4 \cdot p^{3 \cdot 4}$ Power of a Product

$= 256p^{12}$ Simplify.

4. $(-2m^7n^6)^5$

$(-2m^7n^6)^5 = (-2)^5 m^{7 \cdot 5} n^{6 \cdot 5}$ Power of a Product

$= -32m^{35}n^{30}$ Simplify.

Got It? Do these problems to find out.

d. $(8b^9)^2$

e. $(6x^5y^{11})^4$

f. $(-5w^2z^8)^3$

d. ______

e. ______

f. ______

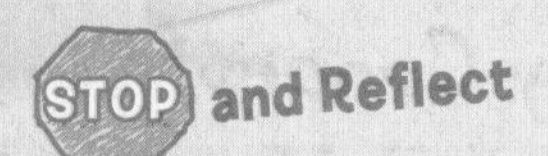

How do you know when an expression is in simplest form? Explain below.

Example

Tutor

5. A magazine offers a special service to its subscribers. If they scan the square logo shown on a smartphone, they can receive special offers from the magazine. Find the area of the logo.

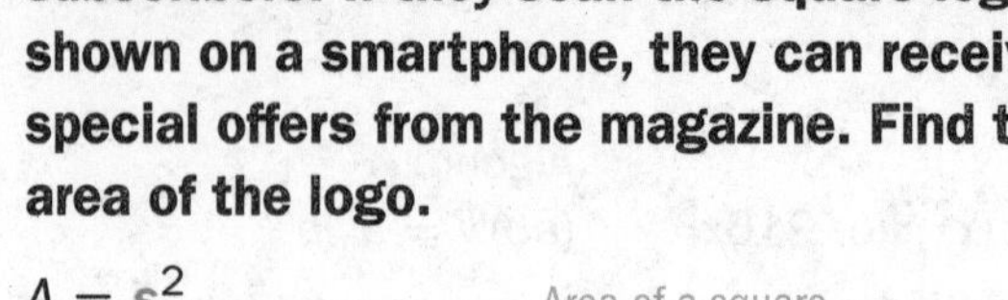

$A = s^2$	Area of a square
$A = (7a^4b)^2$	Replace s with $7a^4b$.
$A = 7^2(a^4)^2(b^1)^2$	Power of a Product
$A = 49a^8b^2$	Simplify.

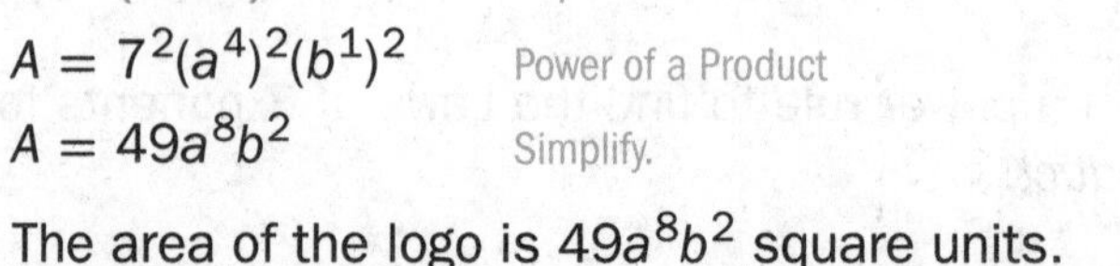

The area of the logo is $49a^8b^2$ square units.

Guided Practice

Simplify using the Laws of Exponents. (Examples 1–4)

Show your work.

1. $(3^2)^5 =$ ______
2. $(h^6)^4 =$ ______
3. $[(2^3)^2]^3 =$ ______
4. $(7w^7)^3 =$ ______
5. $(5g^8k^{12})^4 =$ ______
6. $(-6r^5s^9)^2 =$ ______

7. The floor of the commons room at King Middle School is in the shape of a square with side lengths of x^2y^3 feet. New tile is going to be put on the floor of the room. Find the area of the floor. (Example 5)

______

8. **Building on the Essential Question** How does the Product of Powers law apply to finding the power of a power?

______

Rate Yourself!

How confident are you about powers of monomials? Check the box that applies.

For more help, go online to access a Personal Tutor.

FOLDABLES Time to update your Foldable!

Name ____________________ My Homework ____________________

Independent Practice

Go online for Step-by-Step Solutions 

Simplify using the Laws of Exponents. (Examples 1–4)

1. $(4^2)^3 =$ ________

2. $(5^3)^3 =$ ________

3. $(d^7)^6 =$ ________

4. $(h^4)^9 =$ ________

5. $[(3^2)^2]^2 =$ ________

6. $[(5^2)^2]^2 =$ ________

7. $(5j^6)^4 =$ ________

8. $(11c^4)^3 =$ ________

9. $(6a^2b^6)^3 =$ ________

10. $(2m^5n^{11})^6 =$ ________

11. $(-3w^3z^8)^5 =$ ________

12. $(-5r^4s^{12})^4 =$ ________

13 A shipping box is in the shape of a cube. Each side measures $3c^6d^2$ inches. Express the volume of the cube as a monomial. (Example 5)

____

14. Tamara is decorating her patio with a planter in the shape of a cube like the one shown. Find the volume of the planter. (Example 5)

____

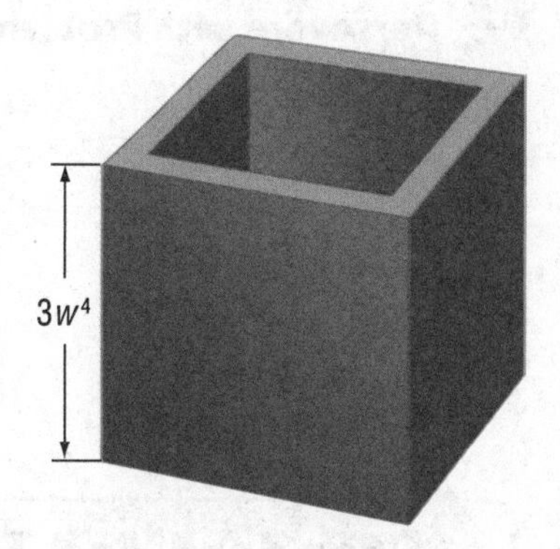

Copy and Solve **Simplify. Show your work on a separate sheet of paper.**

15. $[(3x^2y^3)^2]^3$

16. $\left(\frac{3}{5}a^6b^9\right)^2$

17 $(-2v^7)^3\ (-4v^2)^4$

18. **Identify Structure** Draw a line connecting the Law(s) of Exponents you would use to simplify each of the expressions. Then simplify each one.

Laws	Expressions
Product of Powers	$(a^9)^3 =$ ______
Quotient of Powers	$(m^8) \div (m^4) =$ ______
Power of a Power	$5x^2 \cdot (-7x^4) =$ ______
Power of a Product	$\frac{(xy^4)^3}{xy} =$ ______
	$(n^6)^8 =$ ______

H.O.T. Problems Higher Order Thinking

19. **Reason Inductively** The table gives the area and volume of a square and cube, respectively, with side lengths shown.

a. Complete the table.

b. Describe how the area and volume are each affected if the side length is doubled. Then describe how they are each affected if the side length is tripled.

Side Length (units)	x	$2x$	$3x$
Area of Square (units²)	x^2		
Volume of Cube (units³)	x^3		

______

______

______

Persevere with Problems Solve each equation for *x*.

20. $(7^x)^3 = 7^{15}$ ______

21. $(-2m^3n^4)^x = -8m^9n^{12}$ ______

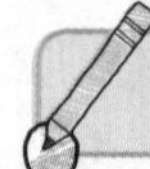

Standardized Test Practice

22. Which expression is equivalent to $(10^4)^8$?

Ⓐ 10^2

Ⓑ 10^4

Ⓒ 10^{12}

Ⓓ 10^{32}

Extra Practice

Simplify using the Laws of Exponents.

23. $(2^2)^7 =$ 2^{14}

$(2^2)^7 = 2^{2 \cdot 7}$

$= 2^{14}$

Homework Help

24. $(8v^9)^5 =$ $32{,}768v^{45}$

$(8v^9)^5 = 8^5 \cdot v^{9 \cdot 5}$

$= 32{,}768v^{45}$

25. $(3^4)^2 =$ ______

26. $(m^8)^5 =$ ______

27. $(z^{11})^5 =$ ______

28. $[(4^3)^2]^2 =$ ______

29. $[(2^3)^3]^2 =$ ______

30. $(14y)^4 =$ ______

Express the area of each square as a monomial.

31. ______

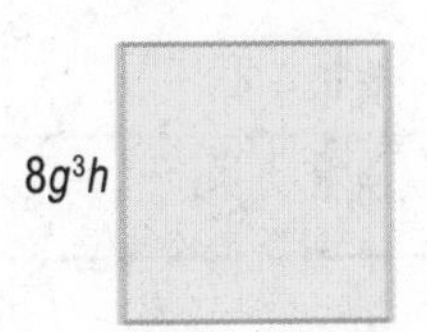

32. ______

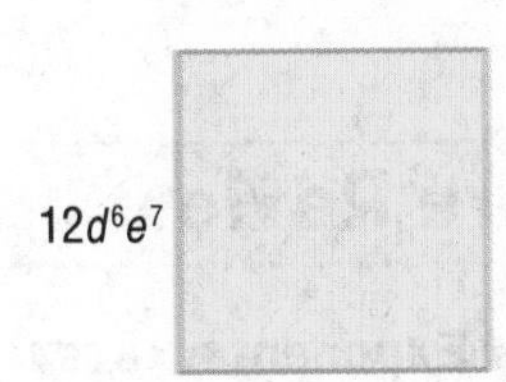

Express the volume of each cube as a monomial.

33. ______

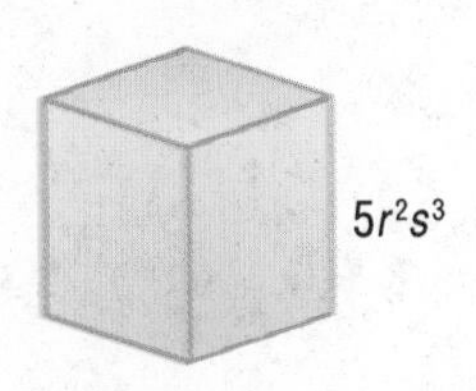

34. ______

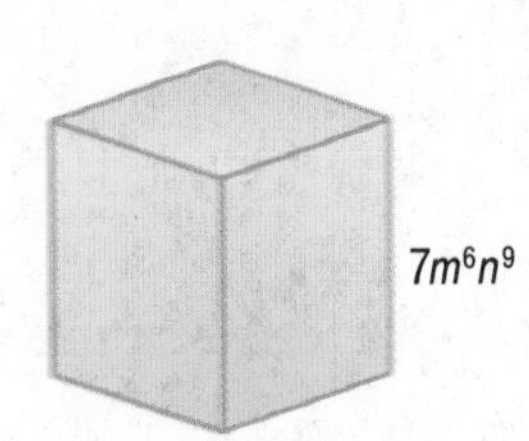

Simplify.

35. $(0.5k^5)^2 =$ ______

36. $(0.3p^7)^3 =$ ______

37. $\left(\frac{1}{4}w^5z^3\right)^2 =$ ______

38. CCSS **Persevere with Problems** A ball is dropped from the top of a building. The expression $4.9x^2$ gives the distance in meters the ball has fallen after x seconds. Write and simplify an expression that gives the distance in meters the ball has fallen after x^2 seconds. after x^3 seconds.

______

Standardized Test Practice

39. What is the volume of the cube shown below?

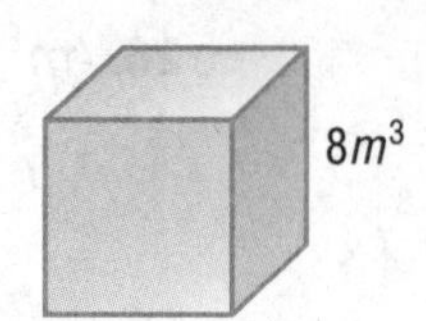

Ⓐ $8m^3$ Ⓒ $64m^9$
Ⓑ $16m^5$ Ⓓ $512m^9$

40. Which expression has the same value as $81h^8k^6$?

Ⓕ $(9h^6k^4)^2$ Ⓗ $(6h^5k^3)^3$
Ⓖ $(9h^4k^3)^2$ Ⓘ $(3h^2k)^6$

41. Which expression is equivalent to $(2x^2)^4(5x^6)$?

Ⓐ $10x^{12}$ Ⓒ $10x^{14}$
Ⓑ $80x^{12}$ Ⓓ $80x^{14}$

42. Short Response Manny has four pieces of carpet in the shape of a square like the one shown. He wants to use them together to carpet a portion of his basement. What is the area of the space he can cover with the carpet? ________

Common Core Review

Simplify using the Laws of Exponents. 8.EE.1

43. $6^4 \cdot 6^7 =$ ________

44. $18^3 \cdot 18^5 =$ ________

45. $(-3x^{11})(-6x^3) =$ ________

46. $(-9a^4)(2a^7) =$ ________

47. The table shows the heights of some United States waterfalls. What is the height of each waterfall? 6.EE.1

____

____

Waterfall	Height (ft)
Bridalveil (California)	$2^2 \cdot 5 \cdot 31$
Fall Creek (Tennessee)	2^8
Shoshone (Idaho)	$2^2 \cdot 53$

Problem-Solving Investigation
The Four-Step Plan

Content Standards
8.EE.1
Mathematical Practices
1, 3, 4

Case #1 Texting Trail

Lillian received a text about a concert. She forwarded the text to two of her friends. They each forwarded it to two more friends, and so on.

How many texts were sent at the 4th stage?

Understand *What are the facts?*

You know that each person at each stage sends a text to two people. You can use counters to represent the trail of texts sent.

Plan *What is your strategy to solve this problem?*

Use red counters to represent the texts in the first stage. Use yellow counters to show the texts sent at the second stage. Continue the pattern. Draw the counters representing the number of texts sent in the 4th stage.

Solve *How can you apply the strategy?*

1st stage
2nd stage
3rd stage
4th stage

There are 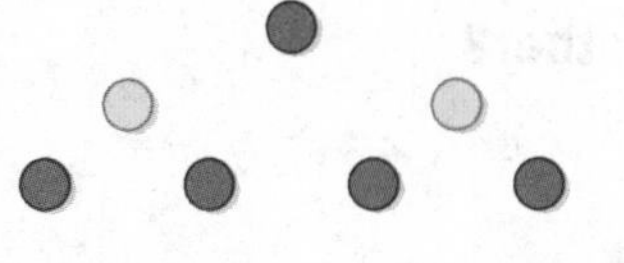 counters in the 4th row. So, 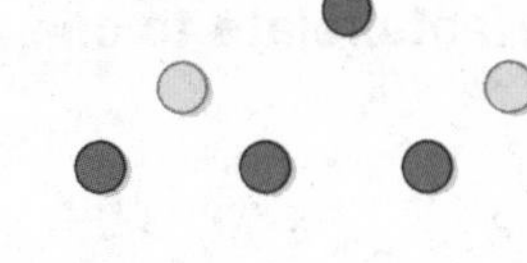 texts were sent during the 4th stage.

Check *Does the answer make sense?*

The number of texts at each stage is a power of 2. So, find 2^4.
Since $2^4 = 16$, the answer is correct. ✓

Analyze the Strategy

Justify Conclusions At what stage would there be more than 1,000 texts sent? Explain.

Case #2 Green Mileage

A test of a hybrid car resulted in 4,840 miles driven using 88 gallons of gas.

At this rate, how many gallons of gas will this vehicle need to travel 1,155 miles?

Understand

Read the problem. What are you being asked to find?

I need to find ______________________.

Underline key words and values in the problem. What information do you know?

The hybrid car can travel ________ miles using ________ gallons of gas.

Is there any information that you do *not* need to know?

I do not need to know ______________________.

Plan

How do the facts relate to one another?

____

Solve

Write and solve a proportion comparing miles to gallons. Let g represent the amount of gas needed to travel 1,155 miles.

$$\frac{\text{miles}}{\text{gallons}} \quad \frac{\square}{\square} = \frac{\square}{\square}$$

How many gallons of gas will the car use to travel 1,155 miles? $\square$

Check

Use information from the problem to check your answer.

____

____

Collaborate Work with a small group to solve the following cases. Show your work on a separate piece of paper.

Case #3 Class Trip

All of Mr. Bassett's science classes are going to the Natural History Museum. A tour guide is needed for each group of eight students. His classes have 28 students, 35 students, 22 students, 33 students, and 22 students.

How many tour guides are needed?

Case #4 Gardening

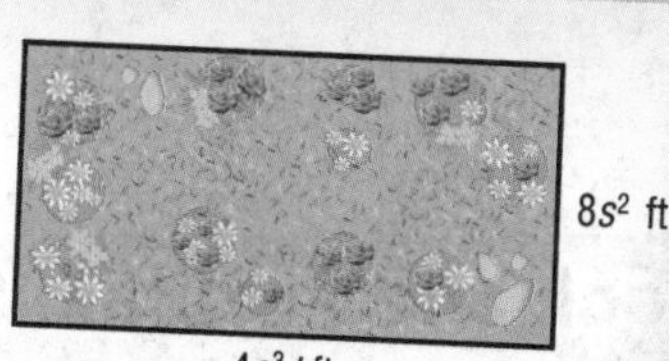

Mrs. Lopez is designing her garden in the shape of a rectangle. The area of her garden is 2 times greater than the area of the rectangle shown.

Write the area of Mrs. Lopez's garden in simplest form.

Case #5 Toothpicks

Figure 1 Figure 2 Figure 3 Figure 4

The figures to the right are made from toothpicks.

How many toothpicks would be needed to make the tenth figure?

Case #6 Number Sense

Study the following sequence:

$1 - \frac{1}{2}, 1 - \frac{1}{2}, 1 - \frac{1}{3}, 1 - \frac{1}{4}, \ldots, 1 - \frac{1}{48}, 1 - \frac{1}{49}$, and $1 - \frac{1}{50}$

What is the product of all of the terms?

Circle a strategy below to solve the problem.

- *Look for a pattern.*
- *Act it out.*
- *Determine reasonable answers.*
- *Make a table.*

Mid-Chapter Check

Vocabulary Check

1. **CCSS Be Precise** Define *power* using the words *base* and *exponent*. Give an example of a power and label the base and exponent. (Lesson 2)

2. Describe the Product of Powers rule. Give an example. (Lesson 3)

Skills Check and Problem Solving

3. Write $1\frac{7}{16}$ as a decimal. (Lesson 1) ______

Show your work.

4. Write $0.\overline{15}$ as a fraction in simplest form. (Lesson 1) ______

5. The mass of a baseball glove is $5 \cdot 5 \cdot 5 \cdot 5$ grams. Write the mass using exponents. Then find the value of the expression. (Lesson 2) ______

Simplify using the Laws of Exponents. (Lessons 3 and 4)

6. $2^3a^7 \cdot 2a^3 =$ ______

7. $\frac{24y^4}{4y^2} =$ ______

8. $(2p^3r^2)^3 =$ ______

9. **Standardized Test Practice** Which expression below has the same value as $5m^2$? (Lesson 2)

Ⓐ $5m$

Ⓑ $5 \cdot m \cdot m$

Ⓒ $5 \cdot 5 \cdot m \cdot m$

Ⓓ $5 \cdot m \cdot m \cdot m$

Lesson 5

Negative Exponents

What You'll Learn

Scan the lesson. Predict two things you will learn about exponents that are not positive.

- ______
- ______

Essential Question

WHY is it helpful to write numbers in different ways?

Common Core State Standards

Content Standards
8.EE.1

Mathematical Practices
1, 3, 4, 7

Real-World Link

Insects The table shows the approximate wing beats per minute for certain insects.

Insect	Wing Beats per Minute
house fly	10,000
small butterfly	100

1. Write a ratio in simplest form that compares the number of wing beats for a butterfly to a housefly. $\frac{\square}{\square}$

2. Write the ratio as a fraction with an exponent in the denominator and as a decimal. $\frac{\square}{\square}$; $\square$

3. Complete the 1st 4 rows of the table showing the exponential and standard forms of power of 10.

4. What operation is performed when you move down the table?
______

5. What happens to the exponent?
______

6. Extend the table to include the next three entries.

Exponential Form	Standard Form
10^3	
$10^{\square}$	100
10^1	
10^0	

Key Concept Zero and Negative Exponents

Words Any nonzero number to the zero power is 1. Any nonzero number to the negative *n* power is the multiplicative inverse of its *n*th power.

Examples

Numbers	Algebra
$5^0 = 1$	$x^0 = 1, x \neq 0$
$7^{-3} = \frac{1}{7} \cdot \frac{1}{7} \cdot \frac{1}{7}$ or $\frac{1}{7^3}$	$x^{-n} = \frac{1}{x^n}, x \neq 0$

Work Zone

Negative Exponents
Remember that 6^{-3} is equal to $\frac{1}{6^3}$, not -216 or -18.

You can use exponents to represent very small numbers. Negative powers are the result of repeated division.

Examples

Tutor

Write each expression using a positive exponent.

1. 6^{-3}

$6^{-3} = \frac{1}{6^3}$ Definition of negative exponent

2. a^{-5}

$a^{-5} = \frac{1}{a^5}$ Definition of negative exponent

Got It? Do these problems to find out.

a. 7^{-2}

b. b^{-4}

c. 5^0

d. m^{-3}

a. ____________

b. ____________

c. ____________

d. ____________

Examples

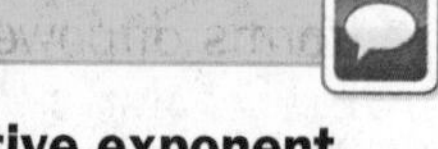

Tutor

Write each fraction as an expression using a negative exponent other than −1.

3. $\frac{1}{5^2}$

$\frac{1}{5^2} = 5^{-2}$ Definition of negative exponent

4. $\frac{1}{36}$

$\frac{1}{36} = \frac{1}{6^2}$ Definition of exponent

$= 6^{-2}$ Definition of negative exponent

Got It? Do these problems to find out.

e. $\frac{1}{8^3}$

f. $\frac{1}{4}$

g. $\frac{1}{c^5}$

h. $\frac{1}{27}$

e. ____________

f. ____________

g. ____________

h. ____________

Example

5. STEM One human hair is about 0.001 inch in diameter. Write the decimal as a power of 10.

$0.001 = \frac{1}{1{,}000}$ Write the decimal as a fraction.

$= \frac{1}{10^3}$ $1{,}000 = 10^3$

$= 10^{-3}$ Definition of negative exponent

A human hair is 10^{-3} inch thick.

Got It? **Do this problem to find out.**

i. STEM A water molecule is about 0.0000000001 meter long. Write the decimal as a power of 10.

and Reflect

Explain below the difference between the expressions $(-4)^2$ and 4^{-2}.

i. ________

Multiply and Divide with Negative Exponents

The Product of Powers and the Quotient of Powers rules can be used to multiply and divide powers with negative exponents.

Examples

Simplify each expression.

6. $5^3 \cdot 5^{-5}$

$5^3 \cdot 5^{-5} = 5^{3 + (-5)}$ Product of Powers

$= 5^{-2}$ Simplify.

$= \frac{1}{5^2}$ or $\frac{1}{25}$ Write using positive exponents. Simplify.

7. $\frac{w^{-1}}{w^{-4}}$

$\frac{w^{-1}}{w^{-4}} = w^{-1 - (-4)}$ Quotient of Powers

$= w^{(-1) + 4}$ or w^3 Subtract the exponents.

Got It? **Do these problems to find out.**

j. $3^{-8} \cdot 3^2$

k. $\frac{11^2}{11^4}$

l. $n^9 \cdot n^{-4}$

m. $\frac{b^{-4}}{b^{-7}}$

j. ________

k. ________

l. ________

m. ________

Guided Practice

Write each expression using a positive exponent. (Examples 1 and 2)

1. $2^{-4} =$ ______

2. $4^{-3} =$ ______

3. $a^{-4} =$ ______

4. $g^{-7} =$ ______

Show your work.

Write each fraction as an expression using a negative exponent other than −1. (Examples 3 and 4)

5. $\frac{1}{3^4} =$ ______

6. $\frac{1}{m^5} =$ ______

7. $\frac{1}{16} =$ ______

8. $\frac{1}{49} =$ ______

9. An American green tree frog tadpole is about 0.00001 kilometer in length when it hatches. Write this decimal as a power of 10.

(Example 5) ______

Simplify. (Examples 6 and 7)

10. $3^{-3} \cdot 3^{-2} =$ ______

11. $r^{-7} \cdot r^{3} =$ ______

12. $\frac{p^{-2}}{p^{-12}} =$ ______

13. **Building on the Essential Question** How are negative exponents and positive exponents related?

______

Rate Yourself!

How well do you understand understand negative exponents? Circle the image that applies.

Clear | Somewhat Clear | Not So Clear

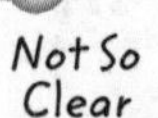

For more help, go online to access a Personal Tutor.

Name ______________________ My Homework ______________________

Independent Practice

Go online for Step-by-Step Solutions

Write each expression using a positive exponent. (Examples 1 and 2)

1. $7^{-10} =$ ______

2. $(-5)^{-4} =$ ______

3. $g^{-7} =$ ______

4. $w^{-13} =$ ______

Write each fraction as an expression using a negative exponent other than −1. (Examples 3 and 4)

5. $\frac{1}{12^4} =$ ______

6. $\frac{1}{(-5)^7} =$ ______

7. $\frac{1}{125} =$ ______

8. $\frac{1}{1{,}024} =$ ______

9. The table shows different metric measurements. Write each decimal as a power of 10. (Example 5) ______

Measurement	Value
Decimeter	0.1
Centimeter	0.01
Millimeter	0.001
Micrometer	0.000001

10. **STEM** An atom is a small unit of matter. A small atom measures about 0.0000000001 meter. Write the decimal as a power of 10. (Example 5)

______

Simplify. (Examples 6 and 7)

11. $2^{-3} \cdot 2^{-4} =$ ______

12. $s^{-5} \cdot s^{-2} =$ ______

13. $y^{-1} \cdot y^{4} =$ ______

14. $(3a)(a^{-3}) =$ ______

15. $\frac{3^{-1}}{3^{-5}} =$ ______

16. $\frac{a^{-4}}{a^{-6}} =$ ______

17. $\frac{y^{-6}}{y^{-10}} =$ ______

18. $\frac{z^{-4}}{z^{-8}} =$ ______

19 **STEM** The mass of a molecule of penicillin is 10^{-18} kilogram and the mass of a molecule of insulin is 10^{-23} kilogram. How many times greater is the mass of a molecule of penicillin than the mass of a molecule of insulin?

20. **CCSS Justify Conclusions** A common flea that is 2^{-4} inch long can jump about 2^3 inches high. About how many times its body size can a flea jump? Explain your reasoning.

H.O.T. Problems Higher Order Thinking

21. **CCSS Identify Structure** Without evaluating, order 11^{-3}, 11^2, and 11^0 from least to greatest. Explain your reasoning.

22. **CCSS Identify Structure** Write an expression with a negative exponent that has a value between 0 and $\frac{1}{2}$.

23. **CCSS Persevere with Problems** Select several fractions between 0 and 1. Find the value of each fraction after it is raised to the −1 power. Explain the relationship between the −1 power and the original fraction.

Standardized Test Practice

24. Which of the following shows the expressions 6^3, 6^0, 6^{-1}, 6^{-2}, and 6^1 in order from least to greatest?

Ⓐ $6^{-2}, 6^{-1}, 6^0, 6^1, 6^3$

Ⓑ $6^{-1}, 6^0, 6^1, 6^{-2}, 6^3$

Ⓒ $6^3, 6^{-2}, 6^{-1}, 6^1, 6^0$

Ⓓ $6^3, 6^1, 6^0, 6^{-1}, 6^{-2}$

Name ______ My Homework ______

Extra Practice

25. Write 3^{-5} using positive exponents. $\frac{1}{3^5}$

$(3)^{-5} = \frac{1}{3^5}$

Homework Help

26. Simplify $(4^{-4})(4^2)$. $\frac{1}{16}$

$(4^{-4})(4^2) = 4^{-4+2}$
$= 4^{-2}$
$= \frac{1}{4^2}$ or $\frac{1}{16}$

Write each expression using a positive exponent.

27. $6^{-8} =$ ______

28. $(-3)^{-5} =$ ______

29. $s^{-9} =$ ______

30. $t^{-11} =$ ______

Simplify.

31. $z^2 \cdot z^{-3} =$ ______

32. $n^{-1} \cdot n^3 =$ ______

33. $\frac{b^{-7}}{b^5} =$ ______

34. $\frac{x^4}{x^{-2}} =$ ______

35. $2^{-4} =$ ______

36. $(-5)^{-4} =$ ______

37. $(-10)^{-4} =$ ______

38. $(0.5)^{-4} =$ ______

CCSS **Persevere with Problems** **Find the missing exponent.**

39. $\frac{17^{\bullet}}{17^4} = 17^8$ ______

40. $\frac{k^6}{k^{\bullet}} = k^2$ ______

41. $\frac{p^{-1}}{p^{\bullet}} = p^{10}$ ______

Standardized Test Practice

42. A blood cell has a diameter of about 5^{-5} inches.

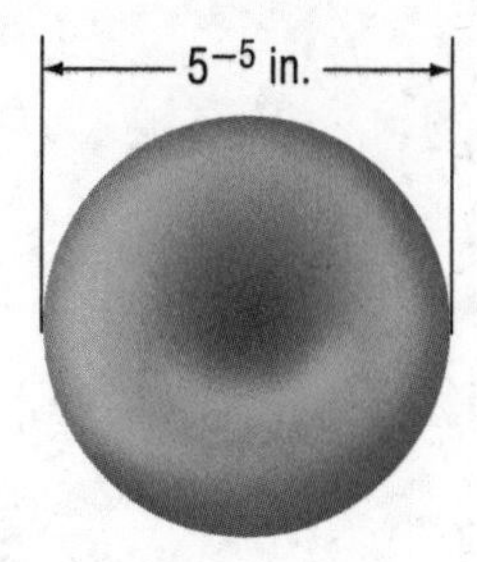

Write 5^{-5} using positive exponents.

Ⓐ 5^5

Ⓑ $\frac{1}{5^{-5}}$

Ⓒ $\frac{5^5}{1}$

Ⓓ $\frac{1}{5^5}$

43. When written without exponents, 10^{-5} is equal to which of the following?

Ⓕ 0.00001

Ⓖ 0.000001

Ⓗ −0.00001

Ⓘ −0.000001

44. Short Response Evaluate 3^{-4}. Write your answer using a positive exponent and as a fraction. ______

Common Core Review

Evaluate. 6.EE.1

45. $10^2 =$ ______

46. $10^3 =$ ______

47. $10^6 =$ ______

48. $10^5 =$ ______

Find each missing value. 6.NS.3

49. 0.003 × ______ = 3

50. 0.079 × ______ = 7.9

51. 0.00041 × ______ = 4.1

52. 987 ÷ ______ = 9.87

53. 3,400 ÷ ______ = 3.4

54. 7,450 ÷ ______ = 745

Lesson 6

Scientific Notation

What You'll Learn

Scan the lesson. List two real-world scenarios in which you would use scientific notation.

-
-

Essential Question

WHY is it helpful to write numbers in different ways?

Vocabulary

scientific notation

Common Core State Standards

Content Standards
8.EE.4

Mathematical Practices
1, 3, 4, 7

Real-World Link

Electronics A single sided, single layer DVD has a storage capacity of 4.7 gigabytes. One gigabyte is equal to 10^9 bytes.

1. Write a multiplication expression that represents how many bytes can be stored on the DVD. ______

2. Complete the table below.

Expression	Product	Expression	Product
$4.7 \times 10^1 = 4.7 \times 10$	47	$4.7 \times 10^{-1} = 4.7 \times \frac{1}{10}$	0.47
$4.7 \times 10^2 = 4.7 \times 100$		$4.7 \times 10^{-2} = 4.7 \times \frac{1}{100}$	
$4.7 \times 10^3 = 4.7 \times 1{,}000$		$4.7 \times 10^{-3} = 4.7 \times \frac{1}{1000}$	
$4.7 \times 10^4 = 4.7 \times$ ______		$4.7 \times 10^{-4} = 4.7 \times$ ______	

3. If 4.7 is multiplied by a positive power of 10, what relationship exists between the decimal point's new position and the exponent?

______

4. When 4.7 is multiplied by a negative power of 10, how does the new position of the decimal point relate to the negative exponent? ______

Key Concept: Scientific Notation

Words **Scientific notation** is when a number is written as the product of a factor and an integer power of 10. The factor must be greater than or equal to 1 and less than 10.

Symbols $a \times 10^n$, where $1 \leq a < 10$ and n is an integer

Example $425{,}000{,}000 = 4.25 \times 10^8$

Work Zone

Use these rules to express a number in scientific notation.

- If the number is greater than or equal to 1, the power of ten is positive.
- If the number is between 0 and 1, the power of ten is negative.

Powers of Ten
Multiplying a factor by a positive power of 10 moves the decimal point right. Multiplying a factor by a negative power of 10 moves the decimal point left.

Examples

Tutor

Write each number in standard form.

1. $\mathbf{5.34 \times 10^4}$

$5.34 \times 10^4 = 53{,}400.$

2. $\mathbf{3.27 \times 10^{-3}}$

$3.27 \times 10^{-3} = 0.00327$

a. ____________

b. ____________

c. ____________

Got It? **Do these problems to find out.**

a. 7.42×10^5 **b.** 6.1×10^{-2} **c.** 3.714×10^2

Examples

Write each number in scientific notation.

3. **3,725,000**

$3{,}725{,}000 = 3.725 \times 1{,}000{,}000$ The decimal point moves 6 places.

$= 3.725 \times 10^6$ Since $3{,}725{,}000 > 1$, the exponent is positive.

4. **0.000316**

$0.000316 = 3.16 \times 0.0001$ The decimal point moves 4 places.

$= 3.16 \times 10^{-4}$ Since $0 < 0.000316 < 1$, the exponent is negative.

Got It? Do these problems to find out.

d. 14,140,000 **e.** 0.00876 **f.** 0.114

d. ____________

e. ____________

f. ____________

Example

5. Refer to the table at the right. Order the countries according to the amount of money visitors spent in the United States from greatest to least.

Dollars Spent by International Visitors in the U.S

Country	Dollars Spent
Canada	1.03×10^7
India	1.83×10^6
Mexico	7.15×10^6
United Kingdom	1.06×10^7

Canada and United Kingdom; Mexico and India

Step 1 $\begin{Bmatrix} 1.06 \times 10^7 \\ 1.03 \times 10^7 \end{Bmatrix} > \begin{Bmatrix} 7.15 \times 10^6 \\ 1.83 \times 10^6 \end{Bmatrix}$ ← Group the numbers by their power of 10.

Step 2 $1.06 > 1.03$ $7.15 > 1.83$ ← Order the decimals.

United Kingdom Canada Mexico India

Got It? Do this problem to find out.

g. Some of the top U.S. cities visited by overseas travelers are shown in the table. Order the cities according to the number of visitors from least to greates

U.S. City	Number of Visitors
Boston	7.21×10^5
Las Vegas	1.3×10^6
Los Angeles	2.2×10^6
Metro D.C. area	9.01×10^5

g. ____________

Example

6. STEM If you could walk at a rate of 2 meters per second, it would take you 1.92×10^8 seconds to walk to the moon. Is it more appropriate to report this time as 1.92×10^8 seconds or 6.09 years? Explain your reasoning.

The measure 6.09 years is more appropriate. The number 1.92×10^8 seconds is very large so choosing a larger unit of measure is more meaningful.

Show your work.

h. ______

Got It? Do this problem to find out.

h. **STEM** In an ocean, the sea floor moved 475 kilometers over 65 million years. Is it more appropriate to report this rate as 7.31×10^{-5} kilometer per year or 7.31 centimeters per year? Explain your reasoning.

Guided Practice

Write each number in standard form. (Examples 1 and 2)

1. $9.931 \times 10^5 =$ ______

2. $6.02 \times 10^{-4} =$ ______

Write each number in scientific notation. (Examples 3 and 4)

3. 8,785,000,000 = ______

4. 0.524 = ______

5. The table lists the total value of music shipments for four years. List the years from least to greatest dollar amount. (Example 5)

Year	Music Shipments ($)
1	1.22×10^{10}
2	1.12×10^{10}
3	7.15×10^{6}
4	1.06×10^{7}

6. **STEM** A plant cell has a diameter of 1.3×10^{-8} kilometer. Is it more appropriate to report the diameter of a plant cell as 1.3×10^{-8} kilometer or 1.3×10^{-2} millimeter? Explain your reasoning. (Example 6)

7. **Building on the Essential Question** How is scientific notation useful in the real world?

Rate Yourself!

☐ I understand how to write numbers in scientific notation.

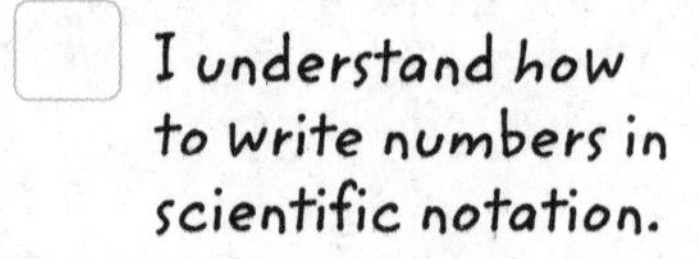

☐ I still have some questions about how to write numbers in scientific notation.

No Problem! Go online to access a Personal Tutor.

Name ________________ My Homework ________________

Independent Practice

Go online for Step-by-Step Solutions 

Write each number in standard form. (Examples 1 and 2)

1. $3.16 \times 10^3 =$ ________

2. $1.1 \times 10^{-4} =$ ________

3. $2.52 \times 10^{-5} =$ ________

Write each number in scientific notation. (Examples 3 and 4)

4. $43{,}000 =$ ________

5. $0.0072 =$ ________

6. $0.0000901 =$ ________

7 The areas of the world's oceans are listed in the table. Order the oceans according to their area from least to greatest. (Example 5)

____

____

World's Oceans

Ocean	Area (mi^2)
Atlantic	2.96×10^7
Arctic	5.43×10^6
Indian	2.65×10^7
Pacific	6×10^7
Southern	7.85×10^6

8. The space shuttle can travel about 8×10^5 centimeters per second. Is it more appropriate to report this rate as 8×10^5 centimeters per second or 8 kilometers per second? Explain. (Example 6)

____

____

____

9. The inside diameter of a certain size of ring is 1.732×10^{-2} meter. Is it more appropriate to report the ring diameter as 1.732×10^{-2} meter or 17.32 millimeters? Explain. (Example 6)

____

____

____

Fill in each ○ with <, >, or = to make a true statement.

10. $678{,}000$ ○ 6.78×10^6

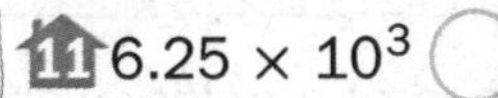

11 6.25×10^3 ○ 6.3×10^3

12. **CCSS Model with Mathematics** Refer to the graphic novel frame below for Exercises a–c.

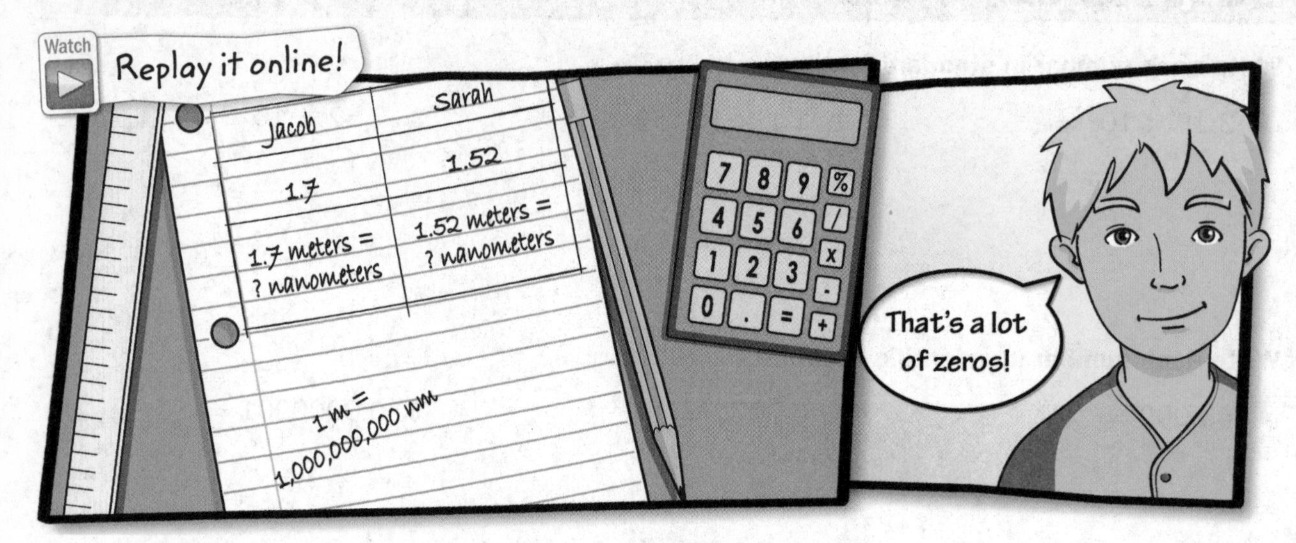

a. Find Jacob's and Sarah's heights in nanometers.

____

b. Write each height using scientific notation.

____

c. Give an example of something that would be appropriately measured by nanometers. ______________________________

____

H.O.T. Problems Higher Order Thinking

13. **CCSS Justify Conclusions** Determine whether 1.2×10^5 or 1.2×10^6 is closer to one million. Explain. ______________________________

____

14. **CCSS Persevere with Problems** Compute and express each value in scientific notation.

a. $\frac{(130{,}000)(0.0057)}{0.0004} =$ ______________________________

b. $\frac{(90{,}000)(0.0016)}{(200{,}000)(30{,}000)(0.00012)} =$ ______________________________

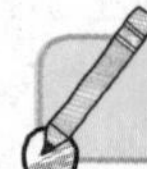

Standardized Test Practice

15. The average width of a strand of a spider web is 7×10^{-6} meter. What is this length expressed in standard notation?

Ⓐ 7,000,000 m

Ⓑ 700,000 m

Ⓒ 0.00007 m

Ⓓ 0.000007 m

Extra Practice

16. Write 7.113×10^7 in standard form.

71,130,000

$7.113 \times 10^7 = 71130000.$ The decimal point moves 7 places right.

Homework Help

17. Write 0.00000707 in scientific notation.

7.07×10^{-6}

$0.00000707 = 7.07 \times 0.000001$
$= 7.07 \times 10^{-6}$

The decimal point moves 6 places. Since $0 < 0.00000707 < 1$, the exponent is negative.

Write each number in standard form.

18. $2.08 \times 10^2 =$ ______

19. $7.8 \times 10^{-3} =$ ______

20. $8.73 \times 10^{-4} =$ ______

Write each number in scientific notation.

21. 6,700 = ______

22. 52,300,000 = ______

23. 0.037 = ______

24. STEM The table shows the mass in grams of one atom of each of several elements. List the elements in order from the least mass to greatest mass per atom.

Element	Mass per Atom
Carbon	1.995×10^{-23} g
Gold	3.272×10^{-22} g
Hydrogen	1.674×10^{-24} g
Oxygen	2.658×10^{-23} g
Silver	1.792×10^{-22} g

CCSS **Identify Structure** **Arrange each set of numbers in increasing order.**

25. 216,000,000, 2.2×10^3, 3.1×10^7, 310,000

26. 4.56×10^{-2}, 4.56×10^3, 4.56×10^2, 4.56×10^{-3}

Standardized Test Practice

27. Short Response By the year 2050, the world population is expected to reach 10 billion people. When 10 billion is written in scientific notation, what is the exponent of the power of ten?

______

28. The thermosphere layer of the atmosphere is between 90 thousand and 110 thousand meters above sea level. What is 110 thousand written in scientific notation?

Ⓐ 1.1×10^5

Ⓑ 1.1×10^4

Ⓒ 1.1×10^{-4}

Ⓓ 1.1×10^{-5}

29. The attendance records for four Major League baseball teams for a recent year are shown below.

Team	Attendance
Florida Marlins	6.76×10^5
Los Angeles Angels	1.87×10^6
Pittsburgh Pirates	9.68×10^5
St. Louis Cardinals	1.98×10^6

Which team had the greatest attendance?

Ⓕ Florida Marlins

Ⓖ Los Angeles Angels

Ⓗ Pittsburgh Pirates

Ⓘ St. Louis Cardinals

CCSS Common Core Review

Find each sum or difference. 6.NS.3

30. $9.7 + 0.532 =$ ______

31. $4.39 - 0.035 =$ ______

32. $679 - 1.4 =$ ______

Find each product or quotient. 6.NS.3

33. $(3.7)(1.2) =$ ______

34. $9.72 \div 1.8 =$ ______

35. $4.64 \div 2.9 =$ ______

Simplify. Express using exponents. 8.EE.1

36. $3a^4 \cdot 12a^2 =$ ______

37. $(5x)^2 \cdot 2x^5 =$ ______

38. $\frac{3^9}{3^2} =$ ______

Lesson 7

Compute with Scientific Notation

What You'll Learn

Scan the lesson. List two real-world scenarios in which you would compute using scientific notation.

-
-

Essential Question

WHY is it helpful to write numbers in different ways?

Common Core State Standards

Content Standards
8.EE.3, 8.EE.4

Mathematical Practices
1, 3, 4

Real-World Link

E-mail Every day, nearly 130 billion spam E-mails are sent worldwide! Use the steps below to find out how many are sent each year. The numbers are too large even for your calculator.

1. Express 130 billion in scientific notation.

2. Round 365 to the nearest hundred and express it in scientific notation.

3. Write a multiplication expression using the number in Exercises 1 and 2 to represent the total number of spam E-mails sent each year.

4. If you use the Commutative Property of Multiplication, you can rewrite the expression in Exercise 3 as $(1.3 \times 4)(10^{11} \times 10^{2})$. Evaluate this expression to find the number of spam E-mails sent in a year. Express the result in both scientific notation and standard form.

Work Zone

Multiplication and Division with Scientific Notation

You can use the Product of Powers and Quotient of Powers properties to multiply and divide numbers written in scientific notation.

Example

Decimal Point
Since, 11.52×10^7 is not written in scientific notation, move the decimal point 1 place to the left and add 1 to the exponent.

1. Evaluate $(7.2 \times 10^3)(1.6 \times 10^4)$. Express the result in scientific notation.

$$
\begin{aligned}
(7.2 \times 10^3)(1.6 \times 10^4) &= (7.2 \times 1.6)(10^3 \times 10^4) && \text{Commutative and Associative Properties} \\
&= (11.52)(10^3 \times 10^4) && \text{Multiply 7.2 by 1.6.} \\
&= 11.52 \times 10^{3+4} && \text{Product of Powers} \\
&= 11.52 \times 10^7 && \text{Add the exponents.} \\
&= 1.152 \times 10^8 && \text{Write in scientific notation.}
\end{aligned}
$$

a. ____________

b. ____________

Got It? Do these problems to find out.

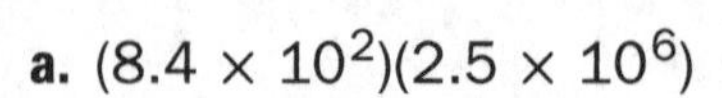

a. $(8.4 \times 10^2)(2.5 \times 10^6)$

b. $(2.63 \times 10^4)(1.2 \times 10^{-3})$

Example

2. In 2010, the world population was about 6,860,000,000. The population of the United States was about 3×10^8. About how many times larger is the world population than the population of the United States?

Estimate the population of the world and write in scientific notation.

$6{,}860{,}000{,}000 \approx 7{,}000{,}000{,}000$ or 7×10^9

Find $\frac{7 \times 10^9}{3 \times 10^8}$.

$$
\begin{aligned}
\frac{7 \times 10^9}{3 \times 10^8} &= \left(\frac{7}{3}\right)\left(\frac{10^9}{10^8}\right) && \text{Associative Property} \\
&\approx 2.3 \times \left(\frac{10^9}{10^8}\right) && \text{Divide 7 by 3. Round to the nearest tenth.} \\
&\approx 2.3 \times 10^{9-8} && \text{Quotient of Powers} \\
&\approx 2.3 \times 10^1 && \text{Subtract the exponents.}
\end{aligned}
$$

So, the population of the world is about 23 times larger than the population of the United States.

Got It? Do this problem to find out.

c. The surface area of Lake Superior, the largest of the Great Lakes, is 8×10^4 square kilometers. The surface area of the smallest Great Lake, Ontario, is 18,160 square kilometers. About how many times as great is the area covered by Lake Superior than Lake Ontario?

c. ____________

Addition and Subtraction with Scientific Notation

When adding or subtracting decimals in standard form, it is necessary to line up the place values. In scientific notation, the place value is represented by the exponent. Before adding or subtracting, both numbers must be expressed in the same form.

Examples

Evaluate each expression. Express the result in scientific notation.

3. $(6.89 \times 10^4) + (9.24 \times 10^5)$

$(6.89 \times 10^4) + (9.24 \times 10^5)$

$= (6.89 \times 10^4) + (92.4 \times 10^4)$ — Write 9.24×10^5 as 92.4×10^4.

$= (6.89 + 92.4) \times 10^4$ — Distributive Property

$= 99.29 \times 10^4$ — Add 6.89 and 92.4.

$= 9.929 \times 10^5$ — Rewrite in scientific notation.

4. $(7.83 \times 10^8) - 11{,}610{,}000$

$(7.83 \times 10^8) - (1.161 \times 10^7)$ — Rewrite 11,610,000 in scientific notation.

$(7.83 \times 10^8) - (1.161 \times 10^7)$

$= (78.3 \times 10^7) - (1.161 \times 10^7)$ — Write 7.83×10^8 as 78.3×10^7.

$= (78.3 - 1.161) \times 10^7$ — Distributive Property

$= 77.139 \times 10^7$ — Subtract 1.161 from 78.3.

$= 7.7139 \times 10^8$ — Rewrite in scientific notation.

and Reflect

Explain below how to estimate the sum of (4.215×10^{-2}) and (3.2×10^{-4}). Then find the estimate.

d. ______

e. ______

f. ______

5. $593{,}000 + (7.89 \times 10^6)$

$593{,}000 + (7.89 \times 10^6)$
$= (5.93 \times 10^5) + (7.89 \times 10^6)$ — Rewrite 593,000 in scientific notation.
$= (0.593 \times 10^6) + (7.89 \times 10^6)$ — Write 5.93×10^5 as 0.593×10^6
$= (0.593 + 7.89) \times 10^6$ — Distributive Property
$= 8.483 \times 10^6$ — Add 0.593 and 7.89.

Got It? Do these problems to find out.

d. $(8.41 \times 10^3) + (9.71 \times 10^4)$

e. $(1.263 \times 10^9) - (1.525 \times 10^7)$

f. $(6.3 \times 10^5) + 2{,}700{,}000$

Guided Practice

Evaluate each expression. Express the result in scientific notation. (Examples 1 and 2)

1. $(2.6 \times 10^5)(1.9 \times 10^2) =$ ______

2. $\dfrac{8.37 \times 10^8}{2.7 \times 10^3} =$ ______

Show your work.

3. In 2005, 8.1×10^{10} text messages were sent in the United States. In 2010, the number of annual text messages had risen to 1,810,000,000,000. About how many times as great was the number of text messages in 2010 than 2005? (Example 2)

______

Evaluate each expression. Express the result in scientific notation. (Examples 3–5)

4. $(8.9 \times 10^9) + (4.2 \times 10^6) =$ ______

5. $(9.64 \times 10^8) - (5.29 \times 10^6) =$ ______

6. $(1.35 \times 10^6) - (117{,}000) =$ ______

7. $5{,}400 + (6.8 \times 10^5) =$ ______

8. **Building on the Essential Question** How does scientific notation make it easier to perform computations with very large or very small numbers? ______

______

______

Rate Yourself!

Are you ready to move on? Shade the section that applies.

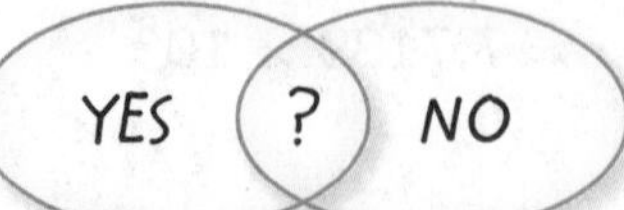

For more help, go online to access a Personal Tutor.

Name ________________ My Homework ________________

Independent Practice

Go online for Step-by-Step Solutions

Evaluate each expression. Express the result in scientific notation. (Examples 1 and 2)

1. $(3.9 \times 10^2)(2.3 \times 10^6) =$ ________

2. $(4.18 \times 10^{-4})(9 \times 10^{-4}) =$ ________

3. $(9.75 \times 10^3)(8.4 \times 10^{-6}) =$ ________

4. $\dfrac{9.45 \times 10^{10}}{1.5 \times 10^6} =$ ________

5. $\dfrac{1.14 \times 10^6}{4.8 \times 10^{-6}} =$ ________

6. $\dfrac{9 \times 10^{-11}}{2.4 \times 10^8} =$ ________

7. **STEM** Neurons are cells in the nervous system that process and transmit information. An average neuron is about 5×10^{-6} meter in diameter. A standard table tennis ball is 0.04 meter in diameter. About how many times as great is the diameter of a ball than a neuron? (Example 2)

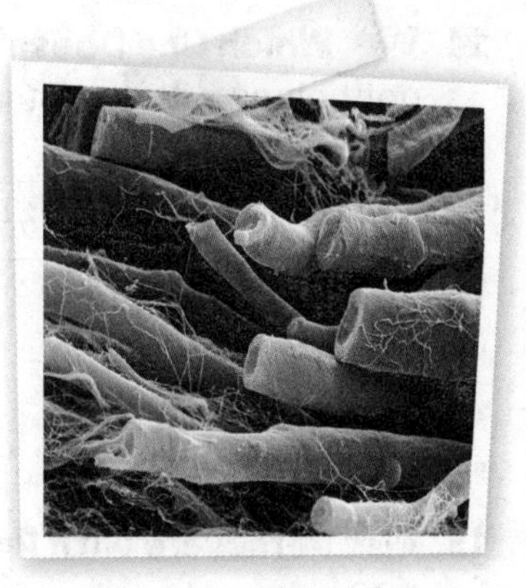

Evaluate each expression. Express the result in scientific notation. (Examples 3–5)

8. $(9.5 \times 10^{11}) + (6.3 \times 10^9) =$ ________

9. $(1.03 \times 10^9) - (4.7 \times 10^7) =$ ________

10. $(1.357 \times 10^9) + 590{,}000 =$ ________

11. $87{,}100 - (6.34 \times 10^1) =$ ________

12. CCSS **Persevere with Problems** Central Park in New York City is rectangular in shape and measures approximately 1.37×10^4 feet by 2.64×10^2 feet. If one acre is equal to 4.356×10^4 square feet, how many acres does Central Park cover? Round to the nearest hundredth.

H.O.T. Problems Higher Order Thinking

13. CCSS **Find the Error** Enrique is finding $\frac{6.63 \times 10^{-6}}{5.1 \times 10^{-2}}$. Circle his mistake and correct it.

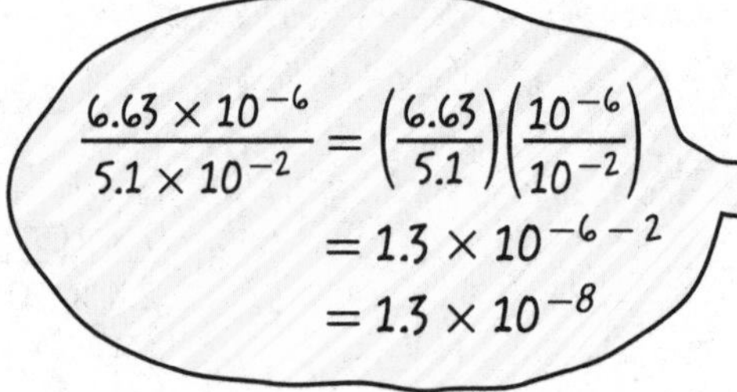

14. CCSS **Which One Doesn't Belong?** Identify the expression that does not belong with the other three. Explain your reasoning.

14.28×10^9	$(3.4 \times 10^6)(4.2 \times 10^3)$	1.4×10^9	$(3.4)(4.2) \times 10^{(6+3)}$

15. CCSS **Persevere with Problems** A *googol* is the number 1 followed by 100 zeros. How many times greater is a googol of meters than a nanometer?

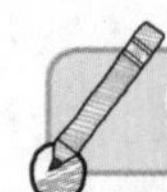

Standardized Test Practice

16. A music download Web site announced that over 4×10^9 songs were downloaded by 5×10^7 registered users. What is the average number of downloads per user?

Ⓐ 8×10^{-1}

Ⓑ 1.25×10^{-2}

Ⓒ 1.25×10^2

Ⓓ 8×10^1

Name ______ My Homework ______

Extra Practice

Evaluate each expression. Express the result in scientific notation.

17. $(3.7 \times 10^{-2})(1.2 \times 10^{3}) =$ 4.44×10^{1}

Homework Help

$$(3.7 \times 10^{-2})(1.2 \times 10^{3}) = (3.7 \times 1.2) \times (10^{-2} \times 10^{3})$$
$$= 4.44 \times 10^{-2+3}$$
$$= 4.44 \times 10^{1}$$

18. $\dfrac{4.64 \times 10^{-4}}{2.9 \times 10^{-6}} =$ 1.6×10^{2}

$$\frac{4.64 \times 10^{-4}}{2.9 \times 10^{-6}} = \frac{4.64}{2.9} \times \frac{10^{-4}}{10^{-6}}$$
$$= 1.6 \times 10^{-4-(-6)}$$
$$= 1.6 \times 10^{2}$$

19. $\dfrac{3.24 \times 10^{-4}}{8.1 \times 10^{-7}} =$ ______

20. $(7.3 \times 10^{5}) + 2{,}400{,}000 =$ ______

21. $(8.64 \times 10^{6}) + (1.334 \times 10^{10}) =$ ______

22. $(1.21 \times 10^{5}) - 9{,}500 =$ ______

23. CCSS **Persevere with Problems** A circular swimming pool holds 1.22×10^{6} cubic inches of water. It is being filled at a rate of 1.5×10^{3} cubic inches per minute. How many hours will it take to fill the swimming pool? ______

24. **Financial Literacy** In 2010, the national debt of the United States was about 14 trillion dollars. In 2003 it was about 7×10^{12} dollars. About how many times larger was the national debt in 2010 than in 2003? ______

Standardized Test Practice

25. The rectangle has an area of 9.14×10^{-7} square kilometers.

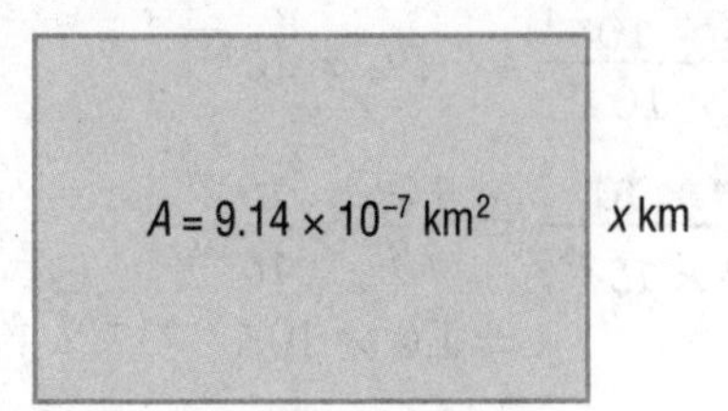

What is the approximate length of the missing side?

Ⓐ 2.74×10^{-6}
Ⓑ 5.52×10^{-4}
Ⓒ 1.656×10^{-3}
Ⓓ 1.51×10^{11}

26. There are approximately 45 hundred species of mammals on Earth and 2.8×10^4 species of fish. What is the difference in the number of species?

Ⓕ 6.2×10^0
Ⓖ 2.35×10^4
Ⓗ 1.6×10^{-1}
Ⓘ 3.25×10^4

27. Alaska is the largest state in the United States with an area of about 1.5×10^6 square kilometers. Rhode Island is the smallest state with an area of about 2,700 square kilometers. About how many times larger is Alaska than Rhode Island?

Ⓐ 5
Ⓑ 50
Ⓒ 500
Ⓓ 5,000

CCSS Common Core Review

28. A cube measures 6.6 inches on each side. **6.G.2**

a. Find the area of one face of the cube. ______________

b. Find the volume of the cube. ______________

29. Complete the table shown. **6.EE.1**

x	x^2	x^3	x	x^2	x^3
1			7		
2			8		
3			9		
4			10		
5			11		
6			12		

Inquiry Lab

Graphing Technology: Scientific Notation Using Technology

WHAT are the similarities and differences between a number written in scientific notation and the calculator notation of the number shown on a screen?

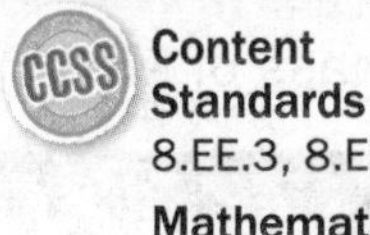

Content Standards
8.EE.3, 8.EE.4

Mathematical Practices
1, 3, 5

Solar System The table shows the mass of some planets in our solar system. What is the mass of Earth written in scientific notation?

Planet	Mass (kg)
Earth	5,973,700,000,000,000,000,000,000
Mars	641,850,000,000,000,000,000,000
Saturn	568,510,000,000,000,000,000,000,000

What do you know? ______

What do you need to find? ______

Investigation 1

You will use a graphing calculator to explore how scientific notation is displayed using technology.

Step 1 Press CLEAR to clear the home screen.

Step 2 Enter the value in standard form for Earth's mass. Press ENTER.

Copy your calculator screen on the blank screen shown.

Step 3 Write the value for Earth's mass using scientific notation.

______

Collaborate

Use Math Tools **Work with a partner. Repeat Steps 1 and 2 for each of the following.**

1. mass of Mars

Show your work.

2. mass of Saturn

3. What does the E symbol represent on the calculator screen?

What does the value after the E symbol represent?

4. Based on your answer for Exercise 1, what is the mass of Mars in scientific notation?

5. Based on your answer for Exercise 2, what is the mass of Saturn in scientific notation?

Analyze

Work with a partner to complete the table.

	Calculator Notation	Scientific Notation	Standard Form
6.	3.1E7		
7.		6.39×10^{10}	
8.			0.02357
9.	1.7E−11		

10. **Reason Inductively** The Moon has a mass of about 73,600,000,000,000,000,000,000 kilograms. Without entering the value in your calculator, predict how the mass of the Moon will be displayed on the calculator screen.

Investigation 2

Measurement A human blood cell is about 1×10^{-6} meter in diameter. The Moon is about 3.476×10^{6} meters in diameter. How many times greater is the diameter of the Moon than the diameter of a blood cell?

Step 1 Press CLEAR to clear the home screen.

Step 2 Perform the following keystrokes:

3.476 2nd [EE] **6** ÷ **1** 2nd [EE] **–6** ENTER

Copy your calculator screen on the blank screen shown.

Step 3 Write the value in standard form.

____

So, the Moon is ______________ times greater than a human blood cell.

Investigation 3

When in "Normal" mode, a calculator will show answers in scientific notation only if they are very large numbers or very small numbers. You can set your calculator to show scientific notation for all numbers by using the "Sci" mode.

Step 1 Press CLEAR to clear the home screen. Put your calculator in scientific mode by pressing MODE ▶ ENTER. Then press CLEAR to return to the home screen.

Step 2 Complete the table by entering the numbers in the first column into your calculator.

Enter	Calculator Notation	Standard Form
14 ÷ 100		
60 – 950		
360 • 15		
1 + 1		

CCSS Use Math Tools Work with a partner. Write down the keystrokes and fill in the calculator screen to find each of the following using a calculator in "Sci" mode. Write your final answer in standard form.

11. $(6.2 \times 10^5)(2.3 \times 10^7)$

Keystrokes: ______________________

____

Answer in standard form: ______________________

12. $(8.5 \times 10^{-3}) - (4.8 \times 10^{-5})$

Keystrokes: ______________________

____

Answer in standard form: ______________________

13. **CCSS Use Math Tools** A *micrometer* is 0.000001 meter. Use your calculator to determine how many micrometers are in each of the following. Write your answer in both calculator and scientific notation.

	Calculator Notation	Scientific Notation
5,000 meters		
4.08E14 meters		
2.9E–10 meter		

14. Inquiry WHAT are the similarities and differences between a number written in scientific notation and the calculator notation of the number shown on a screen? ______________________

____

____

Lesson 8
Roots

What You'll Learn

Scan the lesson. Predict two things you will learn about square roots and cube roots.

-
-

Essential Question

WHY is it helpful to write numbers in different ways?

Vocabulary

square root
perfect square
radical sign
cube root
perfect cube

Common Core State Standards

Content Standards
8.EE.2

Mathematical Practices
1, 3, 4

Vocabulary Start-Up

A **square root** of a number is one of its two equal factors. Numbers such as 1, 4, 9, 16, and 25 are called **perfect squares** because they are squares of integers.

Complete the graphic organizer.

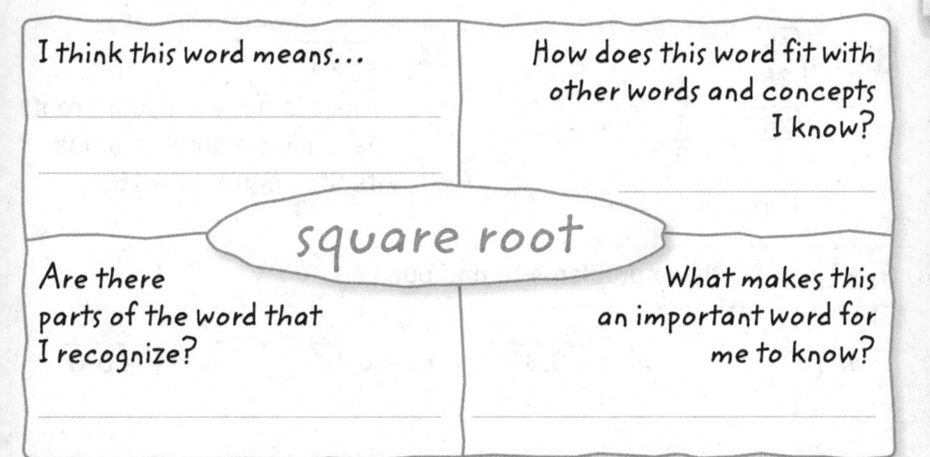

What is the relationship between squaring a number and finding the square root? ______

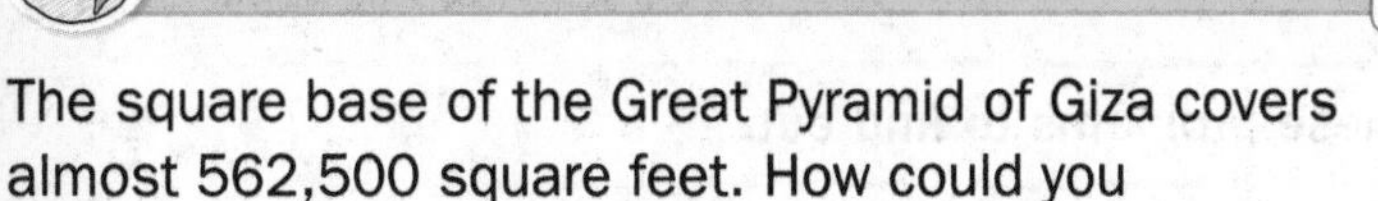

The square base of the Great Pyramid of Giza covers almost 562,500 square feet. How could you determine the length of each side of the base?

Key Concept Square Root

Words A square root of a number is one of its two equal factors.

Symbols If $x^2 = y$, then x is a square root of y.

Example $5^2 = 25$ so 5 is a square root of 25.

Work Zone

Every positive number has *both* a positive and negative square root. In most real-world situations, only the positive or *principal* square root is considered. A **radical sign**, $\sqrt{}$, is used to indicate the principal square root. If $n^2 = a$, then $n = \pm\sqrt{a}$.

Examples

Find each square root.

1. $\sqrt{64}$

$\sqrt{64} = 8$ Find the positive square root of 64; $8^2 = 64$.

2. $\pm\sqrt{1.21}$

$\pm\sqrt{1.21} = \pm 1.1$ Find both square roots of 1.21; $1.1^2 = 1.21$.

3. $-\sqrt{\frac{25}{36}}$

$-\sqrt{\frac{25}{36}} = -\frac{5}{6}$ Find the negative square root of $\frac{25}{36}$; $\left(\frac{5}{6}\right)^2 = \frac{25}{36}$.

4. $\sqrt{-16}$

There is no real square root because no number times itself is equal to −16.

Show your work.

Got It? Do these problems to find out.

a. $\sqrt{\frac{9}{16}}$ **b.** $\pm\sqrt{0.81}$ **c.** $-\sqrt{49}$ **d.** $\sqrt{-100}$

a. ________

b. ________

c. ________

d. ________

Example

5. Solve $t^2 = 169$. Check your solution(s).

$t^2 = 169$ Write the equation.

$t = \pm\sqrt{169}$ Definition of square root

$t = 13$ and -13 Check $13 \cdot 13 = 169$ and $(-13)(-13) = 169$ ✓

Got It? Do these problems to find out.

e. $289 = a^2$ **f.** $m^2 = 0.09$ **g.** $y^2 = \frac{4}{25}$

e. ________

f. ________

g. ________

Cube Roots

Key Concept

Words A **cube root** of a number is one of its three equal factors.

Symbols If $x^3 = y$, then x is the cube root of y.

Numbers such as 8, 27, and 64 are **perfect cubes** because they are the cubes of integers.

$8 = 2 \cdot 2 \cdot 2$ or 2^3 $\quad 27 = 3 \cdot 3 \cdot 3$ or 3^3 $\quad 64 = 4 \cdot 4 \cdot 4$ or 4^3

The symbol $\sqrt[3]{}$ is used to indicate a cube root of a number.

If $n^3 = a$, then $n = \sqrt[3]{a}$. You can use this relationship to solve equations that involve cubes.

Examples

Find each cube root.

6. $\sqrt[3]{125}$

$\sqrt[3]{125} = 5$ $\quad 5^3 = 5 \cdot 5 \cdot 5$ or 125

7. $\sqrt[3]{-27}$

$\sqrt[3]{-27} = -3$ $\quad (-3)^3 = (-3) \cdot (-3) \cdot (-3)$ or -27

Cube Roots
While $\sqrt{-16}$ is not a real number, $\sqrt[3]{-27}$ is a real number. $-3 \cdot -3 \cdot -3 = -27$

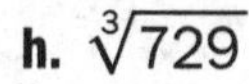

Do these problems to find out.

h. $\sqrt[3]{729}$ **i.** $\sqrt[3]{-64}$ **j.** $\sqrt[3]{1{,}000}$

h. ________

i. ________

j. ________

Example

8. Dylan has a planter in the shape of a cube that holds 8 cubic feet of potting soil. Solve the equation $8 = s^3$ to find the side length s of the container.

$8 = s^3$ Write the equation.

$\sqrt[3]{8} = s$ Take the cube root of each side.

$2 = s$ Definition of cube root

So, each side of the container is 2 feet.

Check $(2)^3 = 8$ ✓

k. ________

Got It? Do this problem to find out.

k. An aquarium in the shape of a cube that will hold 25 gallons of water has a volume of 3.375 cubic feet. Solve $s^3 = 3.375$ to find the length of one side of the aquarium.

Guided Practice

Find each square root. (Examples 1–4)

Show your work.

1. $-\sqrt{1.69} =$ ________

2. $\pm\sqrt{\frac{49}{144}} =$ ________

3. $\sqrt{-1.44} =$ ________

Solve each equation. Check your solution(s). (Example 5)

4. $p^2 = 36$ ________

5. $t^2 = \frac{1}{9}$ ________

6. $6.25 = r^2$ ________

Find each cube root. (Examples 6 and 7)

7. $\sqrt[3]{216} =$ ________

8. $\sqrt[3]{-125} =$ ________

9. $\sqrt[3]{-8} =$ ________

10. A cube-shaped packing box can hold 729 cubic inches of packing material. Solve $729 = s^3$ to find the length of one side of the box. (Example 8) ________

11. **Building on the Essential Question** Why would I need to use square roots and cube roots?

____

____

____

____

Rate Yourself!

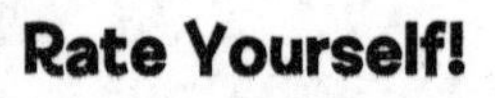

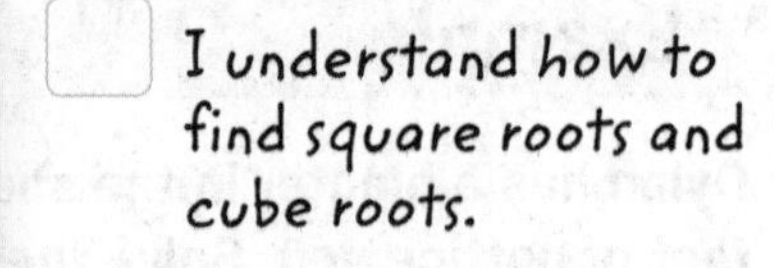

Great! You're ready to move on!

I still have some questions about finding square roots and cube roots.

No Problem! Go online to access a Personal Tutor.

Name ______________________ My Homework ______________________

Independent Practice

Go online for Step-by-Step Solutions

Find each square root. (Examples 1–4)

1. $\sqrt{16} =$ ______

2. $-\sqrt{484} =$ ______

3. $\sqrt{-36} =$ ______

4. $\pm\sqrt{\frac{9}{49}} =$ ______

5. $-\sqrt{2.56} =$ ______

6. $\sqrt{-0.25} =$ ______

Solve each equation. Check your solution(s). (Example 5)

7. $v^2 = 81$ ______

8. $w^2 = \frac{36}{100}$ ______

9. $0.0169 = c^2$ ______

Find each cube root. (Examples 6 and 7)

10. $\sqrt[3]{1{,}728} =$ ______

11. $\sqrt[3]{-0.125} =$ ______

12. $\sqrt[3]{\frac{27}{125}} =$ ______

13. A group of 169 students needs to be seated in a square formation for a yearbook photo. Solve the equation $169 = s^2$ to find how many students should be in each row. (Example 8) ______

14. Chloe wants to build a storage container in the shape of a cube to hold 15.625 cubic meters of hay for her horse. Solve the equation $15.625 = s^3$ to find the length of one side of the container. (Example 8)

______

Persevere with Problems **Given the area of each square, find the perimeter.**

15. Area = 121 square inches ______

16. Area = 25 square feet ______

17. Area = 36 square meters ______

H.O.T. Problems Higher Order Thinking

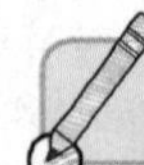

Persevere with Problems **Find each value.**

18. $(\sqrt{36})^2 =$ ______

19. $\left(\sqrt{\frac{25}{81}}\right)^2 =$ ______

20. $(\sqrt{199})^2 =$ ______

21. $(\sqrt{x})^2 =$ ______

22. **Construct an Argument** Explain why the square root of 64 has a positive and a negative value. ______

Standardized Test Practice

23. The area of each square is 16 square units. Find the perimeter of the figure.

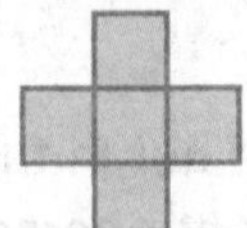

Ⓐ 16 units

Ⓑ 32 units

Ⓒ 40 units

Ⓓ 48 units

Name ____________ My Homework ____________

Extra Practice

Find each square root.

24. $-\sqrt{81} =$ -9

9 • 9 = 81

So, $-\sqrt{81} = -9$.

25. $-\sqrt{\frac{64}{225}} =$ ______

26. $-\sqrt{\frac{16}{25}} =$ ______

27. $\pm\sqrt{1.44} =$ ______

Find each cube root.

28. $\sqrt[3]{-216} =$ ______

29. $\sqrt[3]{-512} =$ ______

30. $\sqrt[3]{-1{,}000} =$ ______

31. $\sqrt[3]{-343} =$ ______

Solve each equation. Check your solution(s).

32. $b^2 = 100$

33. $\frac{9}{64} = c^2$

34. $a^2 = 1.21$

35. $\frac{1}{8} = z^3$

36. $1.331 = c^3$

37. $m^3 = 8{,}000$

38. $\sqrt{x} = 5$

39. $\sqrt{y} = 20$

40. $\sqrt{z} = 10.5$

41. CCSS **Persevere with Problems** A concert crew needs to set up some chairs on the floor level. The chairs are to be placed in a square pattern consisting of four square sections. If one of the square sections holds 900 chairs, how many chairs will there be along each length of the larger square? ______

Standardized Test Practice

42. A marching band wants to form a square in the middle of the field. If there are 100 members in the band, how many should be in each row?

Ⓐ 4
Ⓑ 10
Ⓒ 25
Ⓓ 50

43. Mr. Freeman's farm has a square cornfield. Find the area of the cornfield if the sides are measured in whole numbers.

Ⓕ 164,000 ft^2
Ⓖ 170,150 ft^2
Ⓗ 170,586 ft^2
Ⓘ 174,724 ft^2

44. Short Response A puzzle cube is shown. The volume of the cube is 512 cubic centimeters. What is the length of one side of the puzzle cube?

___

Common Core Review

Evaluate each expression. 8.EE.2

45. $13^3 =$ ___

46. $25^2 =$ ___

47. $15^3 =$ ___

48. $34^2 =$ ___

49. $5 \cdot \sqrt{121} =$ ___

50. $-6 \cdot \sqrt{36} =$ ___

51. $10 \cdot \sqrt[3]{8} =$ ___

52. $-4 \cdot \sqrt{144} =$ ___

Express the volume of each cube as a monomial. 8.EE.1

53. ___

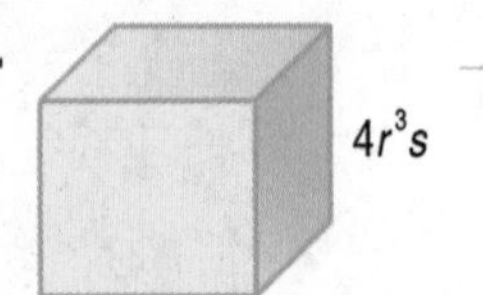

54. ___

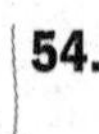

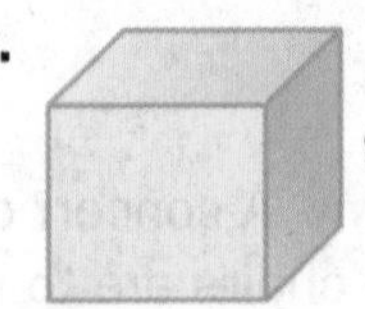

Inquiry Lab

Roots of Non-Perfect Squares

HOW can you estimate the square root of a non-perfect square number?

Content Standards
8.NS.2, 8.EE.2

Mathematical Practices
1, 3, 4, 5

Crafts Mindi is making a quilting piece from a square pattern as shown. Each of the dotted lines is 6 inches. What is the approximate length of one side of the square?

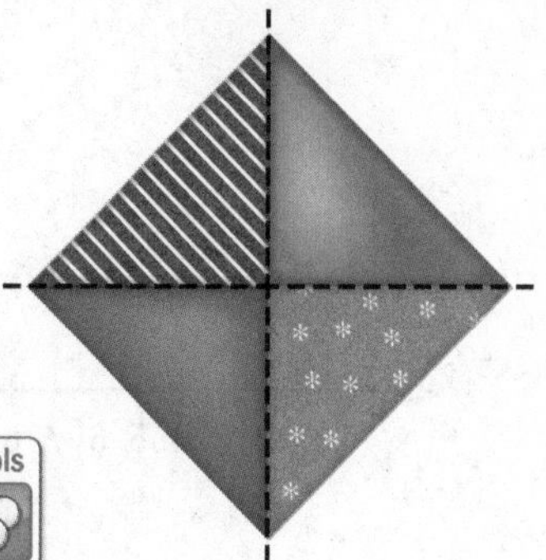

What do you know? ______________________________

What do you need to find? ______________________________

Investigation

Step 1 The outline of the square on dot paper is shown. Draw dotted lines connecting opposite vertices.

When you draw the lines, four triangles that are the same shape and size are formed. What are the dimensions of the triangles?

base = ☐ units   height = ☐ units

The area of one triangle is ☐ square units.

The area of the square is ☐ square units.

Step 2 Copy and cut out the square in Step 1 on another sheet of paper.

Step 3 Place one side of your square on the number line. Between what two consecutive whole numbers is $\sqrt{18}$, the side length of the square,

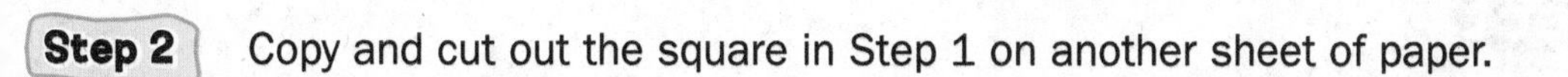

located? ______________________________

The side of the square is closer to which one of the two whole numbers? ________ Estimate $\sqrt{18}$. ________

So, one side of the square is about ☐ units long.

Use Math Tools **Work with a partner. Determine the two consecutive whole numbers the side length of each square is located between using the method shown in the Investigation.**

1. ________ **2.** ________ **3.** ________

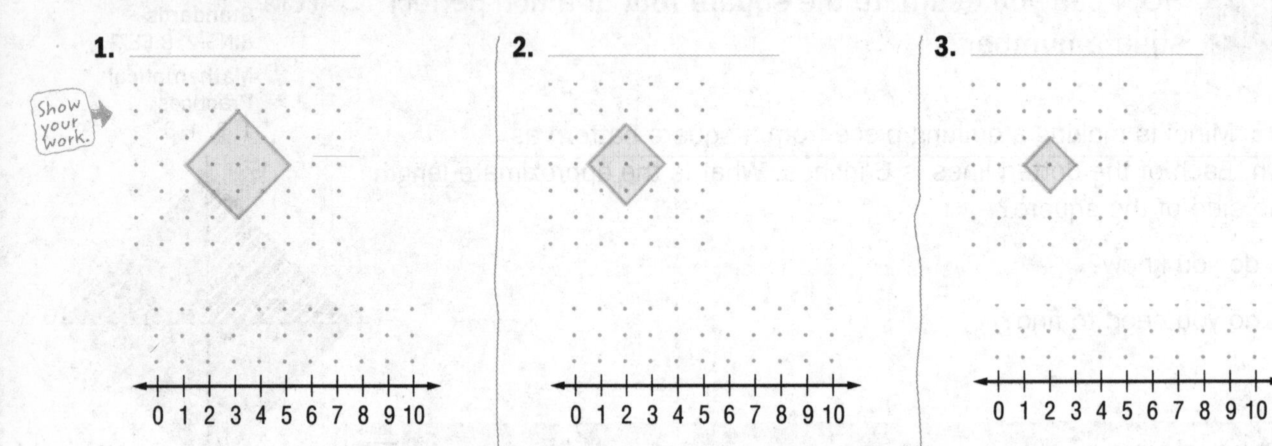

Use Math Tools **Estimate the side length of each square in Exercises 1–3. Verify your estimate by using a calculator.**

4. Exercise 1

Estimate ________

Calculator ________

5. Exercise 2

Estimate ________

Calculator ________

6. Exercise 3

Estimate ________

Calculator ________

7. **Reason Inductively** How does the area of a square relate to the square of a number? ________

8. **Model with Mathematics** How does using a square model help you find the square root of a non-perfect square? ________

9. **Inquiry** HOW can you estimate the square root of a non-perfect square?

Lesson 9

Estimate Roots

What You'll Learn

Scan the lesson. List two headings you would use to make an outline of the lesson.

-
-

Essential Question

WHY is it helpful to write numbers in different ways?

Common Core State Standards

Content Standards
8.NS.2, 8.EE.2

Mathematical Practices
1, 3, 4

Real-World Link

Gravity Legend states that while sitting in his garden one day, Sir Isaac Newton was struck on the head by an apple. Suppose the apple was 25 feet above his head. Use the following steps to find how long it took the apple to fall.

1. What is the square root of 25?

2. The formula $t = \frac{\sqrt{h}}{4}$ can be used to find the time t in seconds it will take an object to fall from a certain height h in feet. How long did it take the apple to fall?

3. Suppose another apple was 13 feet above the ground. Use the formula to write an equation representing the time it would have taken for the apple to hit the ground.

4. Can you write $\frac{\sqrt{13}}{4}$ without a radical sign? Explain.

Work Zone

Estimate Square and Cube Roots

You know that $\sqrt{8}$ is not a whole number because 8 is not a perfect square.

The number line below shows that $\sqrt{8}$ is between 2 and 3. Since 8 is closer to 9 than 4, the best whole number estimate for $\sqrt{8}$ is 3.

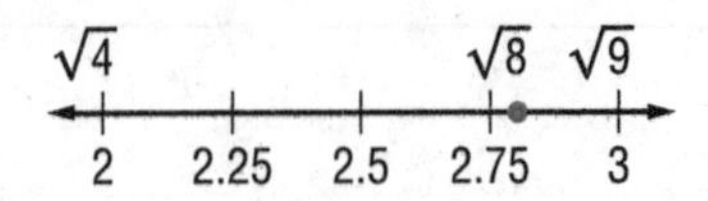

Examples

1. Estimate $\sqrt{83}$ to the nearest integer.

- The largest perfect square less than 83 is 81. $\sqrt{81} = 9$
- The smallest perfect square greater than 83 is 100. $\sqrt{100} = 10$

Plot each square root on a number line. Then estimate $\sqrt{83}$.

$\sqrt{81}$ $\sqrt{83}$ $\sqrt{100}$

9 9.25 9.5 9.75 10

$81 < 83 < 100$ Write an inequality.

$9^2 < 83 < 10^2$ $81 = 9^2$ and $100 = 10^2$

$\sqrt{9^2} < \sqrt{83} < \sqrt{10^2}$ Find the square root of each number.

$9 < \sqrt{83} < 10$ Simplify.

So, $\sqrt{83}$ is between 9 and 10. Since $\sqrt{83}$ is closer to $\sqrt{81}$ than $\sqrt{100}$, the best integer estimate for $\sqrt{83}$ is 9.

Inequalities
$81 < 83 < 100$ is read 81 is less than 83 which is less than 100, or 83 is between 81 and 100.

2. Estimate $\sqrt[3]{320}$ to the nearest integer.

- The largest perfect cube less than 320 is 216. $\sqrt[3]{216} = 6$
- The smallest perfect cube greater than 320 is 343. $\sqrt[3]{343} = 7$

$216 < 320 < 343$ Write an inequality.

$6^3 < 320 < 7^3$ $216 = 6^3$ and $343 = 7^3$

$\sqrt[3]{6^3} < \sqrt[3]{320} < \sqrt[3]{7^3}$ Find the cube root of each number.

$6 < \sqrt[3]{320} < 7$ Simplify.

So, $\sqrt[3]{320}$ is between 6 and 7. Since 320 is closer to 343 than 216, the best integer estimate for $\sqrt[3]{320}$ is 7.

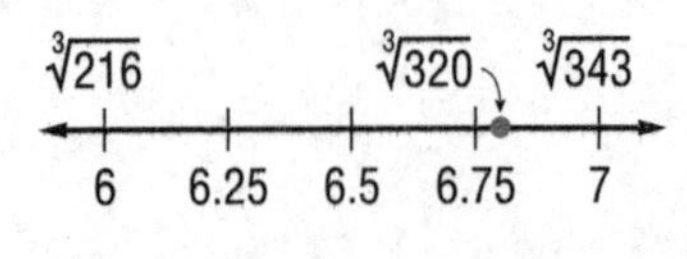

Got It? Do these problems to find out.

a. $\sqrt{35}$ **b.** $\sqrt{170}$ **c.** $\sqrt{44.8}$

d. $\sqrt[3]{62}$ **e.** $\sqrt[3]{25}$ **f.** $\sqrt[3]{129.6}$

a. ________

b. ________

c. ________

d. ________

e. ________

f. ________

Example

3. Wyatt wants to fence in a square portion of the yard to make a play area for his new puppy. The area covered is 2 square meters. How much fencing should Wyatt buy?

$\sqrt{2}$ m

2 m² $\sqrt{2}$ m

Wyatt will need $4 \cdot \sqrt{2}$ meters of fencing. The square root of 2 is between 1 and 2 so $4 \cdot \sqrt{2}$ is between 4 and 8. Is this the best approximation? You can truncate the decimal expansion of $\sqrt{2}$ to find better approximations.

Estimate $\sqrt{2}$ by truncating, or dropping, the digits after the first decimal place, then after the second decimal place, and so on until an appropriate approximation is reached.

$\sqrt{2} \approx 1.414213562$ Use a calculator.

$\sqrt{2} \approx 1.4\sout{14213562}$ Truncate, or drop, the digits after the first decimal place. $\sqrt{2}$ is between 1.4 and 1.5.

$5.6 < 4\sqrt{2} < 6.0$ $4 \cdot 1.4 = 5.6$ and $4 \cdot 1.5 = 6.0$

To find a closer approximation, expand $\sqrt{2}$ then truncate the decimal expansion after the first two decimal places.

$\sqrt{2} \approx 1.41\sout{4213562}$ $\sqrt{2}$ is between 1.41 and 1.42.

$5.64 < 4\sqrt{2} < 5.68$ $4 \cdot 1.41 = 5.64$ and $4 \cdot 1.42 = 5.68$

The approximations indicate that Wyatt should buy 6 meters of fencing.

STOP and Reflect

What is the difference between an exact value and an approximate value when finding square roots of numbers that are not perfect squares? Explain below.

Got It? Do this problem to find out.

g. Kelly needs to put trim around a circular tablecloth with a diameter of 36 inches. Use the equation $C = \pi d$ to find three sets of approximations for the amount of trim she will need. Truncate the value of π to the ones, tenths, and hundredths place. Then determine how much trim she should buy.

g. 

Example

4. The golden rectangle is found frequently in the nautilus shell. The length of the longer side divided by the length of the shorter side is equal to $\frac{1 + \sqrt{5}}{2}$. Estimate this value.

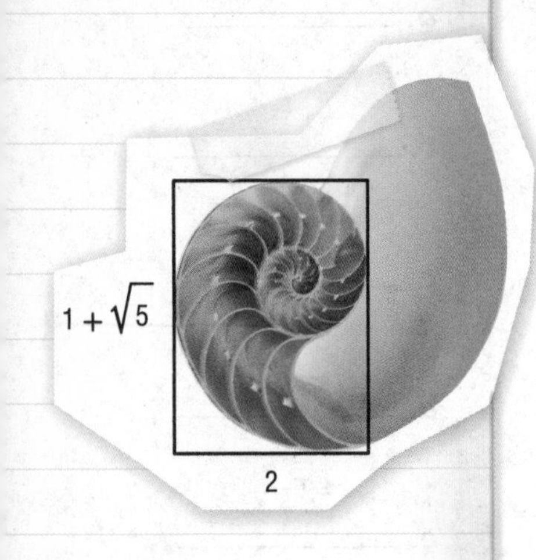

First estimate the value of $\sqrt{5}$.

$4 < 5 < 9$ — 4 and 9 are the closest perfect squares.

$2^2 < 5 < 3^2$ — $4 = 2^2$ and $9 = 3^2$

$\sqrt{2^2} < \sqrt{5} < \sqrt{3^2}$ — Find the square root of each number.

$2 < \sqrt{5} < 3$ — Simplify.

Since 5 is closer to 4 than 9, the best integer estimate for $\sqrt{5}$ is 2. Use this value to evaluate the expression.

$\frac{1 + \sqrt{5}}{2} \approx \frac{1 + 2}{2}$ or 1.5

Guided Practice

Estimate to the nearest integer. (Examples 1 and 2)

1. $\sqrt{28} \approx$ ______

2. $\sqrt{135} \approx$ ______

3. $\sqrt{38.7} \approx$ ______

Show your work.

4. $\sqrt[3]{51} \approx$ ______

5. $\sqrt[3]{200} \approx$ ______

6. $\sqrt[3]{95} \approx$ ______

7. **STEM** Tobias dropped a tennis ball from a height of 60 meters. The time in seconds it takes for the ball to fall 60 feet is $0.25(\sqrt{60})$. Find three sets of approximations for the amount of time it will take. Then determine how long it will take for the ball to hit the ground. (Example 3)

8. The number of swings back and forth of a pendulum of length L in inches each minute is $\frac{375}{\sqrt{L}}$. About how many swings will a 40-inch pendulum make each minute? (Example 4)

9. **Building on the Essential Question** How can I estimate the square root of a non-perfect square?

Rate Yourself!

How confident are you about finding the square root of a non-perfect square? Mark an X in the section that applies.

For more help, go online to access a Personal Tutor.

Name ______ My Homework ______

Independent Practice

Go online for Step-by-Step Solutions

Estimate to the nearest integer. (Examples 1 and 2)

1. $\sqrt{23} \approx$ ______

2. $\sqrt{197} \approx$ ______

3. $\sqrt{15.6} \approx$ ______

4. $\sqrt{85.1} \approx$ ______

5. $\sqrt[3]{22} \approx$ ______

6. $\sqrt[3]{34} \approx$ ______

7. $\sqrt[3]{989} \approx$ ______

8. $\sqrt[3]{250} \approx$ ______

9. The area of Kaitlyn's square garden is 345 square feet. One side of the garden is next to a shed. She wants to put a fence around the other three sides of the garden. Find three sets of approximations for the amount of fence it will take. Then determine how much fence she should buy.

(Example 3) ______

10. In Little League, the bases are squares with sides of 14 inches. The expression $\sqrt{(s^2 + s^2)}$ represents the distance *diagonally across* a square of side length s. Estimate the diagonal distance across a base to the nearest inch. (Example 4) ______

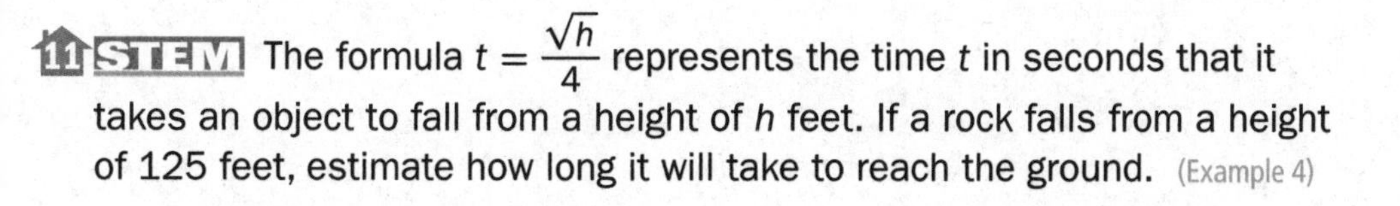

11. **STEM** The formula $t = \frac{\sqrt{h}}{4}$ represents the time t in seconds that it takes an object to fall from a height of h feet. If a rock falls from a height of 125 feet, estimate how long it will take to reach the ground. (Example 4)

______

Order each set of numbers from least to greatest.

12. $\{7, 9, \sqrt{50}, \sqrt{85}\}$ ______

13. $\{\sqrt[3]{105}, 7, 5, \sqrt{38}\}$ ______

14. **CCSS Persevere with Problems** Amanda purchased a storage cube that has a volume of 4 cubic feet. She wants to put it on a bookshelf that is 12 inches tall. Will the cube fit? Explain. ______

15. Without a calculator, determine which is greater, $\sqrt{94}$ or 10. Explain your reasoning. ______

H.O.T. Problems Higher Order Thinking

16. **CCSS Persevere with Problems** Find two numbers that have square roots between 7 and 8. One number should have a square root closer to 7 and the other number should have a square root closer to 8. Justify your answer.

______

17. **CCSS Find the Error** Jasmine is estimating $\sqrt{200}$. Find her mistake and correct it. ______

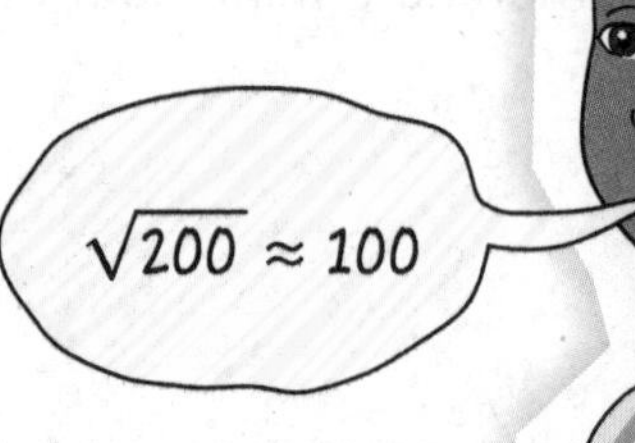

18. **CCSS Construct an Argument** If $x^4 = y$, then x is the fourth root of y. Explain how to estimate the fourth root of 30. Find the fourth root of 30 to the nearest whole number.

______

Standardized Test Practice

19. Leah found the side of a square to be $\sqrt{30}$ inches. Which point is closest to $\sqrt{30}$ on the number line?

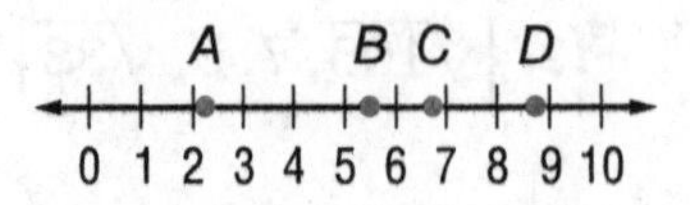

Ⓐ point *A*
Ⓑ point *B*
Ⓒ point *C*
Ⓓ point *D*

Name ______ My Homework ______

Extra Practice

Estimate to the nearest integer.

20. $\sqrt{44} \approx$ 7

36 < 44 < 49

$6^2 < 44 < 7^2$

$\sqrt{6^2} < \sqrt{44} < \sqrt{7^2}$

$\sqrt{44}$ is closer to $\sqrt{49}$ or 7.

Homework Help

21. $\sqrt[3]{199} \approx$ 6

125 < 199 < 216

$5^3 < \sqrt[3]{199} < 6^3$

$\sqrt{5^3} < \sqrt[3]{199} < \sqrt{6^3}$

$\sqrt[3]{199}$ is closer to $\sqrt{216}$ or 6.

22. $\sqrt{125} \approx$ ______

23. $\sqrt{23.5} \approx$ ______

24. $\sqrt[3]{59} \approx$ ______

25. $\sqrt[3]{430} \approx$ ______

Estimate the solution of each equation to the nearest integer.

26. $y^2 = 55$

27. $d^2 = 95$

28. $p^2 = 6.8$

The volume of each cube is given. Estimate the side length of the cube to the nearest integer. Use the formula $V = s^3$.

29. ______

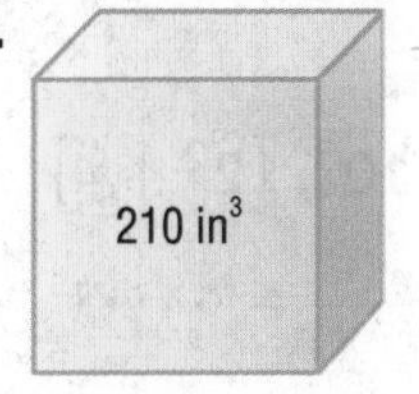

30. ______

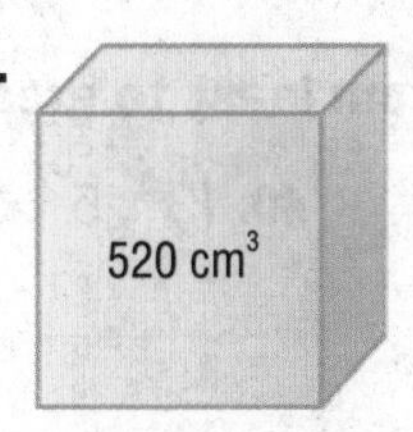

31. CCSS **Use Math Tools** Jacob is buying a bag of grass seed. The two-pound bag will cover 1,000 square feet of lawn. Estimate the side length of the largest square Jacob could seed if he purchases 5 bags.

______

Standardized Test Practice

32. The radius of a circle with area A is approximately $\sqrt{\frac{A}{3}}$. If a pizza has an area of 78 square inches, which of the following is the best approximation of the radius of the pizza?

Ⓐ 3 inches Ⓒ 8 inches

Ⓑ 5 inches Ⓓ 26 inches

33. The new library at Walnut Hills Middle School has a carpeted floor in the shape of a square. If the area of the floor is 52,000 square feet, what is the approximate length in feet of one side of the square floor?

Ⓕ 26,000 Ⓗ 1,500

Ⓖ 13,000 Ⓘ 225

34. Short Response After an accident, officials use the formula below to estimate the speed the car was traveling based on the length of the car's skid marks.

$$s = \sqrt{24m}$$

In the formula, s represents the speed in miles per hour and m is the length of the skid marks in feet. If a car leaves a skid mark of 50 feet, what was its approximate speed? Show all work necessary to justify your answer.

______

Common Core Review

Write each of the following as a fraction in simplest form. 7.NS.2.d

35. -36 = ______

36. 1.7 = ______

37. -0.048 = ______

38. 98% = ______

Order each set of numbers from least to greatest. 6.NS.7

39. $\{4(8), 3^3, 5^2\}$ ______

40. $\{3^4, 5^2, 2^5\}$ ______

41. $\{25^2 \cdot 3, 10^3, 12^2 + 4\}$ ______

42. Of the 150 students in Mr. Bacon's classes, 16% play soccer, $\frac{9}{25}$ play basketball, 3^3 play football and 14 do not play a sport at all. Write the number of students in order from least to greatest. 6.NS.7

______

Lesson 10

Compare Real Numbers

What You'll Learn

Scan the lesson. Write the definitions of irrational number and real number.

- ____
- ____

Essential Question

WHY is it helpful to write numbers in different ways?

Vocabulary

irrational number
real number

Common Core State Standards

Content Standards
8.NS.1, 8.NS.2, 8.EE.2

Mathematical Practices
1, 3, 4, 6

Real-World Link

Sports Major League baseball has rules for the dimensions of the baseball diamond. A model of the diamond is shown.

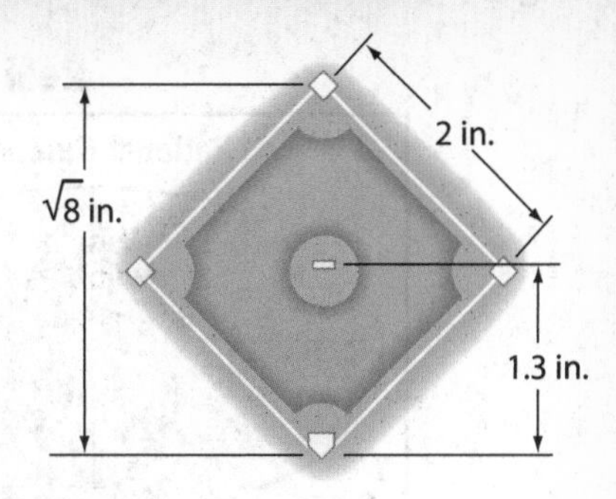

1. On the model, the distance from the pitching mound to home plate is 1.3 inches. Is 1.3 a rational number? Explain.

____

2. On the model, the distance from first base to second base is 2 inches. Is 2 a rational number? Explain.

____

3. The distance from home plate to second base is $\sqrt{8}$ inches. Using a calculator, find $\sqrt{8}$. Does it appear to terminate or repeat?

____

____

4. To determine if the number terminates, on your calculator, multiply your answer to $\sqrt{8}$ by itself. Do not use the x^2 button.

Is the answer 8? ______________________________

5. Based on your results, can you classify $\sqrt{8}$ as a rational number? Explain.

____

____

____

Key Concept Real Numbers

Work Zone

STOP and Reflect

Explain below how you know that $\sqrt{2}$ is an irrational number.

	Rational Number	Irrational Number
Words	A rational number is a number that can be expressed as the ratio $\frac{a}{b}$, where a and b are integers and $b \neq 0$.	An **irrational number** is a number that *cannot* be expressed as the ratio $\frac{a}{b}$, where a and b are integers and $b \neq 0$.
Examples	$-2, 5, 3.\overline{76}, -12\frac{7}{8}$	$\sqrt{2} \approx 1.414213562...$

Numbers that are not rational are called irrational numbers. The square root of any number that is not a perfect square number is irrational. The set of rational numbers and the set of irrational numbers together make up the set of **real numbers**. Study the Venn diagram below.

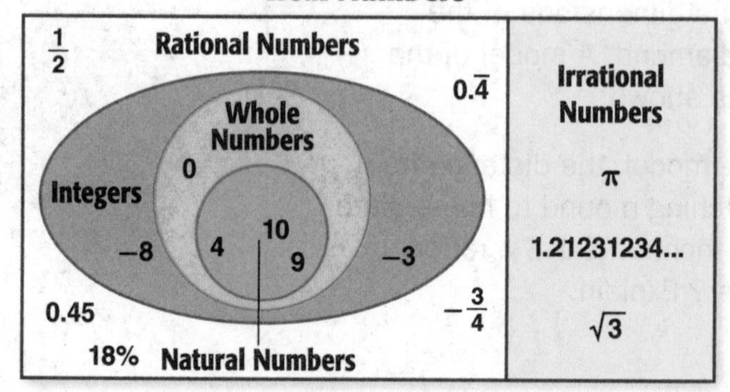

Examples

Name all sets of numbers to which each real number belongs.

1. **0.2525...** The decimal ends in a repeating pattern. It is a rational number because it is equivalent to $\frac{25}{99}$.

2. $\sqrt{36}$ Since $\sqrt{36} = 6$, it is a natural number, a whole number, an integer, and a rational number.

3. $-\sqrt{7}$ $-\sqrt{7} \approx -2.645751311...$ The decimal does not terminate nor repeat, so it is an irrational number.

Show your work.

a. ______

b. ______

c. ______

Got It? Do these problems to find out.

a. $\sqrt{10}$ **b.** $-2\frac{2}{5}$ **c.** $\sqrt{100}$

Compare and Order Real Numbers

You can compare and order real numbers by writing them in the same notation. Write the numbers in decimal notation before comparing or ordering them.

Examples

Fill in each ◯ with <, >, or = to make a true statement.

4. $\sqrt{7}$ ◯ $2\frac{2}{3}$

$\sqrt{7} \approx 2.645751311...$

$2\frac{2}{3} = 2.666666666...$

Since 2.645751311... is less than 2.66666666..., $\sqrt{7} < 2\frac{2}{3}$.

5. **15.7%** ◯ $\sqrt{0.02}$

$15.7\% = 0.157$

$\sqrt{0.02} \approx 0.141$

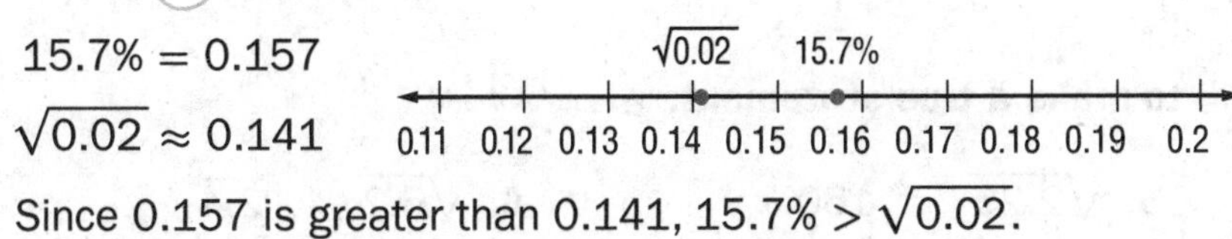

Since 0.157 is greater than 0.141, $15.7\% > \sqrt{0.02}$.

6. **Order the set $\left\{\sqrt{30}, 6, 5\frac{4}{5}, 5.3\overline{6}\right\}$ from least to greatest. Verify your answer by graphing on a number line.**

Write each number as a decimal. Then order the decimals.

$\sqrt{30} \approx 5.48$

$6 = 6.00$

$5\frac{4}{5} = 5.80$

$5.3\overline{6} \approx 5.37$

5.3̄6 √30 5 4/5 6

5.1 5.2 5.3 5.4 5.5 5.6 5.7 5.8 5.9 6

From least to greatest, the order is $5.3\overline{6}$, $\sqrt{30}$, $5\frac{4}{5}$, and 6.

Got It? Do these problems to find out.

Show your work.

d. $\sqrt{11}$ ◯ $3\frac{1}{3}$ **e.** $\sqrt{17}$ ◯ 4.03 **f.** $\sqrt{6.25}$ ◯ 250%

g. Order the set $\left\{-7, -\sqrt{60}, -7\frac{7}{10}, -\frac{66}{9}\right\}$ from least to greatest. Verify your answer by graphing on the number line below.

g. ____________

−7.9 −7.8 −7.7 −7.6 −7.5 −7.4 −7.3 −7.2 −7.1 −7

Real World Example

7. On a clear day, the number of miles a person can see to the horizon is about 1.23 times the square root of his or her distance from the ground in feet. Suppose Frida is at the Empire Building observation deck at 1,250 feet and Kia is at the Freedom Tower observation deck at 1,362 feet. How much farther can Kia see than Frida?

Use a calculator to approximate the distance each person can see.

Frida: $1.23 \cdot \sqrt{1{,}250} \approx 43.49$ Kia: $1.23 \cdot \sqrt{1{,}362} \approx 45.39$

Kia can see 45.39 − 43.49 or 1.90 miles farther than Frida.

Guided Practice

Name all sets of numbers to which each real number belongs. (Examples 1–3)

1. 0.050505... ______
2. $-\sqrt{64}$ ______
3. $\sqrt{17}$ ______

Fill in each ◯ with <, >, or = to make a true statement. (Examples 4 and 5)

4. $\sqrt{15}$ ◯ 3.5
5. $\sqrt{2.25}$ ◯ 150%
6. $\sqrt{6.2}$ ◯ $2.\overline{4}$

7. Order the set $\{\sqrt{5}, 220\%, 2.25, 2.\overline{2}\}$ from least to greatest. Verify your answer by graphing on a number line. (Example 6)

2.19 2.2 2.21 2.22 2.23 2.24 2.25 2.26 2.27 2.28 2.29 2.3

______

8. The formula $A = \sqrt{s(s-a)(s-b)(s-c)}$ can be used to find the area *A* of a triangle. The variables *a*, *b*, and *c* are the side measures and *s* is one half the perimeter. Use the formula to find the area of a triangle with side lengths of 7 centimeters, 9 centimeters, and 10 centimeters. (Example 7) ______

9. **Building on the Essential Question** How are real numbers different from irrational numbers?

______

______

______

Rate Yourself!

How well do you understand real numbers? Circle the image that applies.

Clear

Somewhat Clear

Not So Clear

For more help, go online to access a Personal Tutor.

Name ______________________ My Homework ______________________

Independent Practice

Go online for Step-by-Step Solutions 

Name all sets of numbers to which each real number belongs. (Examples 1–3)

1. $\frac{2}{3}$ ______

2. $-\sqrt{20}$ ______

3. $7.\overline{2}$ ______

4. $\frac{12}{4}$ ______

Fill in each ◯ with <, >, or = to make a true statement. (Examples 4 and 5)

5. $\sqrt{10}$ ◯ 3.2

6. $5\frac{1}{6}$ ◯ $5.1\overline{6}$

7. $2.\overline{21}$ ◯ $\sqrt{5.2}$

Order each set of numbers from least to greatest. Verify your answer by graphing on a number line. (Example 6)

8. $-415\%, -\sqrt{17}, -4.\overline{1}, -4.08$

–4.18 –4.16 –4.14 –4.12 –4.1 –4.08

____

9. $\sqrt{5}, \sqrt{6}, 2.5, 2.55, \frac{7}{3}$

2.1 2.2 2.3 2.4 2.5 2.6 2.7

____

10. The equation $s = \sqrt{30fd}$ can be used to find a car's speed *s* in miles per hour given the length *d* in feet of a skid mark and the friction factor *f* of the road. Police measured a skid mark of 90 feet on a dry concrete road. If the speed limit is 35 mph, was the car speeding? Explain. (Example 7)

Friction Factor		
Road	**Concrete**	**Tar**
Wet	0.4	0.5
Dry	0.8	1.0

____

11. The surface area in square meters of the human body can be found using the expression $\sqrt{\frac{hm}{3{,}600}}$ where *h* is the height in centimeters and *m* is the mass in kilograms. Find the surface area of a 15-year-old boy with a height of 183 centimeters and a mass of 74 kilograms. (Example 7)

____

____

12. CCSS **Be Precise** Write a brief description and give an example of each type of number in the graphic organizer shown.

natural	whole	integer	rational	irrational

Use estimation to fill in each $\bigcirc$ with <, >, or = to make a true statement.

13. $3\pi \bigcirc \sqrt{78}$

14. $\pi^2 \bigcirc 3 \cdot \sqrt{15}$

15. $\sqrt{980} \bigcirc 4\pi^2$

H.O.T. Problems Higher Order Thinking

16. CCSS **Use a Counterexample** Give a counterexample for the statement *All square roots are irrational numbers.* Explain your reasoning.

CCSS **Persevere with Problems Tell whether the following statements are *always, sometimes,* or *never* true. If a statement is not always true, explain.**

17. Integers are rational numbers.

18. Rational numbers are integers.

19. The product of a non-zero rational number and an irrational number is irrational.

Standardized Test Practice

20. Which of the following is an irrational number?

Ⓐ -6

Ⓑ $\frac{2}{3}$

Ⓒ $\sqrt{9}$

Ⓓ $\sqrt{3}$

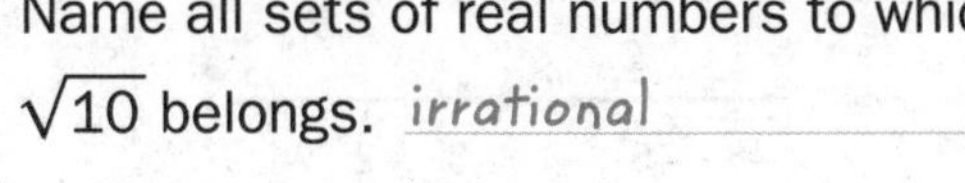

Extra Practice

21. Name all sets of real numbers to which $\sqrt{10}$ belongs. *irrational*

$\sqrt{10} \approx 3.16227766\ldots$ Since the decimal does not terminate nor repeat, it is an irrational number.

22. Fill in ◯ with <, >, or = to make $5.1\overline{5}$ (>) $\sqrt{26}$ a true statement.

Write each number as a decimal.
$5.1\overline{5} = 5.155555\ldots$
$\sqrt{26} \approx 5.099019\ldots$
Since $5.155555\ldots$ is greater than $5.099019\ldots$, $5.1\overline{5} > \sqrt{26}$.

Name all sets of numbers to which each real number belongs.

23. 14

24. $-\sqrt{16}$

25. $-\sqrt[3]{90}$

Fill in each ◯ with <, >, or = to make a true statement.

26. $\sqrt{12}$ ◯ 3.5

27. $6\frac{1}{3}$ ◯ $\sqrt[3]{240}$

28. 240% ◯ $\sqrt{5.76}$

29. About how much greater is the perimeter of a square with area 250 square meters than a square with an area of 125 square meters?

30. CCSS **Persevere with Problems** In the sequence 4, 12, ■, 108, 324, the missing number can be found by simplifying $\sqrt{ab}$ where a and b are the numbers on either side of the missing number. Find the missing number.

Fill in each ◯ with <, >, or = to make a true statement.

31. $3 + \sqrt{7}$ ◯ 6

32. $4 - \sqrt{10}$ ◯ $\sqrt{2}$

33. 13 ◯ $8 + \sqrt{20}$

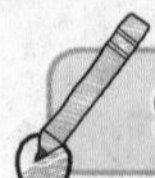

Standardized Test Practice

34. Anna wants to plant a circular garden in her back yard like the one shown.

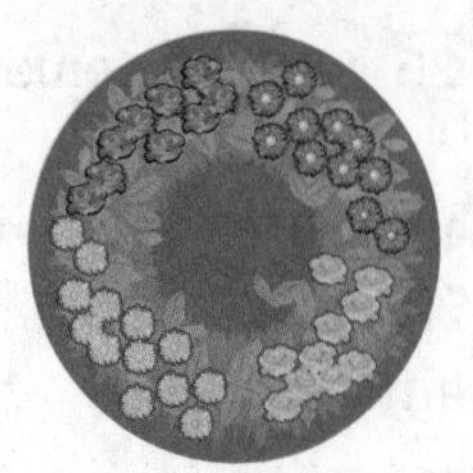

$A = 150 \text{ ft}^2$

The formula $r = \sqrt{\frac{A}{\pi}}$ gives the value of the radius r given the area A. What is the best approximation for the radius of Anna's garden?

Ⓐ 4 feet
Ⓑ 7 feet
Ⓒ 8 feet
Ⓓ 14 feet

35. Which number represents the point graphed on the number line?

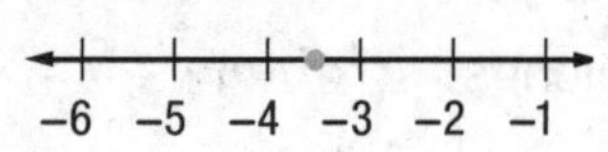

Ⓕ $-\sqrt{12}$
Ⓖ $-\sqrt{10}$
Ⓗ $-\sqrt{15}$
Ⓘ $-\sqrt{8}$

36. Short Response Which of the two real numbers below is greater?

$\sqrt{3}$ $\frac{1}{3}$

______

Common Core Review

37. Order the set $\{7, \sqrt{53}, \sqrt{32}, 6\}$ from least to greatest. 8.EE.2

______

Solve each equation. 8.EE.2

38. $t^2 = 25$ ______

39. $y^2 = \frac{1}{49}$ ______

40. $0.64 = a^2$ ______

Evaluate each expression. Express the result in scientific notation. 8.EE.4

41. $(7.2 \times 10^4)(1.1 \times 10^{-6}) =$ ______

42. $(3.6 \times 10^3) + (5.7 \times 10^5) =$ ______

43. The table shows the approximate population of several countries. Order the countries from the greatest population to the least population. 8.EE.4

______

Country	Population
China	1.3×10^9
India	1.2×10^9
Indonesia	2.3×10^8
United States	3.1×10^8

21ST CENTURY CAREER in Engineering

Robotics Engineer

Are you mechanically inclined? Do you like to find new ways to solve problems? If so, a career as a robotics engineer is something you should consider. Robotics engineers design and build robots to perform tasks that are difficult, dangerous, or tedious for humans. For example, a robotic insect was developed based on a real insect. Its purpose was to travel over water surfaces, take measurements, and monitor water quality.

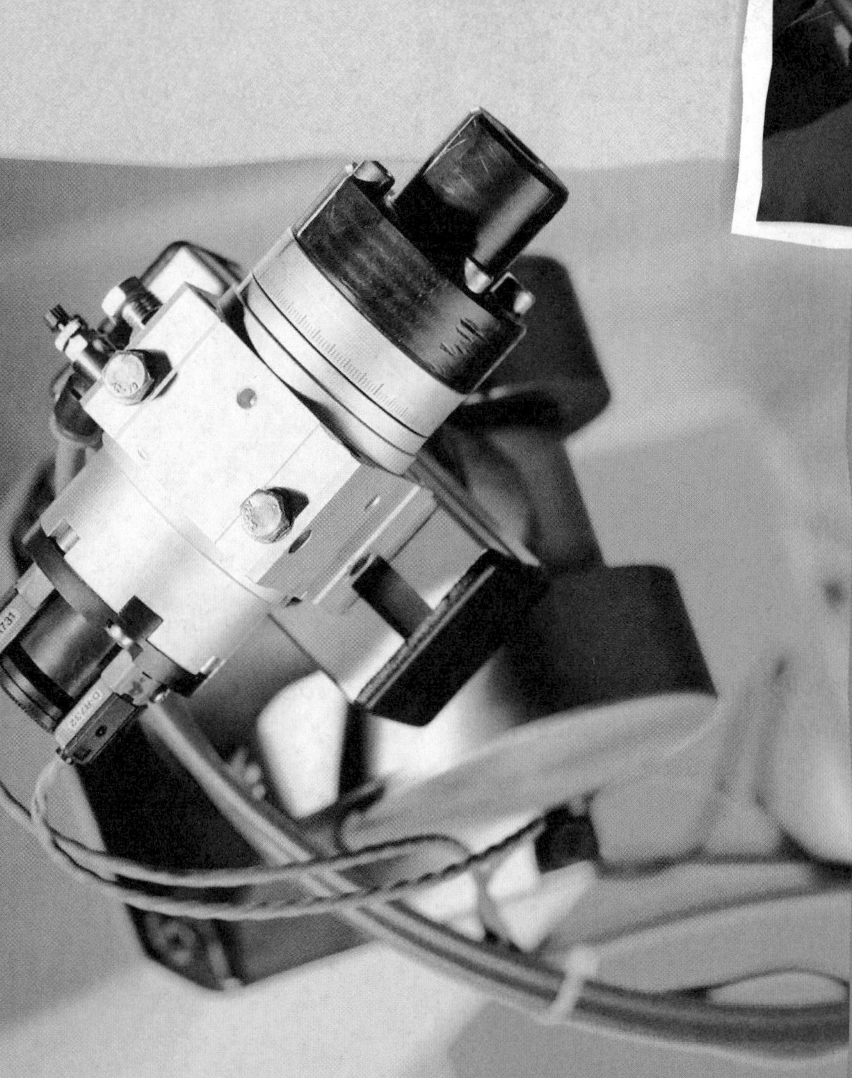

College & Career READINESS

Explore college and careers at ccr.mcgraw-hill.com

Is This the Career for You?

Are you interested in a career as a robotics engineer? Take some of the following courses in high school.

- Calculus
- Electro-Mechanical Systems
- Fundamentals of Robotics
- Physics

Turn the page to find out how math relates to a career in Engineering.

Relying on Robots

Use the information in the table to solve each problem.

1. Write the mass of the robot in standard form. ________
2. Write the length of the robot in scientific notation. ________
3. Write the leg diameter of the robot in scientific notation. ________
4. What is the mass in milligrams? Write in standard form. ________
5. Real insects called water striders can travel 8.3 times faster than the robot. Write the speed of water striders in scientific notation.

____

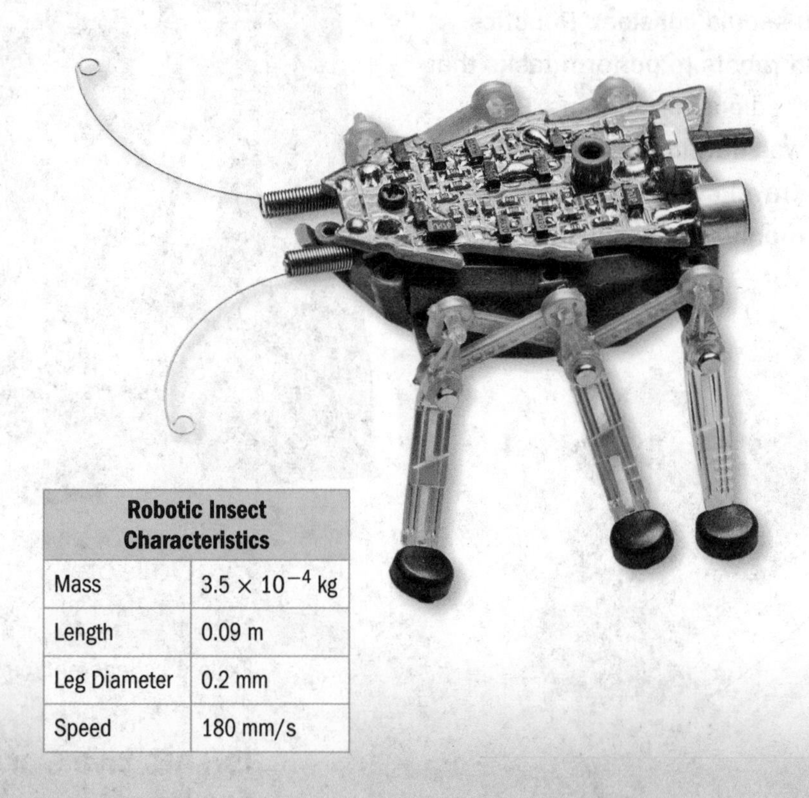

Robotic Insect Characteristics	
Mass	3.5×10^{-4} kg
Length	0.09 m
Leg Diameter	0.2 mm
Speed	180 mm/s

Career Project

It's time to update your career portfolio! Investigate the education and training requirements for a career in robotics engineering.

____

____

____

____

____

____

What skills would you need to improve to succeed in this career?

- ________
- ________
- ________
- ________
- ________

Chapter Review

Vocabulary Check

Complete the crossword puzzle using the vocabulary list at the beginning of the chapter.

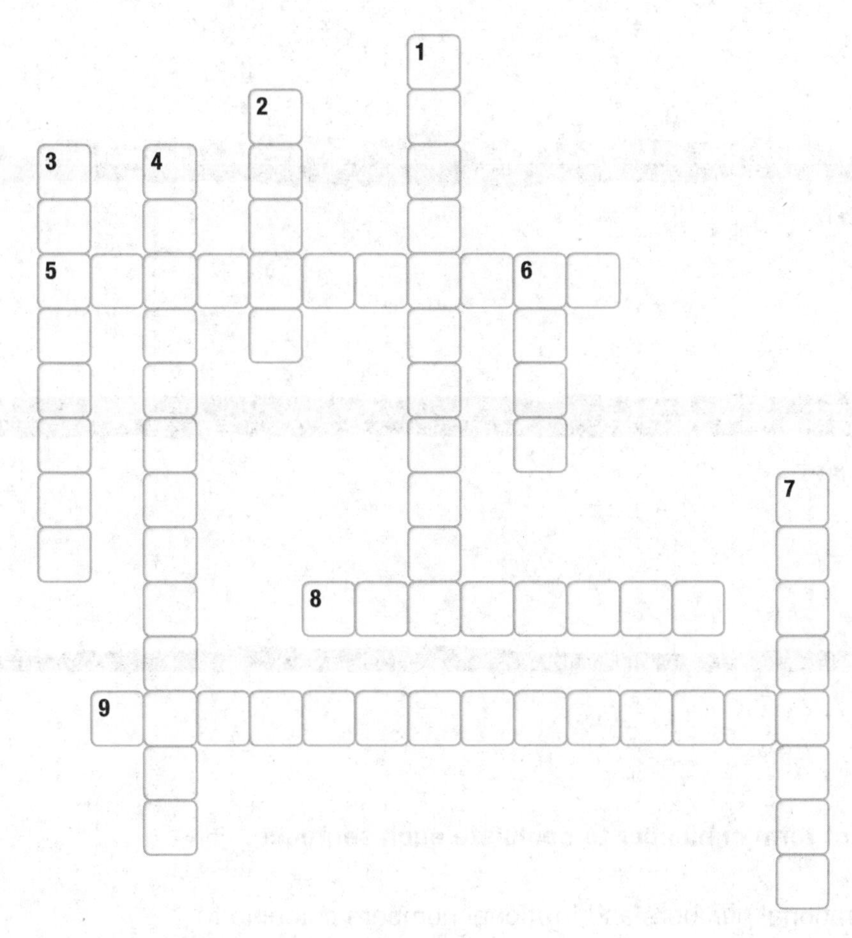

Across

5. a rational number whose cube root is a whole number
8. a number, a variable, or a product of a number and one or more variables
9. numbers that can be written as a comparison of two integers, expressed as a fraction

Down

1. the symbol used to indicate a positive square root
2. a product of repeated factors using a base and exponent
3. this tells how many times a number is used as a factor
4. a rational number whose square root is a whole number
6. in a power, the number that is the common factor
7. one of a number's three equal factors

Key Concept Check

Use Your FOLDABLES

Use your Foldable to help review the chapter.

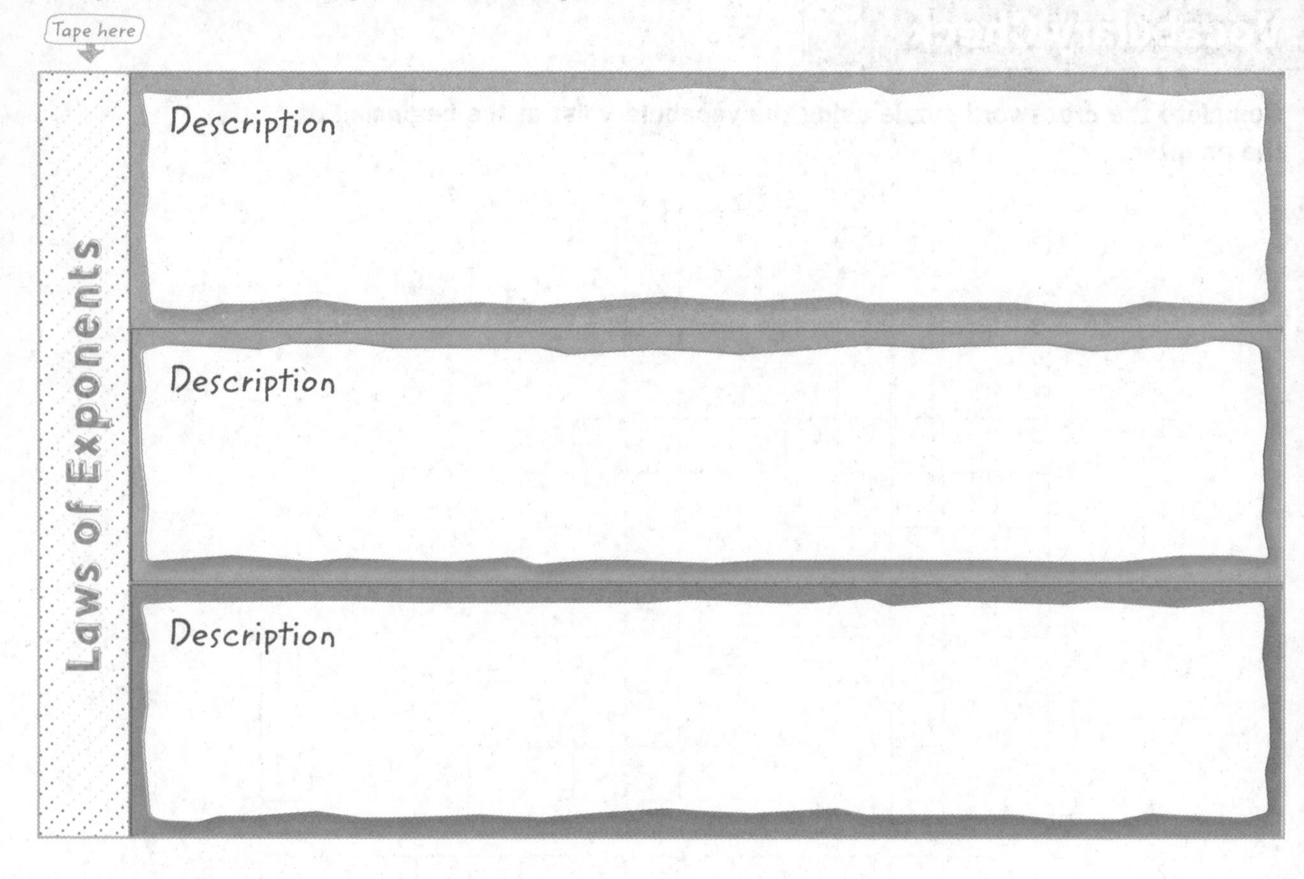

Got it?

Circle the correct term or number to complete each sentence.

1. The sets of rational numbers and irrational numbers combine to make the (whole, real) numbers.
2. The product of $3a^2b$ and $-3a^2b$ is $(-9a^4b^2, a^4b)$.
3. You would use the (Power of a Product rule, Product of Powers rule) to simplify the expression $(p^2r)^4$.
4. The expression $\frac{6^2 \cdot 2^6 \cdot 8^4}{6 \cdot 2^3 \cdot 8^3}$ is equal to (384, 288).
5. Another way to write $(9^4)^7$ is $(9^{11}, 9^{28})$.
6. 3^{-4} is equal to $\left(-81, \frac{1}{81}\right)$.
7. Scientific notation is when a number is written as a product of a power of 10 and a factor greater than or equal to 1 and (less than, less than or equal to) 10.

Problem Solving

1. To close school, the principal calls six teachers, who in turn call six more. If each of those teachers call six more, how many calls will be made by the teachers in this last group? (Lesson 3) ______

2. In a house, the area of the den is 3^4 square feet. The area of the kitchen is 3^3 square feet. How many times larger is the den than the kitchen? (Lesson 3) ______

3. The smallest mammal is Kitti's hog-nosed bat weighing about 4.375×10^{-3} pound. Write this weight in standard form. (Lesson 6)

______

4. The Guadalupe River is 2.56×10^2 miles long. The Amazon River is 4.096×10^3 miles long. How many times longer is the Amazon River than the Guadalupe River? (Lesson 7) ______

5. Jodie took the measurements of her favorite jewelry box. Order the set of measurements $\left\{2.\overline{2}, 2\frac{1}{5}, 2.25, \sqrt{5}\right\}$ from least to greatest. (Lesson 10)

______

6. A quilter made 256 small squares for a large quilt. If the quilt is shaped like a square, how many small squares will she use on each side? (Lesson 8)

______

Reflect

Answering the Essential Question

Use what you learned about numbers to complete the graphic organizer. For each category, describe why you would use that form for the number $35{,}036\frac{1}{3}$. Then write the number in that form. If you would not use the number in that form, explain why.

Decimal	Power
Fraction	**Scientific Notation**

WHY is it helpful to write numbers in different ways?

Answer the Essential Question. WHY is it helpful to write numbers in different ways?

UNIT PROJECT

Watch

Music to My Ears When you listen to music, you may not be aware of the math used to create it. In this project you will:

- **Collaborate** with your classmates as you research the connections between math and music.
- **Share** the results of your research in a creative way.
- **Reflect** on how mathematical ideas can be represented.

By the end of this project, you just might be ready to write a hit song!

Collaborate

Go Online Work with your group to research and complete each activity. You will use your results in the Share section on the following page.

1. Choose a song on a CD or on your MP3 player. Listen to the song and describe the beats or rhythm using repeating numbers. For example, a song may have a rhythm that can be described by 1-2-3-1-2-3-... .

2. Research and describe the different types of musical notes. Make sure to use rational numbers. Include a drawing of each note on sheet music.

3. Research Pythagoras' findings about music, notes and frequency, and harmony. Write a few paragraphs about what you found and create a list of the types of numbers you find in your research.

4. Describe the Fibonacci Sequence. Then give some examples of how Fibonacci numbers are found in music.

5. Find the digital music sales in a recent year. Write this number in both standard form and scientific notation. Then compare the digital music sales to CD music sales for the same year. Create a display to show what you find.

With your group, decide on a way to share what you have learned about math and music. Some suggestions are listed below, but you can also think of other creative ways to present your information. Remember to show how you used mathematics to complete each of the activities in this project!

- Create your own short piece of music based on your knowledge of notes and frequency. Make a recording of the music and explain how it is harmonious.
- Use presentation software to demonstrate some ways math and music are connected.

Check out the note on the right to connect this project with other subjects.

with Health

Health Literacy Many studies have been done that show a positive connection between music and good health. Research the Internet to find information about one such study.

6. **Answer the Essential Question** HOW can mathematical ideas be represented?

   a. How did you use what you learned about real numbers in this chapter to represent mathematical ideas in this project?

   ___

   ___

   ___

   b. In this project, you discovered how mathematical ideas are represented in music. Explain how mathematical ideas are represented in other parts of culture.

   ___

   ___

   ___

UNIT 2

CCSS

Expressions and Equations

Essential Question

HOW can you communicate mathematical ideas effectively?

Chapter 2
Equations in One Variable

Linear equations in one variable can have one solution, infinitely many solutions, or no solutions. In this chapter, you will write and solve two-step equations and solve equations with variables on both sides.

Chapter 3
Equations in Two Variables

In a proportional relationship, the unit rate is the slope of the graph. In this chapter, you will graph equations of the form $y = mx$ and $y = mx + b$. You will then solve systems of equations algebraically and by graphing.

Unit Project Preview

Web Design 101 A Web page is a useful way of presenting a summary of facts and statistics about a subject.

To design a Web page, you must first collect all the information that you want to include on the page. You will also need to decide how to balance the text and graphics. This will ensure that your Web page not only looks good but is functional too.

At the end of Chapter 3, you'll complete a project in which you will plan the design of a Web page about your favorite insect or other animal. But for now, it's time to do an activity in your book. Choose a subject that interests you. In the space provided, make an outline of all the items you would include if you were to design a Web page about that subject.

My Web page about ______________________ would include:

Chapter 2 Equations in One Variable

Essential Question

WHAT is equivalence?

Common Core State Standards

Content Standards
8.EE.7, 8.EE.7a, 8.EE.7b

Mathematical Practices
1, 2, 3, 4, 5, 7

Math in the Real World

Tips Maria and her family had salads and dinner at a local restaurant. Her mother wants to leave an 18% tip. The proportion $\frac{p}{\$35.60} = \frac{18}{100}$ can be used to find the amount of tip she should leave.

Use the proportion to find the amount of tip Maria's mother should leave. Then find the total amount.

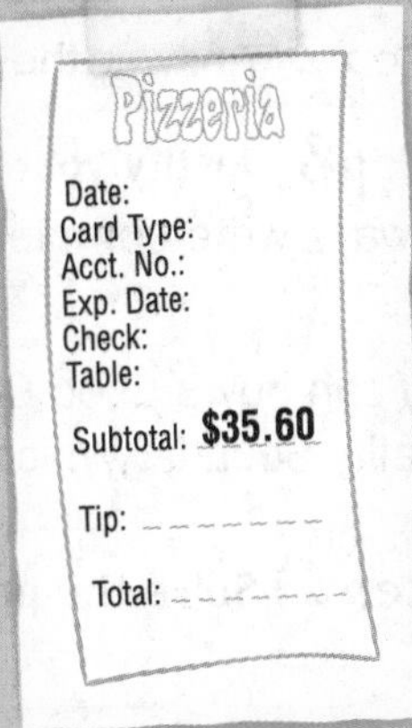

FOLDABLES® Study Organizer

1. Cut out the Foldable on page FL5 of this book.

2. Place your Foldable on page 164.

3. Use the Foldable throughout this chapter to help you learn about solving equations.

What Tools Do You Need?

Vocabulary

coefficient
identity
multiplicative inverse
null set
properties
two-step equation

Study Skill: Writing Math

Justify Your Answer When you justify your answer, you give *reasons* why your answer is correct.

The different plans an online movie rental company offers are shown. Mariah wants to purchase the 2 DVDs at-a-time plan. This month, the plans are advertised at $\frac{1}{4}$ off. If she has $10.00 to spend, does she have enough? Justify your answer.

Online DVD Rental	
Plan	Monthly Price ($)
3 DVDs at-a-time	14.50
2 DVDs at-a-time	12.00
1 DVD at-a-time	8.00

Step 1 **Solve the problem.**	Find the discount. $\frac{1}{4}$ of $12 = 3$ The discount is $3.00. Find the discounted price. $12 – $3 = $9
Step 2 **Answer the question.**	Mariah does have enough money.
Step 3 **Justify your answer.** Always write complete sentences.	Mariah has enough money because $9.00 is less than $10.00.

You can buy 3 used CDs at The Music Shoppe for $12.99, or you can buy 5 for $19.99 at Quality Sounds. Which is the better buy? Justify your answer.

Step 1 **Solve the problem.**	
Step 2 **Answer the question.**	
Step 3 **Justify your answer.** Always write complete sentences.	

When Will You Use This?

Your Turn! **You will solve this problem in the chapter.**

Try the Quick Check below.
Or, take the Online Readiness Quiz.

Quick Review

Common Core Review 7.EE.4

Example 1

Solve $44 = k - 7$.

$$\begin{aligned} 44 &= k - 7 \\ +7 &= +7 \\ \hline 51 &= k \end{aligned}$$

Write the equation.
Addition Property of Equality

Example 2

Solve $18m = -360$.

$18m = -360$ Write the equation.

$\frac{18m}{18} = \frac{-360}{18}$ Division Property of Equality

$m = -20$ Simplify.

Quick Check

One-Step Equations **Solve each equation. Check your solution.**

1. $n + 8 = -9$

2. $4 = p + 19$

3. $-4 + a = 15$

4. $3c = -18$

5. $-42 = -6b$

6. $\frac{w}{4} = -8$

7. Barry has 18 more marbles than Heidi. If Barry has 92 marbles, write and solve an equation to determine the number of marbles Heidi has.

Which problems did you answer correctly in the Quick Check? Shade those exercise numbers below.

1 2 3 4 5 6 7

Lesson 1

Solve Equations with Rational Coefficients

Essential Question

WHAT is equivalence?

Vocabulary

multiplicative inverse
coefficient

Common Core State Standards

Content Standards
8.EE.7, 8.EE.7a, 8.EE.7b

Mathematical Practices
1, 3, 4, 7

Scan the lesson. List two headings you would use to make an outline of the lesson.

-
-

Vocabulary Start-Up

Two numbers with a product of 1, such as $\frac{3}{4}$ and $\frac{4}{3}$, are called reciprocals or **multiplicative inverses**.

Complete the graphic organizer.

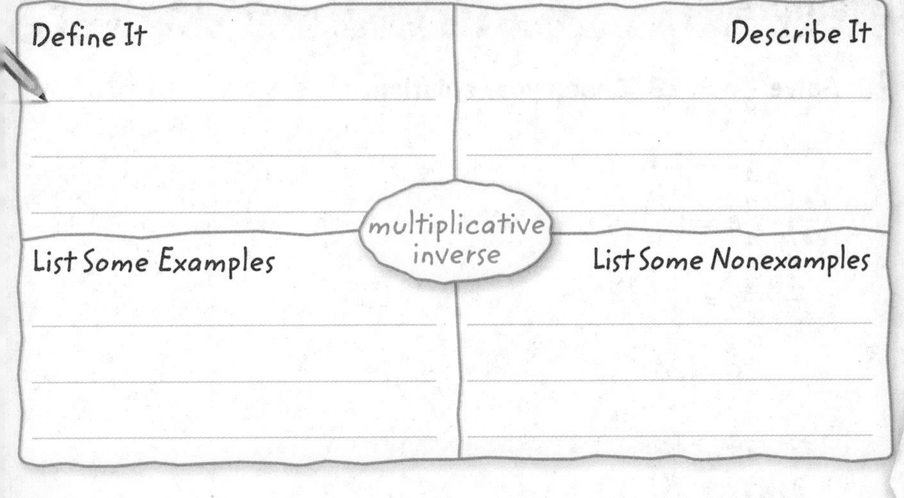

Describe how a multiplicative inverse is used in division of fractions.

Real-World Link

How can the action of the motorcyclist in the photo help you remember what the multiplicative inverse is?

Yikes!

Key Concept Inverse Property of Multiplication

Words The product of a number and its multiplicative inverse is 1.

Numbers $\frac{7}{8} \times \frac{8}{7} = 1$ $-\frac{3}{2} \times -\frac{2}{3} = 1$

Symbols $\frac{a}{b} \cdot \frac{b}{a} = 1$, where a and $b \neq 0$

Work Zone

STOP and Reflect

What is the multiplicative inverse of $-\frac{3}{2}$?

The numerical factor of a term that contains a variable is called the **coefficient** of the variable.

In the equation $\frac{3}{4}c = 18$, the coefficient of c is a rational number. To solve an equation when the coefficient is a fraction, multiply each side by the multiplicative inverse of the fraction.

Example

1. Solve $\frac{3}{4}c = 18$. Check your solution.

$\frac{3}{4}c = 18$ Write the equation.

$\left(\frac{4}{3}\right) \cdot \frac{3}{4}c = \left(\frac{4}{3}\right) \cdot 18$ Multiply each side by the multiplicative inverse of $\frac{3}{4}, \frac{4}{3}$.

$\frac{\overset{1}{\cancel{4}}}{\underset{1}{\cancel{3}}} \cdot \frac{\overset{1}{\cancel{3}}}{\underset{1}{\cancel{4}}}c = \frac{4}{\underset{1}{\cancel{3}}} \cdot \frac{\overset{6}{\cancel{18}}}{1}$ Write 18 as $\frac{18}{1}$. Divide by common factors.

$c = 24$ Simplify.

Check $\frac{3}{4}c = 18$ Write the original equation.

$\frac{3}{4}(24) \stackrel{?}{=} 18$ Replace c with 24.

$\frac{3}{\underset{1}{\cancel{4}}} \cdot \frac{\overset{6}{\cancel{24}}}{1} \stackrel{?}{=} 18$ Write 24 as $\frac{24}{1}$. Divide by common factors.

$18 = 18$ ✓ This sentence is true.

Show your work.

a. ____________

b. ____________

c. ____________

d. ____________

Got It? **Do these problems to find out.**

a. $\frac{1}{5}x = 12$ **b.** $-\frac{2}{9}d = 4$

c. $15 = \frac{5}{3}n$ **d.** $-24 = -\frac{6}{7}p$

Example

2. Solve $1\frac{1}{2}s = 16\frac{1}{2}$. Check your solution.

$1\frac{1}{2}s = 16\frac{1}{2}$ Write the equation.

$\frac{3}{2}s = \frac{33}{2}$ Rename $1\frac{1}{2}$ as $\frac{3}{2}$ and $16\frac{1}{2}$ as $\frac{33}{2}$.

$\left(\frac{2}{3}\right) \cdot \frac{3}{2}s = \left(\frac{2}{3}\right) \cdot \frac{33}{2}$ Multiply each side by the multiplicative inverse of $\frac{3}{2}$, $\frac{2}{3}$.

$\frac{\overset{1}{\cancel{2}}}{\underset{1}{\cancel{3}}} \cdot \frac{\overset{1}{\cancel{3}}}{\underset{1}{\cancel{2}}}s = \frac{\overset{1}{\cancel{2}}}{\underset{1}{\cancel{3}}} \cdot \frac{\overset{11}{\cancel{33}}}{\underset{1}{\cancel{2}}}$ Divide by common factors.

$s = 11$ Simplify.

Got It? **Do these problems to find out.**

d. $4\frac{1}{6} = 3\frac{1}{3}c$ **e.** $-9\frac{5}{8}w = 108$ **f.** $1\frac{7}{8}y = 4\frac{1}{2}$

Show your work.

d. ________

e. ________

f. ________

Solve Equations with Decimal Coefficients

In the equation $3.15 = 0.45n$ the coefficient of n is a decimal. To solve an equation with a decimal coefficient, divide each side of the equation by the coefficient.

Quick Review

Division

$0.45\overline{)3.15}$ = 7; -315; 0

Example

3. Solve $3.15 = 0.45n$. Check your solution.

$3.15 = 0.45n$ Write the equation.

$\frac{3.15}{0.45} = \frac{0.45n}{0.45}$ Division Property of Equality

$7 = n$ Simplify.

Check $3.15 = 0.45n$ Write the original equation.

$3.15 = 0.45(7)$ Replace n with 7.

$3.15 = 3.15$ ✓ The sentence is true.

Got It? **Do these problems to find out.**

g. $4.9 = 0.7t$ **h.** $-1.4m = 2.1$ **i.** $-5.6k = -12.88$

g. ________

h. ________

i. ________

Example

4. Latoya's softball team won 75%, or 18, of its games. Define a variable. Then write and solve an equation to determine the number of games the team played.

Quick Review
To write a percent as a decimal, move the decimal point two places to the left. Add zeros, if necessary. For example, 3% = 0.03 and 75% = 0.75.

Latoya's softball team won 18 games, which was 75% of the games played. Let n represent the number of games played. Write and solve an equation.

$0.75n = 18$ Write the equation. Write 75% as 0.75.

$\frac{0.75n}{0.75} = \frac{18}{0.75}$ Division Property of Equality

$n = 24$ Simplify.

Latoya's softball team played 24 games.

Guided Practice

Solve each equation. Check your solution. (Examples 1–3)

1. $60 = \frac{3}{4}p$

2. $-\frac{27}{25}x = -\frac{9}{5}$

3. $-2.7t = 810$

4. Paula has read 70% of the total pages in a book she is reading for English class. Paula has read 84 pages. Define a variable. Then write and solve an equation to determine how many pages are in the book. (Example 4)

5. **Building on the Essential Question** How is the multiplicative inverse used to solve an equation that has a rational coefficient?

Name ______________________ My Homework ______________________

Independent Practice

Go online for Step-by-Step Solutions

Solve each equation. Check your solution. (Examples 1–3)

1. $6 = \frac{1}{12}v$

2. $-\frac{2}{3}w = 60$

3. $-\frac{7}{8}k = -21$

4. $9.6 = 1.2b$

5. $0.75a = -9$

6. $-413.4 = -15.9n$

7. $3\frac{1}{10}s = 6\frac{1}{5}$

8. $2\frac{2}{9} = -\frac{4}{5}m$

9. $-2\frac{4}{5} = -3\frac{1}{2}n$

Define a variable. Then write and solve an equation for each situation. (Example 4)

10. The Parker family drove a total of 180 miles on their road trip. This distance is 1.5 times the distance they drove on the first day. How many miles did the Parker family drive on the first day?

____

11. José correctly answered 80% of the questions on a language arts quiz. If he answered 16 questions correctly, how many questions were on the language arts quiz?

____

12. **Financial Literacy** Demitrius deposited 60% of his paycheck into his savings account. What was the amount of his paycheck?

____

Savings Deposit Slip
Demetrius Matthews
Name
Amount Deposited **$41.67**

13. **Identify Structure** Suppose the numbers, $1\frac{1}{3}$, 0.2, −5, $-\frac{1}{2}$, are each coefficients in separate equations. Choose whether you would solve the equation by multiplying each side by the multiplicative inverse of the coefficient or by dividing each side by the coefficient. Write the number in the appropriate space.

Multiplicative Inverse

Division

H.O.T. Problems Higher Order Thinking

14. **Model with Mathematics** Write a real-world problem that can be represented by the equation $\frac{3}{4}c = 21$.

Persevere with Problems **Determine whether each statement is *true* or *false*. Explain your reasoning.**

15. The product of a fraction and its multiplicative inverse is 1.

16. To solve an equation with a coefficient that is a fraction, divide each side of the equation by the reciprocal of the fraction.

17. **Reason Inductively** Complete the statement: If $10 = \frac{1}{5}x$, then $x + 3 = ■$. Explain your reasoning.

Standardized Test Practice

18. A store is having a sale on notebook computers. Which equation can be used to find the regular price x of a notebook computer that is on sale for $799?

Ⓐ $799x = 0.5$

Ⓑ $0.5x = 799$

Ⓒ $\frac{1}{799}x = 0.5$

Ⓓ $\frac{1}{0.5}x = 799$

Name ________________ My Homework ________________

Extra Practice

Solve each equation. Check your solution.

19. $\frac{1}{2} = \frac{2}{5}z$

Homework Help

$\frac{5}{2} \cdot \frac{1}{2} = \frac{5}{2} \cdot \frac{2}{5}z$

$\frac{5}{4} = 1z$

$1\frac{1}{4} = z$

20. $-\frac{3}{4}t = 5$

21. $-\frac{2}{9}g = -\frac{7}{9}$

22. $0.6w = 0.48$

23. $-226.8 = 21.6y$

24. $-30 = 1.25c$

25. $1\frac{1}{2}x = 9\frac{9}{20}$

26. $-12\frac{2}{3} = -1\frac{1}{9}y$

27. $1\frac{5}{7} = 1\frac{13}{14}a$

28. One third of the bagels in a bakery are sesame bagels. There are 72 sesame bagels. Define a variable. Then write and solve an equation to find how many bagels there were in the bakery.

____

29. CCSS **Find the Error** Sarah is solving the equation $-\frac{7}{8}x = 24$. Circle her mistake and correct it.

____

____

____

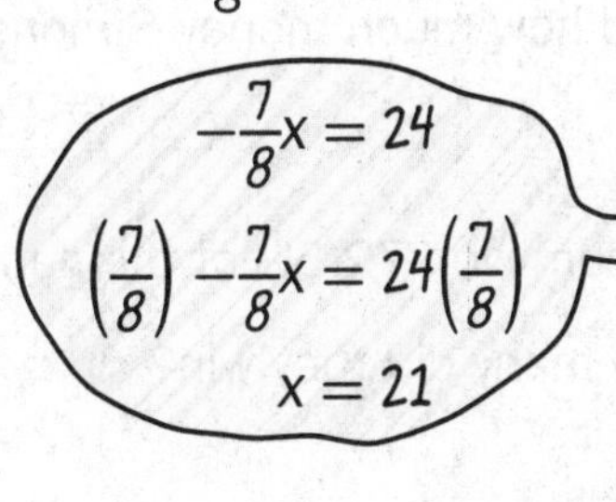

Standardized Test Practice

30. What is the reciprocal of $-\frac{4}{3}$?

Ⓐ $-1\frac{1}{3}$

Ⓑ $-\frac{3}{4}$

Ⓒ $\frac{3}{4}$

Ⓓ $1\frac{1}{3}$

31. An airplane travels 100 miles in 0.4 hour. Which speed represents the rate of the airplane?

Ⓕ 50 miles per hour

Ⓖ 100 miles per hour

Ⓗ 250 miles per hour

Ⓘ 500 miles per hour

32. **Short Response** To train for a marathon, Uyen ran a total of 71 miles in one month. This distance is $2\frac{1}{2}$ times the distance that she ran in the first week. How many miles did Uyen run in the first week?

Common Core Review

Solve each equation. Check your solution. 7.EE.4

33. $w + 5 = -20$

34. $x - 17 = -32$

35. $t + 7.2 = 1.65$

36. $-0.4 = g - 4.9$

37. $y - \frac{2}{5} = 1\frac{3}{5}$

38. $-5\frac{1}{6} = 2\frac{1}{3} + p$

39. **Financial Literacy** Simone saved $65.35 more than her brother Dan and $37.50 less than her sister Carly. Carly saved $127.75. Write and solve equations to find how much money Simone and Dan saved. 6.EE.7

40. A college basketball team won 72.5% of the games they played. If they played 80 games, how many did they win? 6.RP.3

Inquiry Lab

Solve Two-Step Equations

HOW does a bar diagram help you solve a real-world problem involving a two-step equation?

Content Standards
8.EE.7, 8.EE.7a
Mathematical Practices
1, 2, 3, 4

Postcards Miranda bought two large postcards and four small postcards at a souvenir shop. Each small postcard costs \$0.50. If Miranda spent \$5.00 on postcards, what is the cost of one large postcard?

What do you know? ______________________________

What do you need to find? ______________________________

Investigation

Step 1 The bar diagram represents the total number of postcards and the total cost. Label the missing parts.

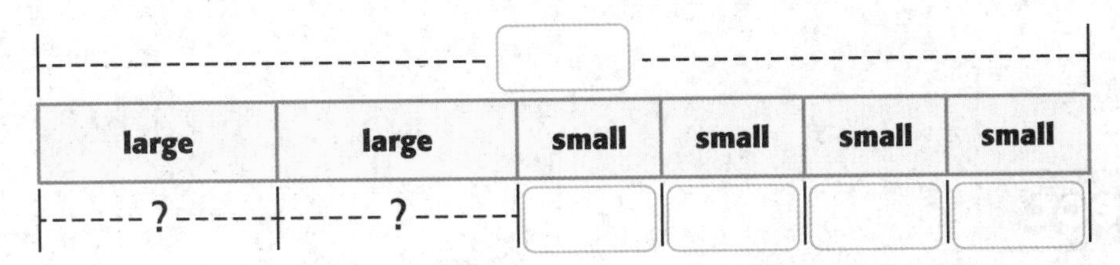

Step 2 Fill in the boxes to write an equation that represents the bar diagram. The cost of a large postcard is the unknown, so it is represented by the variable p.

$2p + \square = \square$

Step 3 Find the cost of the large postcards by working backward.

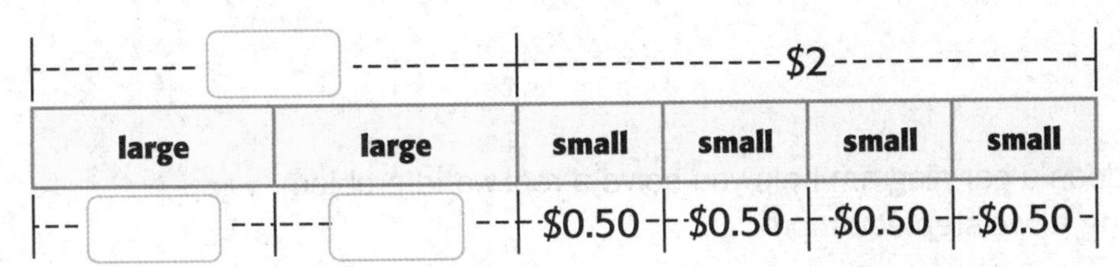

The cost of one large postcard is ______________________.

Collaborate

CCSS Reason Abstractly **Work with a partner. Use a bar diagram to write and solve an equation for each exercise.**

1. Brooke and two friends went to the movies and spent a total of $42. The movie tickets were $5 each and they each bought a popcorn combo. What is the cost of one popcorn combo?

2. Four medium postcards and 4 small postcards cost $5. What is the cost of one medium postcard?

$5							
medium	medium	medium	medium	small	small	small	small
				0.50	0.50	0.50	0.50

Reflect

3. **CCSS Model with Mathematics** Write and solve a word problem that could represent the bar diagram shown.

4. **Inquiry** HOW does a bar diagram help you solve a real-world problem involving a two-step equation?

Lesson 2

Solve Two-Step Equations

What You'll Learn

Scan the lesson. List two real-world scenarios in which you would solve two-step equations.

- ______________________________

- ______________________________

Essential Question

WHAT is equivalence?

Vocabulary

properties
two-step equation

Common Core State Standards

Content Standards
8.EE.7, 8.EE.7a, 8.EE.7b

Mathematical Practices
1, 2, 3, 4

Vocabulary Start-Up

Recall that in mathematics, **properties** are statements that are true for any number.

Complete the graphic organizer by matching the Property of Equality with the correct example.

Property	Example
Addition Property of Equality	$\frac{1}{2}x = 10$ $2 \cdot \frac{1}{2}x = 10 \cdot 2$
Division Property of Equality	$3x = 9$ $\frac{3x}{3} = \frac{9}{3}$
Multiplication Property of Equality	$x + 3 = 1$ $x + 3 - 3 = 1 - 3$
Subtraction Property of Equality	$x - 5 = 6$ $x - 5 + 5 = 6 + 5$

Real-World Link

A property in science is a trait of matter that is always true under a given set of conditions. For example, pure water freezes at 0°C. How is the definition of *property* similar in science and math?

____

Work Zone

Solve Two-Step Equations

A **two-step equation** contains two operations. In the equation $2x + 3 = 7$, x is multiplied by 2 and then 3 is added. To solve two-step equations, undo each operation in reverse order.

Example

1. Solve $2x + 3 = 7$.

Method 1 **Use a model.**

Remove three 1-tiles from each mat.

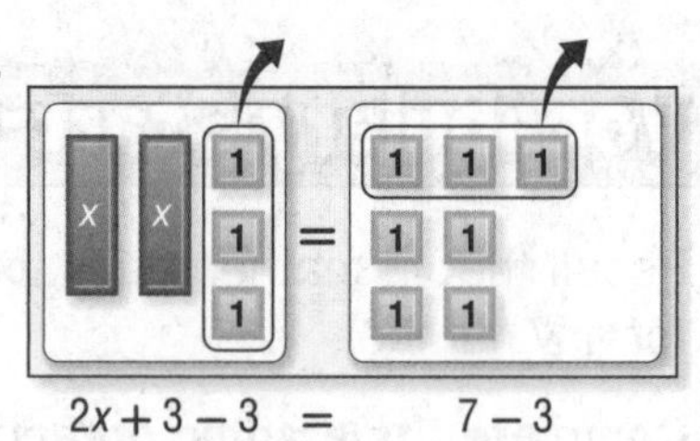

Separate the remaining tiles into 2 equal groups.

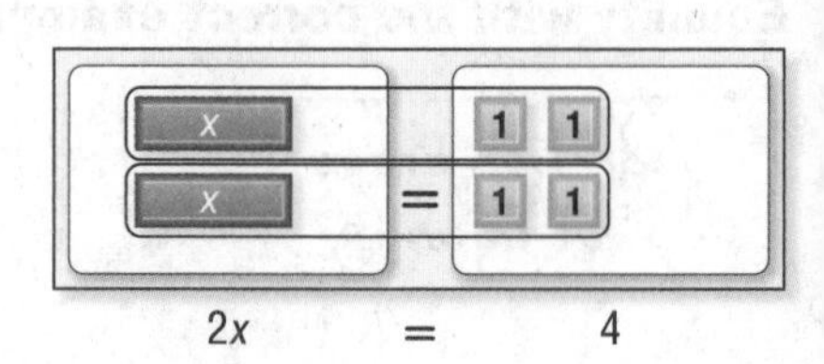

There are two 1-tiles in each group, so $x = 2$.

Method 2 **Use symbols.**

$2x + 3 = 7$	Write the equation.
$-3 = -3$	Subtraction Property of Equality
$2x = 4$	
$\frac{2x}{2} = \frac{4}{2}$	Division Property of Equality
$x = 2$	Simplify.

Using either method, the solution is 2.

a. ______________

b. ______________

Got It? **Do these problems to find out.**

a. $3x + 2 = 20$

b. $5 + 2n = -1$

Example

2. Solve $25 = \frac{1}{4}n - 3$.

$$25 = \frac{1}{4}n - 3$$ Write the equation.

$$+3 = \quad +3$$ Addition Property of Equality

$$28 = \frac{1}{4}n$$ Simplify.

$$4 \cdot 28 = 4 \cdot \frac{1}{4}n$$ Multiplication Property of Equality

$$112 = n$$

The solution is 112.

Show your work.

c. ______

d. ______

Got It? Do these problems to find out.

c. $-1 = \frac{1}{2}a + 9$ **d.** $\frac{2}{5}r - 5 = 7$

Example

3. Solve $6 - 3x = 21$.

$$6 - 3x = 21$$ Write the equation.

$$6 + (-3x) = 21$$ Rewrite the left side as addition.

$$-6 \quad = -6$$ Subtraction Property of Equality

$$-3x = 15$$ Simplify.

$$\frac{-3x}{-3} = \frac{15}{-3}$$ Division Property of Equality

$$x = -5$$ Simplify.

The solution is –5.

Check $6 - 3x = 21$ Write the equation.

$$6 - 3(-5) \stackrel{?}{=} 21$$ Replace x with –5.

$$6 - (-15) \stackrel{?}{=} 21$$ Multiply.

$$6 + 15 \stackrel{?}{=} 21$$ To subtract a negative number, add its opposite.

$$21 = 21 \checkmark$$ The sentence is true.

Common Error

A common mistake when solving the equation in Example 3 is to divide each side by 3 instead of –3. Since $6 - 3x = 6 + (-3x)$, the coefficient is –3.

Got It? Do these problems to find out.

e. $10 - \frac{2}{3}p = 52$ **f.** $-19 = -3x + 2$ **g.** $\frac{n}{-3} - 2 = -18$

e. ______

f. ______

g. ______

Example

4. **STEM** **Chicago's lowest recorded temperature in degrees Fahrenheit is −27°. Solve the equation $-27 = 1.8C + 32$ to convert to degrees Celsius.**

$-27 = 1.8C + 32$	Write the equation.
$\underline{-32 = \qquad -32}$	Subtraction Property of Equality
$-59 = 1.8C$	Simplify.
$\frac{-59}{1.8} = \frac{1.8C}{1.8}$	Division Property of Equality
$-32.8 \approx C$	Simplify. Check the solution.

So, Chicago's lowest recorded temperature is about −32.8 degrees Celsius.

Guided Practice

Solve each equation. Check your solution. (Examples 1–3)

1. $6x + 5 = 29$

2. $3 - 5y = -37$

3. $\frac{2}{3}x - 5 = 7$

4. Cassidy went to the movies with some of her friends. The tickets cost \$6.50 each, and they spent \$17.50 on snacks. The total amount paid was \$63.00. Solve the equation $63 = 6.50p + 17.50$ to determine how many people went to the movies. (Example 4)

5. **Building on the Essential Question** How can you use the *work backward* problem-solving strategy to solve a two-step equation?

Rate Yourself!

How confident are you about solving equations? Check the box square that applies.

For more help, go online to access a Personal Tutor.

Name ______________________ My Homework ______________________

Independent Practice

Go online for Step-by-Step Solutions eHelp

Solve each equation. Check your solution. (Examples 1–3)

1. $5 = 4a - 7$

2. $16 = 5x - 9$

3. $3 - 8c = 35$

4. $-\frac{1}{2}x - 7 = -11$

5. $15 - \frac{w}{4} = 28$

6. $-3 - 6x = 9$

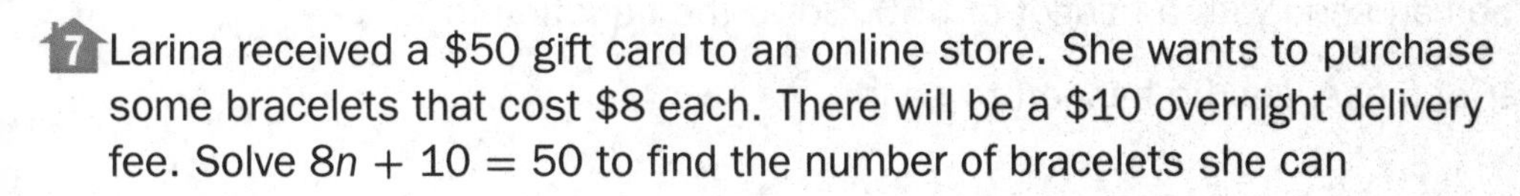

7. Larina received a \$50 gift card to an online store. She wants to purchase some bracelets that cost \$8 each. There will be a \$10 overnight delivery fee. Solve $8n + 10 = 50$ to find the number of bracelets she can purchase. (Example 4) ______________________

8. LaTasha paid \$75 to join a summer golf program. The course where she plays charges \$30 per round. Since she is a student, she receives a \$10 discount per round. If LaTasha spent \$375, use the equation $375 = 20g + 75$ to find how many rounds of golf LaTasha played. (Example 4) ______________________

Copy and Solve **Solve each equation. Show your work on a separate piece of paper.**

9. $\frac{a-4}{5} = 12$

10. $\frac{n+3}{8} = -4$

11. $\frac{6+z}{10} = -2$

12. CCSS **Reason Abstractly** If Mr. Arenth wants to put new carpeting in the room shown, how many square feet should he order?

____

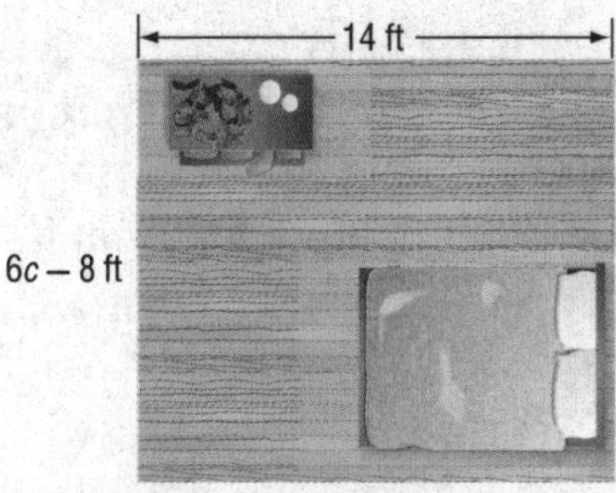

13. **Model with Mathematics** Refer to the graphic novel frame below for Exercises a–b.

a. The equation $50 = 28.10 + 0.15m$ represents the number of messages Jacob can send with a budget of $50. Solve the equation to find the number of messages he has left to send. ________

b. The equation $50 = 36.50 + 0.10m$ represents the number of messages Roberto can send with a budget of $50. Solve the equation to find the number of messages he has left to send. ________

H.O.T. Problems Higher Order Thinking

14. **Persevere with Problems** Solve $(x + 5)(x + 5) = 49$. (*Hint:* There are two solutions.)

____

15. **Model with Mathematics** Write a real-world problem that could be solved by using the equation $3x - 25 = 125$. Then solve the equation.

____

____

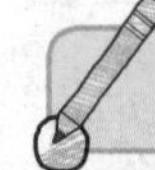

Standardized Test Practice

16. What is the value of m if $-6m + 4 = -32$?

Ⓐ 6
Ⓑ $4\frac{2}{3}$
Ⓒ $2\frac{1}{3}$
Ⓓ -6

Extra Practice

Solve each equation. Check your solution.

17. $2h + 9 = 21$

$$2h + 9 = 21$$
$$-9 = -9$$
$$\frac{2h}{2} = \frac{12}{2}$$
$$h = 6$$

Homework Help

18. $12 - \frac{3}{5}p = -27$

$$12 - \frac{3}{5}p = -27$$
$$-12 \quad = -12$$
$$-\frac{3}{5}p = -39$$
$$\left(-\frac{5}{3}\right)\left(-\frac{3}{5}p\right) = -39\left(-\frac{5}{3}\right)$$
$$p = 65$$

19. $11 = 2b + 17$

20. $-17 = 6p - 5$

21. $2g - 3 = -19$

22. $13 = \frac{g}{3} + 4$

23. $13 - 3d = -8$

24. $-\frac{2}{3}m - 4 = 10$

25. $-5y - 25 = 25$

26. Some friends decide to go to the aquarium together. Each person pays $7.50 to get in. They spend a total of $40 for the shark exhibit. The total cost is $70. Solve $7.5x + 40 = 70$ to find how many people went to the aquarium. ______

27. CCSS **Identify Structure** Brent had $26 when he went to the fair. After playing 7 games, he had $15.50 left. Solve $15.50 = 26 - 7p$ to find the price for each game. Then list the Properties of Equality you used to solve the equation.

______

Standardized Test Practice

28. The width of the rectangle below can be found by solving the equation $6w + 6 = 36$.

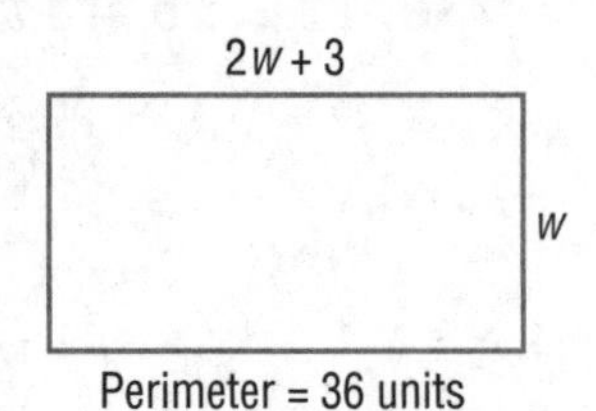

What is the width of the rectangle?

Ⓐ 4 units
Ⓑ 5 units
Ⓒ 6 units
Ⓓ 7 units

29. **Short Response** What value of y makes the equation true? ______

$$\frac{y}{4} - 7 = 3$$

30. What is the vaue of x in the following equation?

$$40 = -11 + 3x$$

Ⓕ -17
Ⓖ $-\frac{29}{3}$
Ⓗ $\frac{29}{3}$
Ⓘ 17

Common Core Review

Solve each equation. Check your solution. 6.EE.7

31. $t - 17 = 5$

32. $a - 5 = 14$

33. $9 = 5 + x$

Write and solve an equation for each of the following. 6.EE.7

34. Solomon is 9 years younger than his brother. His brother is 21. How old is Solomon? ______

35. Kelly spent \$45 more on boots than she did on a pair of jeans. She spent \$79.50 on the boots. How much did she spend on the jeans?

______

36. The product of two integers is 72. If one integer is 18, what is the other integer? ______

Lesson 3

Write Two-Step Equations

What You'll Learn

Scan the lesson. List two headings you would use to make an outline of the lesson.

- _______________
- _______________

Essential Question

WHAT is equivalence?

Common Core State Standards

Content Standards
8.EE.7, 8.EE.7a, 8.EE.7b

Mathematical Practices
1, 2, 3, 4

Real-World Link

Robotics You want to attend a two-week robotic day camp that costs $700. Your parents will pay the deposit of $400 if you pay the rest in weekly payments of $15. Use the questions below to help you find the number of weeks you will need to make payments.

1. Complete the table below. How much is paid after 2, 3, and 4 weeks?

Payments	Amount Paid
0	$400 + 15(0) = 400$
1	$400 + 15(1) = 415$
2	
3	
4	

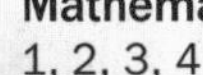

2. It will take a long time to solve the problem with a table. Instead, write and solve an equation to find the number of payments p you will need to make.

____

3. How many payments will you make? ☐

4. Suppose you received $75 in birthday money that you want to use towards the camp. Write and solve an equation to find the number of payments p you will need to make. _______________

____

Work Zone

Translate Sentences into Equations

There are three steps to writing a two-step equation.

Words	Describe the situation. Use only the most important words.
Variable	Define a variable to represent the unknown quantity.
Equation	Translate your verbal model into an algebraic equation.

You know how to write verbal sentences as one-step equations. Some verbal sentences translate into two-step equations.

Examples

Translate each sentence into an equation.

1. Eight less than three times a number is −23.

Words	Eight less than three times a number is −23.
Variable	Let n represent the number.
Equation	$3n - 8 = -23$

2. Thirteen is 7 more than one-fifth of a number.

Words	Thirteen is 7 more than one-fifth of a number.
Variable	Let n represent the number.
Equation	$13 = \frac{1}{5}n + 7$

STOP and Reflect

Name 3 words that indicate an addition statement.

Got It? Do these problems to find out.

a. Fifteen equals three more than six times a number.

b. Ten increased by the quotient of a number and 6 is 5.

c. The difference between 12 and $\frac{2}{3}$ of a number is 18.

a. ______

b. ______

c. ______

Examples

3. You buy 3 books that each cost the same amount and a magazine, all for $55.99. You know that the magazine costs $1.99. How much does each book cost?

Words Three books and a magazine cost $55.99.

Variable Let b represent the cost of one book.

Equation $3b + 1.99 = 55.99$

$$\begin{aligned} 3b + 1.99 &= 55.99 && \text{Write the equation.} \\ -1.99 &= -1.99 && \text{Subtraction Property of Equality} \\ 3b &= 54.00 && \text{Simplify.} \\ \frac{3b}{3} &= \frac{54.00}{3} && \text{Division Property of Equality} \\ b &= 18 && \text{Simplify.} \end{aligned}$$

So, the books each cost $18.

4. A personal trainer buys a weight bench for $500 and *w* weights for $24.99 each. The total cost of the purchase is $849.86. How many weights were purchased?

Words Bench plus $24.99 per weight equals $849.86

Variable Let w represent the number of weights.

Equation $500 + 24.99 \cdot w = 849.86$

$$\begin{aligned} 500 + 24.99w &= 849.86 && \text{Write the equation.} \\ -500 &= -500 && \text{Subtraction Property of Equality} \\ 24.99w &= 349.86 && \text{Simplify.} \\ \frac{24.99w}{24.99} &= \frac{349.86}{24.99} && \text{Division Property of Equality} \\ w &= 14 && \text{Simplify.} \end{aligned}$$

So, 14 weights were purchased.

Got It? **Do this problem to find out.**

d. The current temperature is 54°F. It is expected to rise 2.5°F each hour. In how many hours will the temperature be 84°F?

d. ______________

Example

5. Your and your friend's lunch cost \$19. Your lunch cost \$3 more than your friend's. How much was your friend's lunch?

Words	Your friend's lunch plus your lunch equals \$19.
Variable	Let f represent the cost of your friend's lunch.
Equation	$f + f + 3 = 19$

$$f + f + 3 = 19 \quad \text{Write the equation.}$$
$$2f + 3 = 19 \quad f + f = 2f$$
$$-3 = -3 \quad \text{Subtraction Property of Equality}$$
$$2f = 16 \quad \text{Simplify.}$$
$$\frac{2f}{2} = \frac{16}{2} \quad \text{Division Property of Equality}$$
$$f = 8 \quad \text{Simplify.}$$

Your friend spent \$8.

Defining the Variable
When the equation is solved, you can refer back to the definition of the variable to see if the question is answered or if additional steps are needed.

Guided Practice

Translate each sentence into an equation. (Examples 1 and 2)

1. One more than three times a number is 7. ______

2. Seven less than one-fourth of a number is −1. ______

3. The quotient of a number and 5, less 10, is 3. ______

4. You already owe \$4.32 in overdue rental fees and are returning a movie that is 4 days late. Now you owe \$6.48. Define a variable. Then write and solve an equation to find the daily fine for an overdue movie. (Examples 3–5)

______

5. **Building on the Essential Question** Why is it important to define a variable before writing an equation? ______

______

______

______

Rate Yourself!

☐ I understand how to write two-step equations.

Great! You're ready to move on!

☐ I still have some questions about writing two-step equations.

No Problem! Go online to access a Personal Tutor.

Independent Practice

eHelp Go online for Step-by-Step Solutions

Translate each sentence into an equation. (Examples 1 and 2)

1. Four less than five times a number is equal to 11. ______

2. Fifteen more than half a number is 9. ______

3. Six less than seven times a number is equal to -20. ______

4. Eight more than four times a number is -12. ______

Define a variable. Then write and solve an equation to solve each problem. (Examples 3–5)

5. **Financial Literacy** The cost for a certain music plan is $9.99 per year plus $0.25 per song you download. If you paid $113.74 one year, find the number of songs you downloaded. ______

6. Amy has saved $725 for a new guitar and lessons. Her guitar costs $475, and guitar lessons are $25 per hour. Determine how many hours of lessons she can afford. ______

7. From ground level to the tip of the torch, the Statue of Liberty and its pedestal are 92.99 meters tall. The pedestal is 0.89 meter taller than the statue. How tall is the Statue of Liberty?

______

8. CCSS **Reason Abstractly** Elsie would like to take snowboarding lessons at Powder Mountain. She has saved $550 for lessons and a junior season pass. How many more semi-private lessons than private lessons can she take? ______

Powder Mountain Ski Resort Snowboarding Lessons	
Semi-Private	$45/lesson
Private	$60/lesson
Junior Season Pass	$315

9. When diving, the peregrine falcon can reach speeds of up to 175 miles per hour. Write and solve equations to find each of the following.

   a. The top speed of a peregrine falcon is 20 miles per hour less than three times the top speed of a cheetah. What is the cheetah's top speed? ______________________

   b. A sailfish can swim up to 1 mile per hour less than one fifth the top speed of a peregrine falcon. Find the top speed that a sailfish can swim. ______________________

   c. The peregrine falcon can reach speeds about 13 miles per hour more than 6 times the speed of the fastest human. What is the approximate top speed of the fastest human? ______________

H.O.T. Problems Higher Order Thinking

10. **CCSS Model with Mathematics** If 12 less than 4 times a number is 8, the number is 5. Write a different sentence where the unknown number is also 5. ______________________

11. **CCSS Persevere with Problems** The ages of three siblings combined is 27. The oldest is twice the age of the youngest. The middle child is 3 years older than the youngest. Write and solve an equation to find the ages of each sibling. ______________________

    ____

12. **CCSS Model with Mathematics** Write about a real-world situation that can be solved using a two-step equation. Then write the equation and solve the problem. ______________________

    ____

    ____

Standardized Test Practice

13. Kimberly needs $45 to go to the amusement park. She has $13. She earns $8 per hour working at her job. The equation $8h + 13 = 45$ shows this relationship. How many hours does Kimberly need to work to earn enough money to go to the park?

    Ⓐ 8　　Ⓒ 6

    Ⓑ 7　　Ⓓ 4

Extra Practice

Translate each sentence into an equation.

14. Twenty-two less than three times a number is −70. $3n - 22 = -70$

Words Twenty-two less than three times a number is −70.

Variable Let n represent the number.

Equation $3n - 22 = -70$

15. The product of a number and 4 increased by 16 is −2. ______

16. Twelve less than the one-fifth of a number is −7. ______

17. Six more than nine times a number is 456. ______

Define a variable. Then write and solve an equation to solve each problem.

18. It costs $13 for admission to an amusement park, plus $1.50 for each ride. If you have a total of $35.50 to spend, what is the greatest number of rides you can go on? ______

19. Trey went to the batting cages to practice hitting. He rented a helmet for $4 and paid $0.75 for each group of 20 pitches. If he spent a total of $7 at the batting cages, how many groups of pitches did he pay for?

______

20. CCSS **Make a Conjecture** Hunter and Amado are each trying to save $600 for a summer trip. Hunter started with $150 and earns $7.50 per hour working at a grocery store. Amado has nothing saved, but he earns $12 per hour painting houses.

a. Make a conjecture about who will take longer to save enough money for the trip. Justify your reasoning. ______

______

______

b. Write and solve two equations to check your conjecture.

______

Standardized Test Practice

21. A company employs 72 workers. It plans to increase the number of employees by 6 per month until it has twice its current workforce. Which equation can be used to determine m, the number of months it will take for the number of employees to double?

Ⓐ $6m + 72m = 144$

Ⓑ $2m + 72 = 144$

Ⓒ $2(6m + 72) = 144$

Ⓓ $6m + 72 = 144$

22. What is the value of x in the following equation?

$$-3x + 4 = -23$$

Ⓕ -9

Ⓖ $-6\frac{1}{3}$

Ⓗ $6\frac{1}{3}$

Ⓘ 9

23. Short Response The table shows the number of baseball cards in two baseball collections. If Marcus and James have 120 cards altogether, write and solve an equation that could be used to find the number of cards in Marcus' collection. ______

Person	Cards
Marcus	m
James	$2m + 6$

Common Core Review

Solve each equation. Check your solution. 7.EE.4

24. $\frac{y}{7} = 22$

25. $\frac{a}{6} = -108$

26. $-6 = \frac{n}{8} + 1$

27. $-15 = -4p + 9$

28. In a recent NFL game, the Green Bay Packers scored 14 points less than the Tennessee Titans. Write and solve an equation to find the total points the Tennessee Titans scored. 6.EE.7 ______

Preseason Week 4	
Team	**Total Points**
Packers	17
Titans	p

Problem-Solving Investigation
Work Backward

Content Standards
8.EE.7
Mathematical Practices
1, 4, 7

Case #1 Game Switcheroo!

Hector and Alex traded video games. Alex gave Hector one fourth of his video games in exchange for 6 video games. Then he sold 3 video games and gave 2 video games to his brother. Alex ended up with 16 video games.

How many video games did Alex have when he started?

Understand *What are the facts?*

- Alex now has 16 games.
- He gave some away, sold some, and traded some.

Plan *What is your strategy to solve this problem?*

Start with the ending number of video games, 16, and work backward.

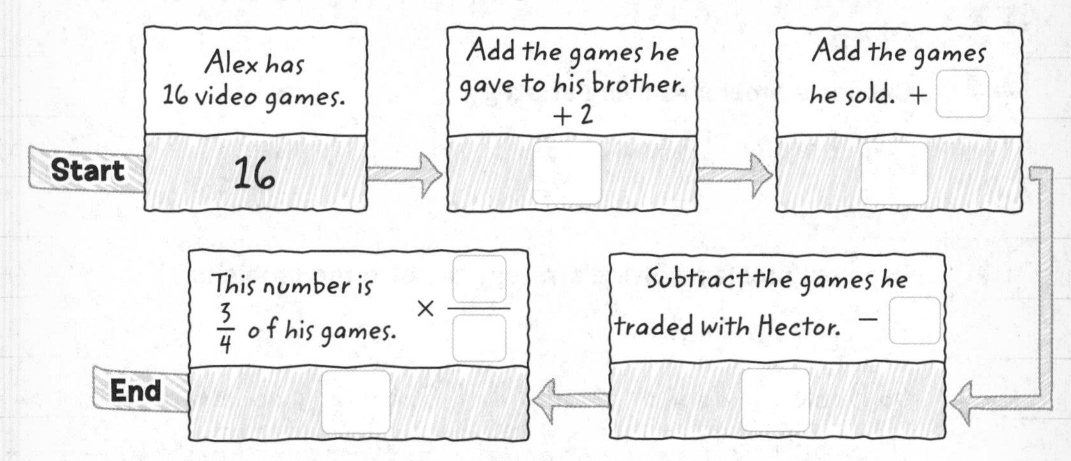

Solve *How can you apply the strategy?*

So, Alex had ______ video games at the beginning.

Check *Does the answer make sense?*

Start with 20. Perform operations in reverse order.

Analyze the Strategy

 **Identify Structure** How is working backward similar to solving an equation?

Case #2 Shoot the Rapids

Aurora raised money for a white water rafting trip. Jacy made the first donation. Guillermo's donation was twice Jacy's donation. Rosa's mother tripled what Aurora had raised so far. Now Aurora has \$120.

How much did Jacy donate?

1 Understand

Read the problem. What are you being asked to find?

I need to find ______________________.

Underline key words and values. What information do you know?

Jacy donated __________. Guillermo ______________________ and Rosa's mother ______________ the whole amount collected.

Is there any information that you do *not* need to know?

I do not need to know ______________________.

2 Plan

Choose a problem-solving strategy.

I will use the ______________________ strategy.

3 Solve

Use your problem-solving strategy to solve the problem.

Aurora raised a total of __________.

Go Back Divide that amount by 4. One part is Guillermo's donation and three parts is the amount donated by Rosa's mother. $\$120 \div 4 =$ __________

Go Back Divide that amount by 3. One part is for Jacy's donation and two parts is the amount that Guillermo donated. ______ $\div 3 =$ __________

Jacy was the first to donate. So, Jacy donated __________.

4 Check

Use information from the problem to check your answer.

Begin with \$10 and perform operations in reverse. ______ $\times 2 = \$20$;

$\$20 + \$10 =$ ______; ______ $\times 3 = \$90$; ______ + ______ = ______.

Collaborate Work with a small group to solve the following cases. Show your work on a separate piece of paper.

Case #3 Shopping

Janelle has $75. She buys jeans that are on sale for half price and then uses an in-store coupon for $10 off. After paying $1.80 in sales tax, she receives $37.20 in change.

What was the original price of the jeans?

Case #4 Schedule

Nyoko needs to be at school at 7:45 A.M. It takes her 25 minutes to walk to school, 25 minutes to eat breakfast, and 35 minutes to get dressed.

What time should Nyoko get up to be at school 5 minutes early?

Case #5 Money

At the end of the month, Mr. Copley had $1,475 in his checking account. His checkbook showed the following transactions.

What was his balance at the beginning of the month?

Chk No.	Date	Payment or Withdrawal		Deposit	
				$150	00
132		$45	00		
		$100	00		
133		$18	50		
				$250	00

Case #6 Geometry

Study the pattern below.

Draw the next two figures in the pattern.

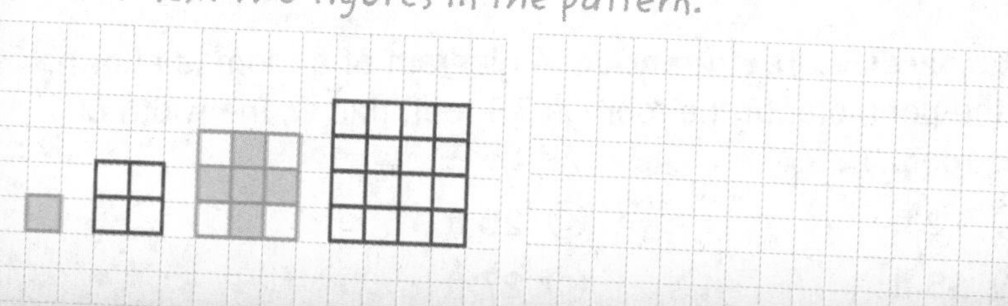

Circle the strategy below to solve the problem.

- Look for a pattern.
- Act it out.
- Use logical reasoning.
- Guess, check, and revise.

Mid-Chapter Check

Vocabulary Check

1. CCSS **Be Precise** Define *multiplicative inverse.* Give an example of a number and its multiplicative inverse. (Lesson 1)

____

____

2. Fill in the blank in the sentence below with the correct term. (Lesson 2)

The first step in solving the equation $3x + 4 = 20$ is to ____________ from each side. This is an example of the ____________ Property of ____________.

Skills Check and Problem Solving

Solve each equation. Check your solution. (Lessons 1–2)

Show your work.

3. $\frac{2}{3}x = -8$

4. $-4.5 = -0.15p$

5. $2\frac{1}{3}c = 2\frac{1}{10}$

6. $3m + 5 = 14$

7. $-2k + 7 = -3$

8. $11 = \frac{1}{3}a + 2$

9. **Standardized Test Practice** A diagram of a room is shown. If the perimeter of the room is 78 feet, find w, the width of the room. (Lesson 3)

Ⓐ 12 ft
Ⓑ 15 ft
Ⓒ 25 ft
Ⓓ 27 ft

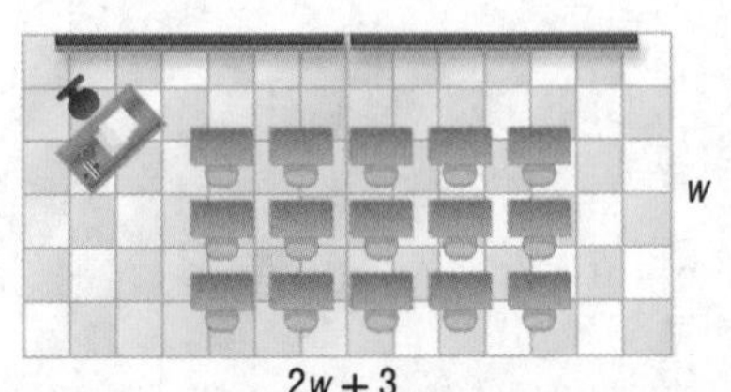

Inquiry Lab

Equations with Variables on Each Side

HOW do you use the Properties of Equality when solving an equation using algebra tiles?

Content Standards
8.EE.7, 8.EE.7a

Mathematical Practices
1, 3, 5

Shopping Leah bought 4 pens and a bottle of nail color. Her sister bought 2 of the same pens and 4 bottles of nail color, and spent the same amount as Leah. The nail color cost $2. Use algebra tiles to find the cost of each pen.

Investigation 1

The equation $4x + 2 = 2x + 8$ represents the situation above. Use algebra tiles to solve the equation.

Step 1 Model the equation.

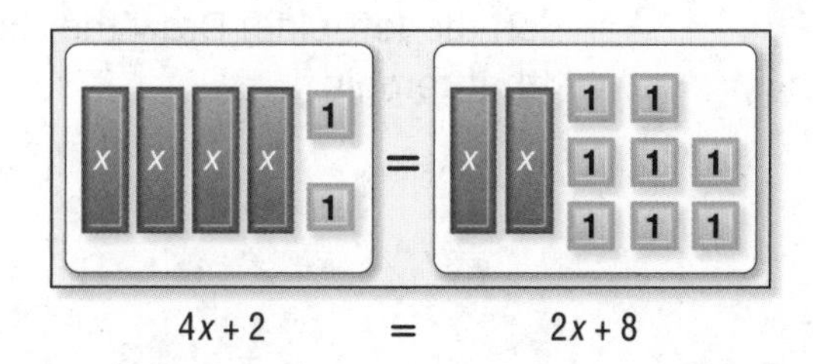

Step 2 Remove ☐ x-tiles from each side of the mat until there are x-tiles on only one side.

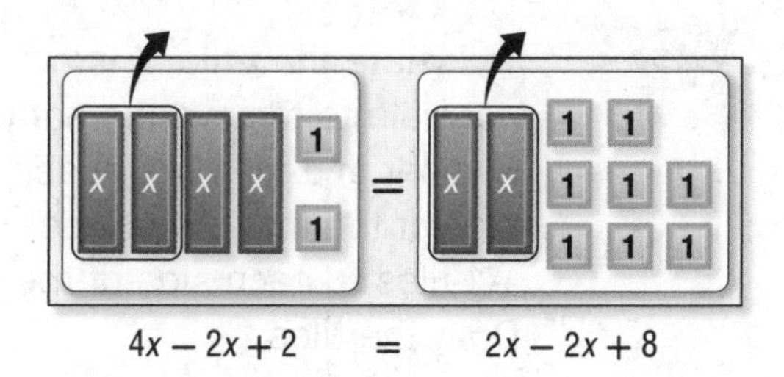

Step 3 Remove ☐ 1-tiles from each side of the mat until the x-tiles are by themselves on one side.

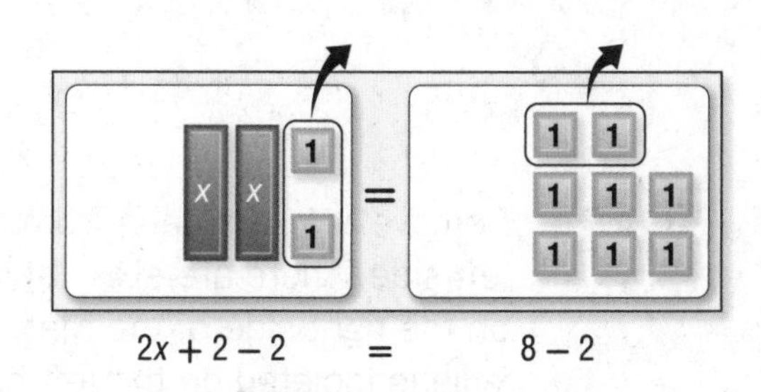

Step 4 Separate the tiles in ☐ equal groups.

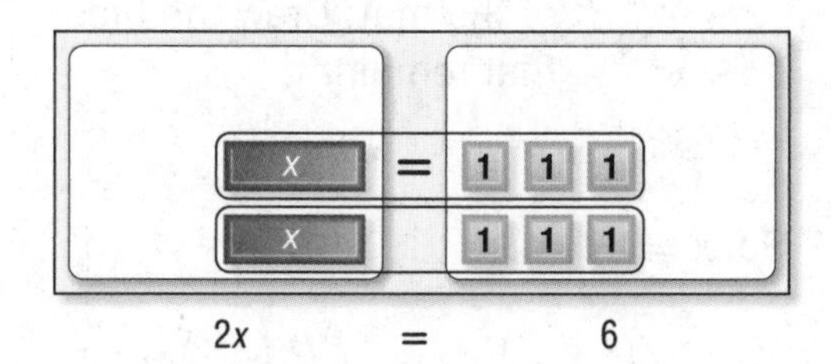

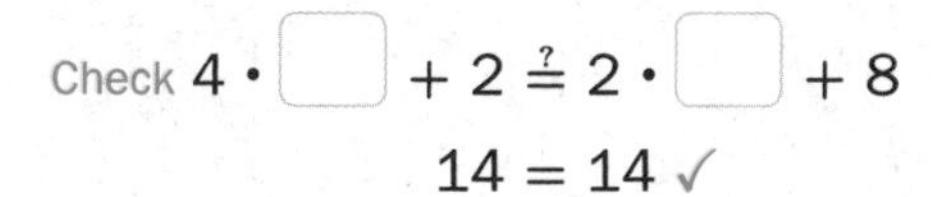

Check $4 \cdot$ ☐ $+ 2 \stackrel{?}{=} 2 \cdot$ ☐ $+ 8$

$14 = 14$ ✓

So, each pen costs $ ☐.

Investigation 2

Use algebra tiles to solve $3x + 3 = 2x - 3$. Draw the tiles in the blank mats shown. The first one is done for you.

Step 1 Model the equation.

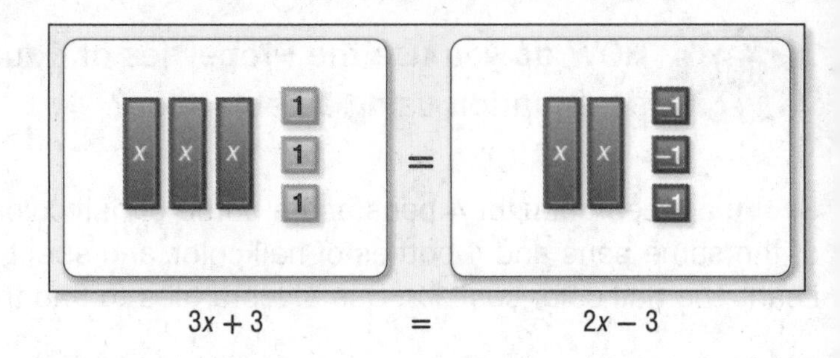

$3x + 3 \quad = \quad 2x - 3$

Step 2 Remove 2 x-tiles from each side of the mat in Step 1 so that there is an x-tile by itself on the left side. Draw the tiles that remain.

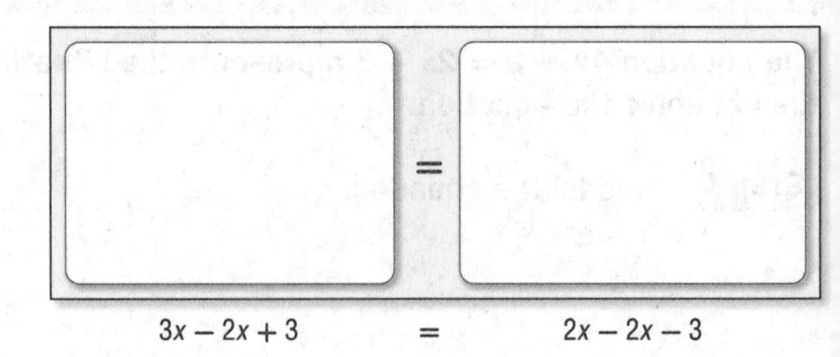

$3x - 2x + 3 \quad = \quad 2x - 2x - 3$

Step 3 To isolate the x-tile, it is not possible to remove the same number of 1-tiles from each side of the mat. Add three −1-tiles to each side of the mat. Draw the tiles.

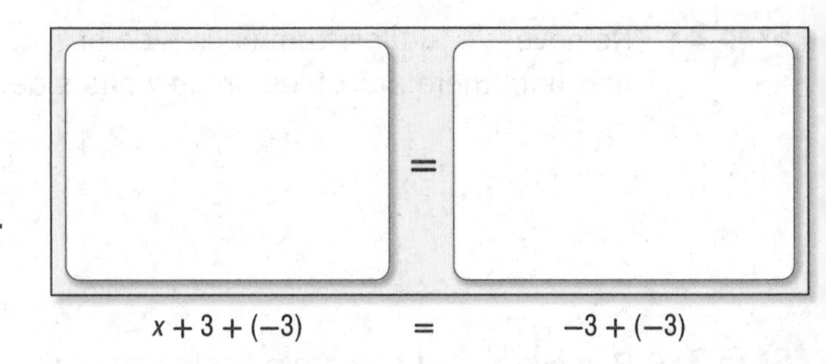

$x + 3 + (-3) \quad = \quad -3 + (-3)$

Step 4 Remove the zero pairs from the left side. There are six −1-tiles on the right side of the mat. The x-tile is isolated on the left side of the mat. Draw the tiles that remain.

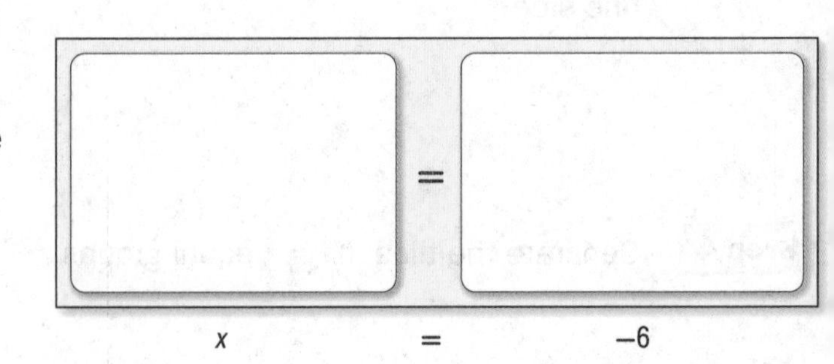

$x \quad = \quad -6$

So, $x =$ ☐.

Check $3(\square) + 3 \stackrel{?}{=} 2(\square) - 3$

$-15 = -15$ ✓ The solution is correct.

CCSS Use Math Tools Work with a partner. Solve each equation. Show your work using drawings. Write the solution below the mat.

1. $x + 2 = 2x + 1$

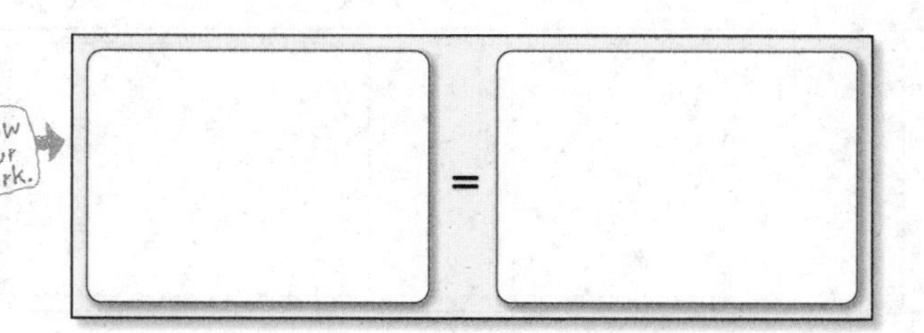

$x =$ ______

2. $2x + 7 = 3x + 4$

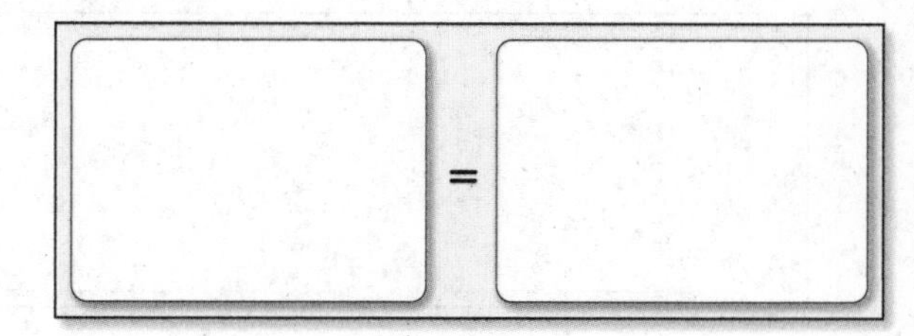

$x =$ ______

3. $2x - 5 = x - 7$

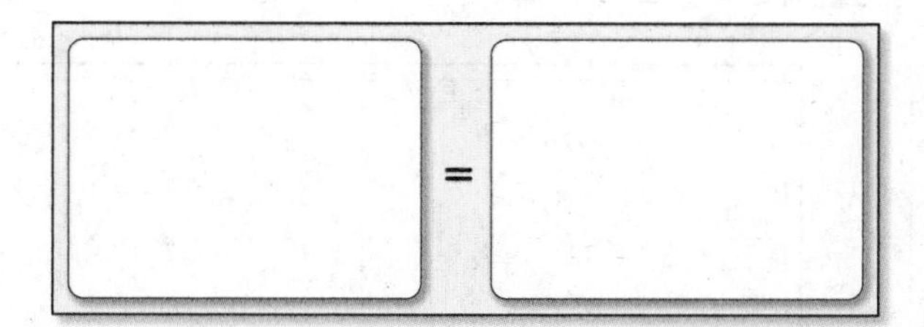

$x =$ ______

4. $x + 6 = 3x - 2$

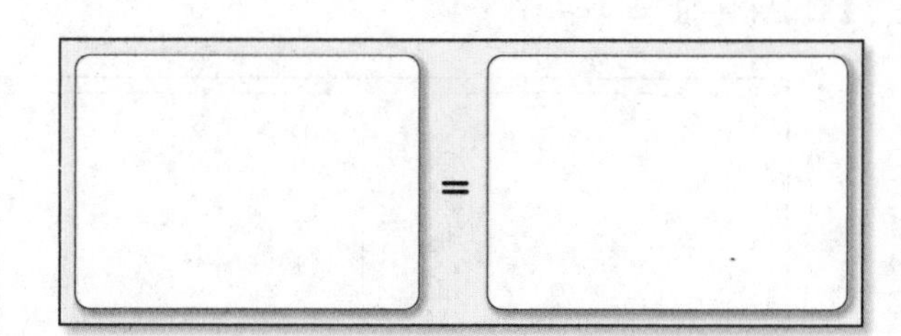

$x =$ ______

5. $8 + x = 3x$

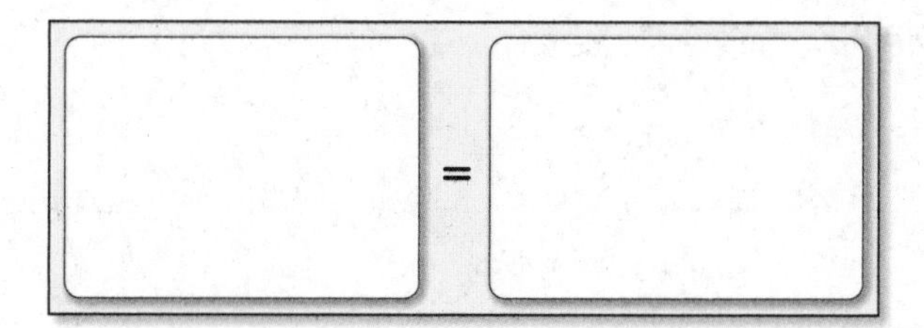

$x =$ ______

6. $3x + 6 = 6x$

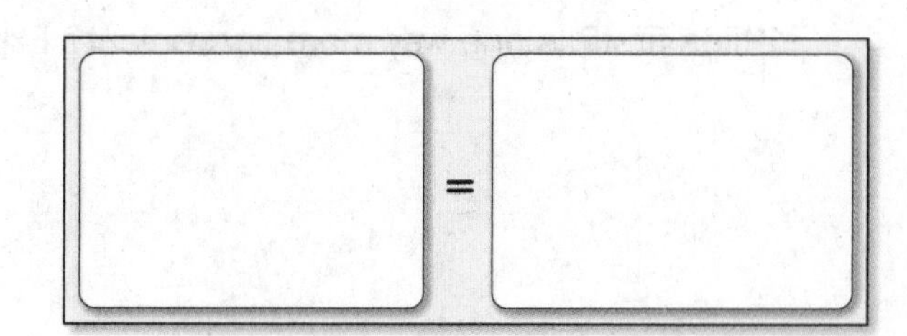

$x =$ ______

7. $3x + 3 = x - 5$

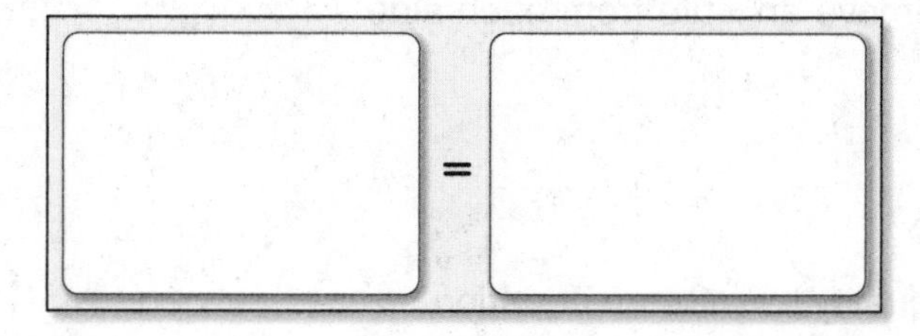

$x =$ ______

8. $2x + 5 = 4x - 1$

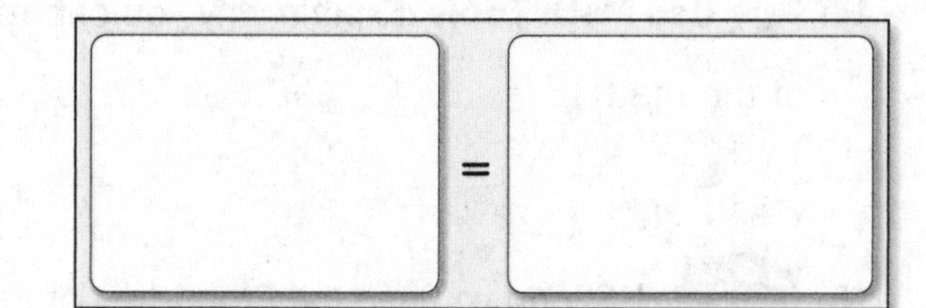

$x =$ ______

Analyze

Work with a partner. One of you should solve the following equations by removing 1-tiles first. The other one should solve the equations by removing x-tiles first. Compare your answers.

9. $x + 4 = 3x - 4$

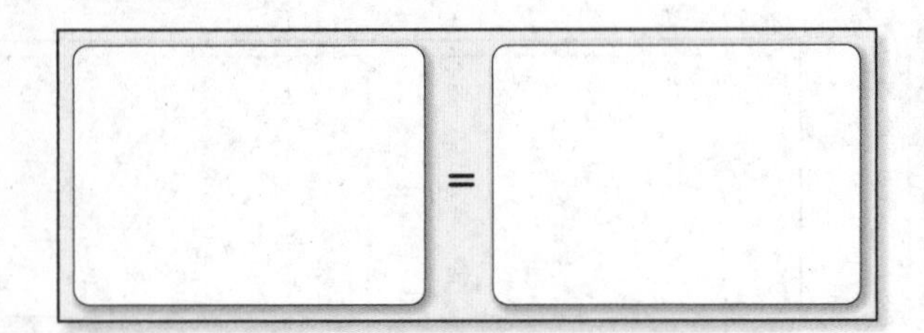

$x =$ ______

10. $x + 2 = 2x - 3$

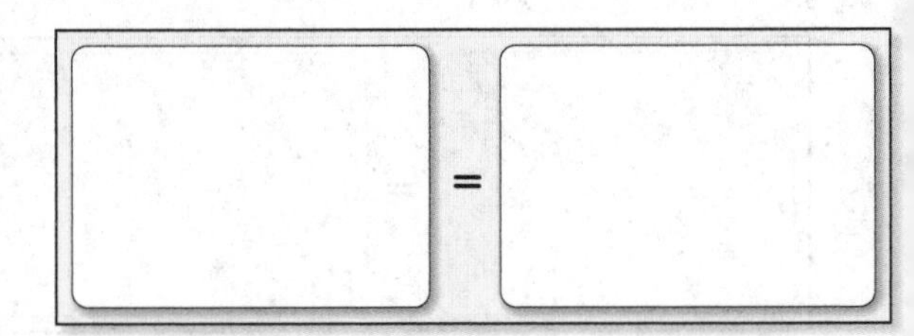

$x =$ ______

11. $2x + 1 = x - 7$

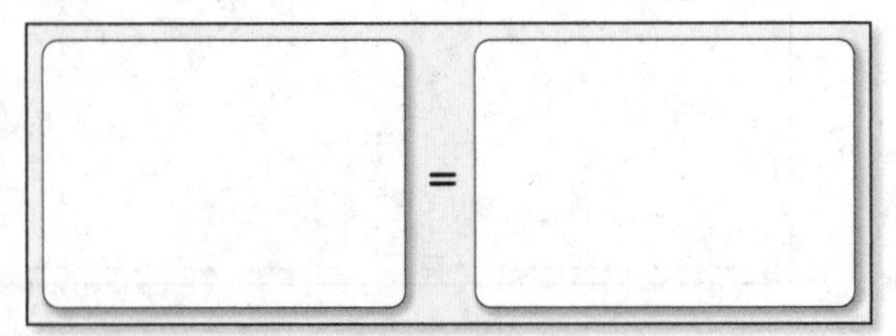

$x =$ ______

12. $4x + 2 = x - 4$

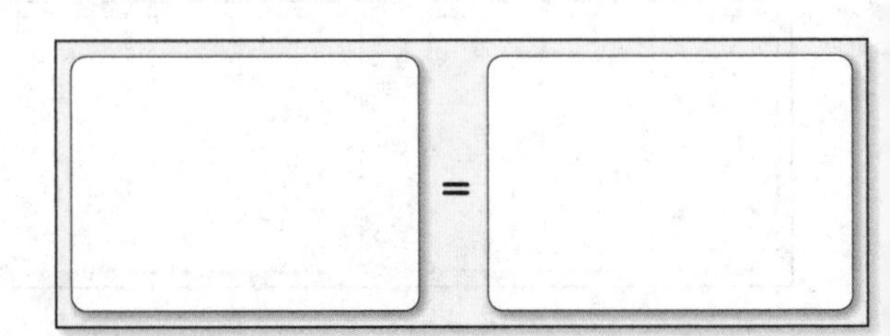

$x =$ ______

13. CCSS **Reason Inductively** Does it matter whether you remove x-tiles or 1-tiles first? Is one way more convenient? Explain. ______

Reflect

14. CCSS **Use Math Tools** Explain why you can remove an x-tile from each side of the mat. ______

15. Inquiry HOW do you use the Properties of Equality to solve an equation using algebra tiles? ______

Lesson 4

Solve Equations with Variables on Each Side

What You'll Learn

Scan the lesson. Predict two things you will learn about solving equations when the variable is on each side.

- ______
- ______

Essential Question

WHAT is equivalence?

Common Core State Standards

Content Standards
8.EE.7, 8.EE.7a, 8.EE.7b

Mathematical Practices
1, 3, 4

Real-World Link

Cell Phones A wireless company offers two cell phone plans. Plan A charges \$24.95 per month plus \$0.10 per minute for calls. Plan B charges \$19.95 per month plus \$0.20 per minute. Use the questions to find when the two plans cost the same.

1. Complete the table.

Minutes (m)	Plan A $24.95 + 0.10m$	Plan B $19.95 + 0.20m$
10		
20		
30		
40		
50		
60		
70		

2. For what value(s) does Plan A cost less?

______

3. For what value(s) does Plan B cost less?

______

4. For what value(s) do both Plans cost the same?

______

Work Zone

Equations with Variables on Each Side

Some equations, like $8 + 4d = 5d$, have variables on each side of the equals sign. To solve, use the properties of equality to write an equivalent equation with the variables on one side of the equals sign. Then solve the equation.

Examples

1. Solve $8 + 4d = 5d$. Check your solution.

$$
\begin{aligned}
8 + 4d &= 5d && \text{Write the equation.}\\
-4d &= -4d && \text{Subtraction Property of Equality}\\
8 &= d && \text{Simplify by combining like terms.}
\end{aligned}
$$

Subtract $4d$ from the left side of the equation to isolate the variable.

Subtract $4d$ from the right side of the equation to keep it balanced.

To check your solution, replace d with 8 in the original equation.

Check
$$
\begin{aligned}
8 + 4d &= 5d && \text{Write the original equation.}\\
8 + 4(8) &\stackrel{?}{=} 5(8) && \text{Replace } d \text{ with 8.}\\
40 &= 40 \checkmark && \text{The sentence is true.}
\end{aligned}
$$

2. Solve $6n - 1 = 4n - 5$.

$$
\begin{aligned}
6n - 1 &= 4n - 5 && \text{Write the equation.}\\
-4n \quad &= -4n && \text{Subtraction Property of Equality}\\
2n - 1 &= -5 && \text{Simplify.}\\
+1 &= +1 && \text{Addition Property of Equality}\\
2n &= -4 && \text{Simplify.}\\
n &= -2 && \text{Mentally divide each side by 2.}
\end{aligned}
$$

Check
$$
\begin{aligned}
6n - 1 &= 4n - 5 && \text{Write the original equation.}\\
6(-2) - 1 &\stackrel{?}{=} 4(-2) - 5 && \text{Replace } n \text{ with } -2.\\
-13 &= -13 \checkmark && \text{The sentence is true.}
\end{aligned}
$$

Got It? Do these problems to find out.

Solve each equation. Check your solution.

a. $8a = 5a + 21$

b. $3x - 7 = 8x + 23$

a. ____________

b. ____________

Example

3. Green's Gym charges a one time fee of $50 plus $30 per session for a personal trainer. A new fitness center charges a yearly fee of $250 plus $10 for each session with a trainer. For how many sessions is the cost of the two plans the same?

Words	fee of $50 plus $30 per session	is the same as	a fee of $250 plus $10 per session.
Variable	Let s represent the number of sessions.		
Equation	$50 + 30s = 250 + 10s$		

$50 + 30s = 250 + 10s$ Write the equation.

$\quad - 10s = \quad - 10s$ Subtraction Property of Equality

$50 + 20s = 250$ Simplify.

$-50 \quad = -50$ Addition Property of Equality

$20s = 200$ Simplify.

$\frac{20s}{20} = \frac{200}{20}$ Division Property of Equality

$s = 10$ Simplify.

So, the cost is the same for 10 personal trainer sessions.

Check

Green's Gym: $50 plus 10 sessions at $30 per session

$50 + 10 \cdot 30 = 50 + 300$

$= \$350$

new fitness center: $250 plus 10 sessions at $10 per session

$250 + 10 \cdot 10 = 250 + 100$

$= \$350$ ✓

Got It? Do this problem to find out.

c. The length of a flag is 0.3 foot less than twice its width. If the perimeter is 14.4 feet longer than the width, find the dimensions of the flag.

c. ______

Equations with Rational Coefficients

In some equations, the coefficients of the variables are rational numbers. Remember when working with fractions, you need to have a common denominator before you add or subtract.

Example

4. Solve $\frac{2}{3}x - 1 = 9 - \frac{1}{6}x$.

$\frac{4}{6}x - 1 = 9 - \frac{1}{6}x$	The common denominator of the coefficients is 6. Rewrite the equation.
$+\frac{1}{6}x \quad = \quad +\frac{1}{6}x$	Addition Property of Equality
$\frac{5}{6}x - 1 = 9$	Simplify.
$+1 = +1$	Addition Property of Equality
$\frac{5}{6}x = 10$	Simplify.
$\left(\frac{6}{5}\right)\frac{5}{6}x = 10\left(\frac{6}{5}\right)$	Multiplication Property of Equality
$x = 12$	Simplify.

Show your work.

e. ______________

f. ______________

Got It? Do these problems to find out.

e. $\frac{1}{2}p + 7 = \frac{3}{4}p + 9$

f. $-\frac{5}{4}c - \frac{1}{2} = -\frac{3}{4} + \frac{5}{8}c$

Guided Practice

Solve each equation. Check your solution. (Examples 1, 2, 4)

1. $5n + 9 = 2n$

2. $7y - 8 = 6y + 1$

3. $\frac{3}{5}x - 15 = \frac{6}{5}x + 12$

4. EZ Car Rental charges \$40 a day plus \$0.25 per mile. Ace Rent-A-Car charges \$25 a day plus \$0.45 per mile. What number of miles results in the same cost for one day? (Example 3) ______________

5. **Building on the Essential Question** How is solving an equation with the variable on each side similar to solving a two-step equation? ______________

____

____

____

Rate Yourself!

How well do you understand how to solve equations? Circle the figure that applies.

Clear — Somewhat Clear — Not So Clear

For more help, go online to access a Personal Tutor.

Name ______________________ My Homework ______________________

Independent Practice

Go online for Step-by-Step Solutions

Solve each equation. Check your solution. (Examples 1, 2, 4)

1. $7a + 10 = 2a$

2. $11x = 24 + 8x$

3. $8y - 3 = 6y + 17$

4. $5p + 2 = 4p - 1$

5. $15 - \frac{1}{6}n = \frac{1}{6}n - 1$

6. $3 - \frac{2}{9}b = \frac{1}{3}b - 7$

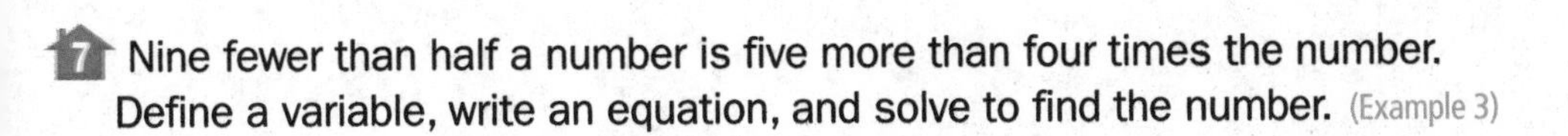

7. Nine fewer than half a number is five more than four times the number. Define a variable, write an equation, and solve to find the number. (Example 3)

8. The table shows ticket prices for the local minor league baseball team for fan club members and non-members. For how many tickets is the cost the same for club members and non-members? (Example 3)

	Ticket Prices	
	Club Members	Non-Club Members
Membership Fee (one-time)	$30	none
Ticket Price	$3	$6

9. **Multiple Representations** Refer to the square at the right.

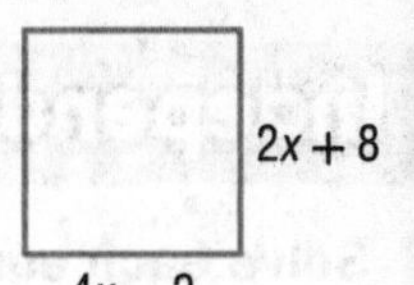

a. **Words** Explain a method you could use to find the value of x.

b. **Symbols** Write an equation to find the side length of the square.

c. **Algebra** What is the side length of the square?

H.O.T. Problems Higher Order Thinking

10. **Find the Error** Alma is solving the equation $4a - 5 = 2a - 3$. Circle her mistake and correct it.

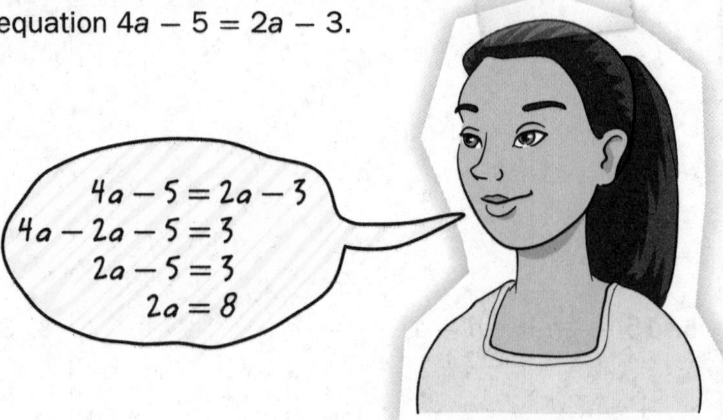

11. **Model with Mathematics** Write a word problem that can be solved using the equation $5x = 3x + 20$.

12. **Persevere with Problems** Find the area of the rectangle at the right.

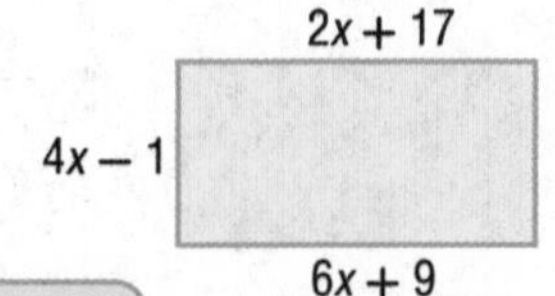

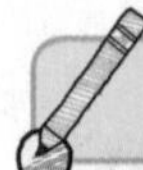

Standardized Test Practice

13. What is the solution of the following equation?

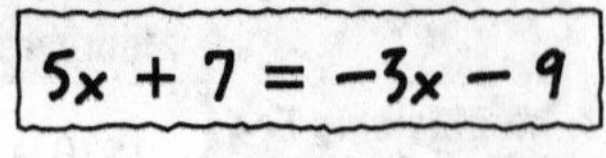

Ⓐ −2
Ⓑ 1
Ⓒ 2
Ⓓ 8

Name ______________________ My Homework ______________________

Extra Practice

Solve each equation. Check your solution.

14. $9g - 14 = 2g$

mework Help

$$\begin{aligned} 9g - 14 &= 2g \\ -9g \quad &= -9g \\ \hline \frac{-14}{-7} &= \frac{-7g}{-7} \\ 2 &= g \end{aligned}$$

15. $-6f + 13 = 2f - 11$

16. $2.5h - 15 = 4h$

17. $2z - 31 = -9z + 24$

18. Will averages 18 points a game and is the all-time scoring leader on his team with 483 points. Tom averages 21 points a game and is currently second on the all-time scorers list with 462 points. If both players continue to play at the same rate, how many more games will it take until Tom and Will have scored the same number of total points?

____

19. Eighteen less than three times a number is twice the number. Define a variable, write an equation, and solve to find the number.

____

CCSS Reason Abstractly **Write an equation to find the value of *x* so that each pair of polygons has the same perimeter. Then solve.**

20.

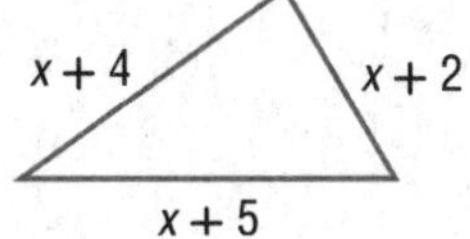

x + 1

x + 3

____

21.

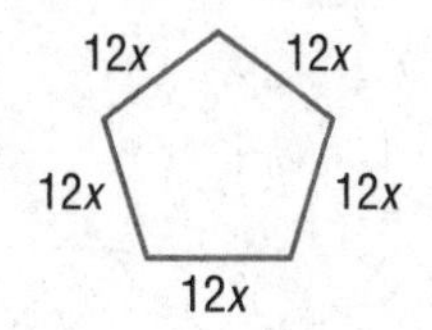

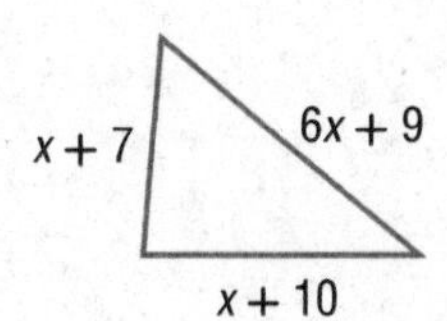

____

Standardized Test Practice

22. Find the value of x so that the polygons have the same perimeter.

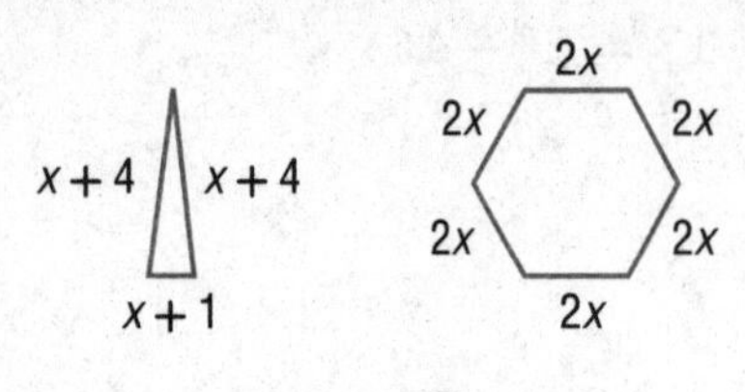

Ⓐ 4
Ⓑ 3
Ⓒ 2
Ⓓ 1

23. Which of the following equations has a solution of 5?

Ⓕ $-12x - 6 = -10x + 4$
Ⓖ $12x - 6 = 10x + 4$
Ⓗ $12x + 6 = 10x - 4$
Ⓘ $12x - 6 = 10x - 4$

24. Carpet cleaner A charges \$28.25 plus \$18 a room. Carpet cleaner B charges \$19.85 plus \$32 a room. Which equation can be used to find the number of rooms for which the total cost of both carpet cleaners is the same?

Ⓐ $28.25x + 18 = 19.85x + 32$
Ⓑ $28.25 + 32x = 19.85 + 18x$
Ⓒ $28.25 + 18x = 19.85 + 32x$
Ⓓ $(28.25 + 18)x = (19.85 + 32)x$

Common Core Review

Write each expression in simplest form. 7.EE.2

25. $5x + 6 - x =$ ______

26. $8 - 3n + 3n =$ ______

27. $7a - 7a - 9 =$ ______

28. $3 - 4y + 9y =$ ______

Use the Distributive Property to write each expression as an equivalent expression. 7.EE.1

29. $6(x + 5) =$ ______

30. $-8(y - 1) =$ ______

31. $-3(-5z + 12) =$ ______

32. $\frac{1}{3}(6z + 10) =$ ______

Lesson 5

Solve Multi-Step Equations

What You'll Learn

Scan the lesson. Write the definitions of null set and identity.

- ____________________
- ____________________

Essential Question

WHAT is equivalence?

Vocabulary

null set
identity

Math Symbols

∅ null set

{ } empty set

Common Core State Standards

Content Standards
8.EE.7, 8.EE.7a, 8.EE.7b

Mathematical Practices
1, 2, 3, 4

Real-World Link

Lacrosse Coach Everly wants to order uniform shirts for all the players p on her women's lacrosse team. Each shirt costs \$20. There is an additional cost d for a player to put her name on the shirt. Use the steps below to write an equation for the total cost c if every player on the team orders a shirt with her name on it.

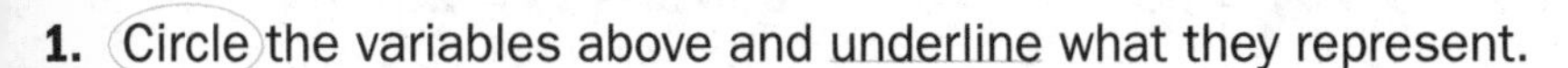

1. Circle the variables above and underline what they represent.

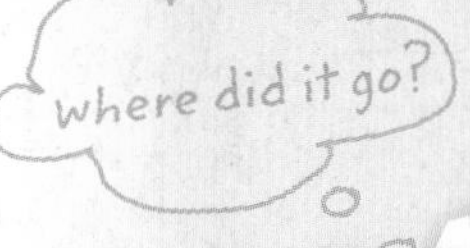

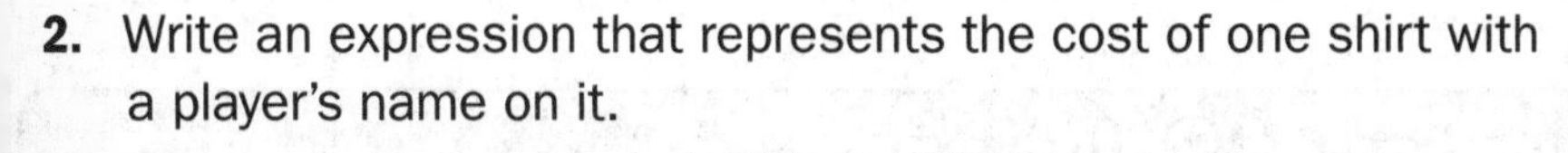

2. Write an expression that represents the cost of one shirt with a player's name on it.

   ☐ + ☐

   cost of shirt + cost of name

3. Use the expression to write an equation that can be used to find the total cost if every player on the team orders a shirt with her name on it.

   ☐ (☐ + ☐) = c

   number of players (cost of shirt + cost of name) = total cost

4. Suppose the total cost for 15 players to buy shirts is \$420. Write an equation to show the total cost of the shirts if all of the players put their names on the shirts.

   ____

Work Zone

Solve Multi-Step Equations

Some equations contain expressions with grouping symbols. To solve these equations, first expand the expression using the Distributive Property. Then collect like terms if needed, and solve the equation using the Properties of Equality.

Example

1. Solve $15(20 + d) = 420$.

$15(20 + d) = 420$	Write the equation.
$300 + 15d = 420$	Distributive Property
$-300 \quad = -300$	Subtraction Property of Equality
$15d = 120$	Simplify.
$\frac{15d}{15} = \frac{120}{15}$	Division Property of Equality
$d = 8$	Simplify.

a. ____________

b. ____________

Got It? **Do these problems to find out.**

a. $-3(9 + x) = 33$

b. $5(a - 7) = 25$

Key Concept

Number of Solutions

	Null Set	One Solution	Identity
Words	no solution	one solution	infinitely many solutions
Symbols	$a = b$	$x = a$	$a = a$
Example	$3x + 4 = 3x$ $4 = 0$ Since $4 \neq 0$, there is no solution.	$2x = 20$ $x = 10$	$4x + 2 = 4x + 2$ $2 = 2$ Since $2 = 2$, the solution is all numbers.

Some equations have no solution. When this occurs, the solution is the **null set** or empty set and is shown by the symbol ∅ or { }. Other equations may have every number as their solution. An equation that is true for every value of the variable is called an **identity**.

Examples

2. Solve $6(x - 3) + 10 = 2(3x - 4)$.

$6(x - 3) + 10 = 2(3x - 4)$	Write the equation.
$6x - 18 + 10 = 6x - 8$	Distributive Property
$6x - 8 = 6x - 8$	Collect like terms.
$+8 = \quad +8$	Addition Property of Equality
$6x = 6x$	Simplify.
$\frac{6x}{6} = \frac{6x}{6}$	Division Property of Equality
$x = x$	Simplify.

The statement $x = x$ is *always* true. The equation is an identity and the solution set is all numbers.

Check		
	$6(x - 3) + 10 = 2(3x - 4)$	Write the original equation.
	$6(5 - 3) + 10 \stackrel{?}{=} 2[3(5) - 4]$	Substitute any value for x.
	$6(2) + 10 \stackrel{?}{=} 2(15 - 4)$	Simplify.
	$22 = 22$ ✓	

3. Solve $8(4 - 2x) = 4(3 - 5x) + 4x$.

$8(4 - 2x) = 4(3 - 5x) + 4x$	Write the equation.
$32 - 16x = 12 - 20x + 4x$	Distributive Property
$32 - 16x = 12 - 16x$	Collect like terms.
$+16x = \quad +16x$	Addition Property of Equality
$32 = 12$	Simplify.

The statement $32 = 12$ is *never* true. The equation has no solution and the solution set is $\varnothing$.

Check		
	$8(4 - 2x) = 4(3 - 5x) + 4x$	Write the equation.
	$8[4 - 2(2)] \stackrel{?}{=} 4[3 - 5(2)] + 4(2)$	Substitute any value for x.
	$8(0) \stackrel{?}{=} 4(-7) + 8$	Simplify.
	$0 \neq -20$ ✓	Since $0 \neq -20$, the equation has no solution.

STOP and Reflect

How do you know if the solution $5 = 0$ indicates no solution, one solution, or infinitely many solutions?

Got It? Do these problems to find out.

c. $3(6 - 4x) = -2(6x - 9)$

d. $2(3x + 5) = 5(2x - 4) - 4x$

Show your work.

c. ____________

d. ____________

Example

4. At the fair, Hunter bought 3 snacks and 10 ride tickets. Each ride ticket costs $1.50 less than a snack. If he spent a total of $24.00, what was the cost of each snack?

Write an equation to represent the problem.

$3s + 10(s - 1.5) = 24$	Write the equation.
$3s + 10s - 15 = 24$	Distributive Property
$13s - 15 = 24$	Collect like terms.
$+15 = +15$	Addition Property of Equality
$13s = 39$	Simplify.
$\frac{13s}{13} = \frac{39}{13}$	Division Property of Equality
$s = 3$	Simplify.

So, the cost of each snack was $3.

Guided Practice

Solve each equation. Check your solution. (Examples 1–3)

1. $-8(w - 6) = 32$

2. $8z - 22 = 3(3z + 11) - z$

3. Mr. Richards's class is holding a canned food drive for charity. Juliet collected 10 more cans than Rosana. Santiago collected twice as many cans as Juliet. If they collected 130 cans altogether, how many cans did Juliet collect? (Example 4) ______

4. **Building on the Essential Question** How many possible solutions are there to a linear equation in one variable? Describe each one.

______

______

______

______

Rate Yourself!

Are you ready to move on? Shade the section that applies.

YES ? NO

For more help, go online to access a Personal Tutor.

FOLDABLES *Time to update your Foldable!*

Independent Practice

Go online for Step-by-Step Solutions

Solve each equation. Check your solution. (Examples 1–3)

1. $-12(k + 4) = 60$

2. $8(3a + 6) = 9(2a - 4)$

3. $\frac{1}{3}h - 4\left(\frac{2}{3}h - 3\right) = \frac{2}{3}h - 6$

4. $8(c - 9) = 6(2c - 12) - 4c$

Copy and Solve **Solve each equation. Show your work on a separate piece of paper.** (Examples 2 and 3)

5. $-10y + 18 = -3(5y - 7) + 5y$

6. $8(t + 2) - 3(t - 4) = 6(t - 7) + 8$

7. $4(5 + 2x) - 5 = 3(3x + 7)$

8. $6(2x - 8) + 3 = 15$

9. The school has budgeted $2,000 for an end-of-year party at the local park. The cost to rent the park shelter is $150. How much can the student council spend per student on food if each of the 225 students receives a $3.50 gift? (Example 4) ______

10. CCSS **Reason Abstractly** The table shows the number of students in each homeroom.

    a. Write an equation to find the number of students in Mr. Boggs's homeroom if the total number of students is 90. ______

    ______

    b. Solve the equation from part **a** to find the number of students in Mr. Boggs's homeroom. ______

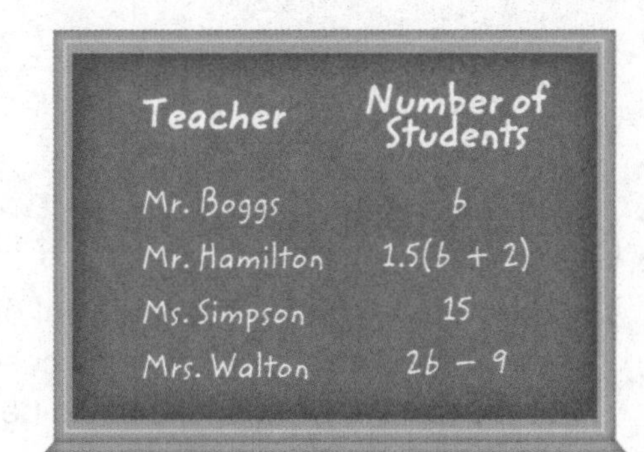

Teacher	Number of Students
Mr. Boggs	b
Mr. Hamilton	$1.5(b + 2)$
Ms. Simpson	15
Mrs. Walton	$2b - 9$

11. **Model with Mathematics** Refer to the graphic novel frame below for Exercises a–b.

a. Write an equation that can be used to determine the number of text messages Jacob and Roberto can send for their plans to cost the same.

b. Solve the equation from part **a** to find the number of text messages each person can send for their costs to be the same.

H.O.T. Problems Higher Order Thinking

12. **Reason Inductively** Does a multi-step equation *always, sometimes,* or *never* have a solution? Explain your reasoning.

13. **Persevere with Problems** The perimeter of a rectangle is $8(2x + 1)$ inches. If the length of the sides of the rectangle are $3x + 4$ inches and $4x + 3$ inches, what is the length of each side of the rectangle?

Standardized Test Practice

14. The Yeoman family spent a total of \$26.75 on lunch. They bought 5 drinks and 3 sandwiches. Each drink costs \$2.50 less than a sandwich. Which of the following equations could be used to find the cost of each sandwich?

Ⓐ \$26.75 = 5(\$2.50) + 3s

Ⓑ \$26.75 = 3(\$2.50) + 5s

Ⓒ \$26.75 = 5s + 3(s + 2.50)

Ⓓ \$26.75 = 3s + 5(s − \$2.50)

Name ______________________ My Homework ______________________

Extra Practice

Solve each equation. Check your solution.

15. $9(j - 4) = 81$

$$9j - 36 = 81$$
$$+36 = +36$$
$$9j = 117$$
$$\frac{9j}{9} = \frac{117}{9}$$
$$j = 13$$

16. $8(4q - 5) - 7q = 5(5q - 8)$

$$32q - 40 - 7q = 25q - 40$$
$$25q - 40 = 25q - 40$$
$$-25q \quad = -25q$$
$$-40 = -40$$

The solution set is all numbers.

17. $\frac{1}{2}r + 2\left(\frac{3}{4}r - 1\right) = \frac{1}{4}r + 6$

18. $-5(3m + 6) = -3(4m - 2)$

19. $-7(k + 9) = 9(k - 5) - 14k$

20. $10p - 2(3p - 6) = 4(3p - 6) - 8p$

21. $12(x + 3) = 4(2x + 9) + 4x$

22. $0.2(x + 50) - 6 = 0.4(3x + 20)$

23. CCSS **Identify Structure** Give an example of a multi-step equation for each of the following solutions.

a. all numbers ______________________

b. null set ______________________

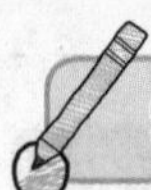

Standardized Test Practice

24. What is the solution of the equation?

$-2(3x + 1) - 2x = -4(2x) - 4$

Ⓐ $x = \frac{1}{2}$

Ⓑ $x = -\frac{1}{2}$

Ⓒ all numbers

Ⓓ no solution

25. What value of x makes the perimeters of the figures below equal?

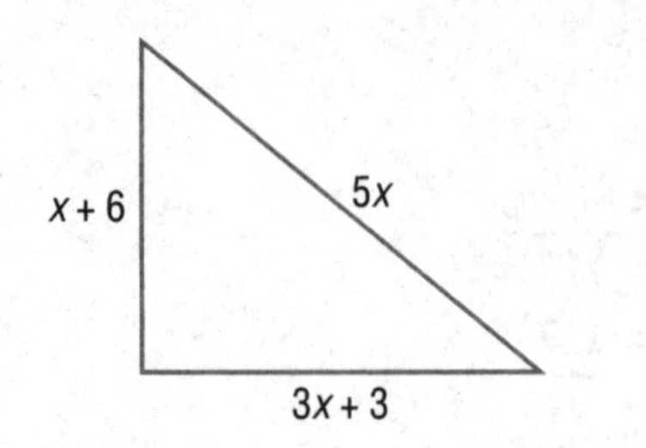

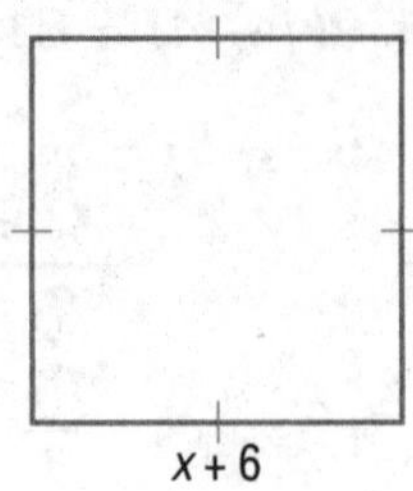

Ⓕ 2
Ⓖ 3
Ⓗ 4
Ⓘ 5

Common Core Review

Solve each inequality. Graph the solution set on a number line. 7.EE.4b

26. $a - 5 < 2$ ______

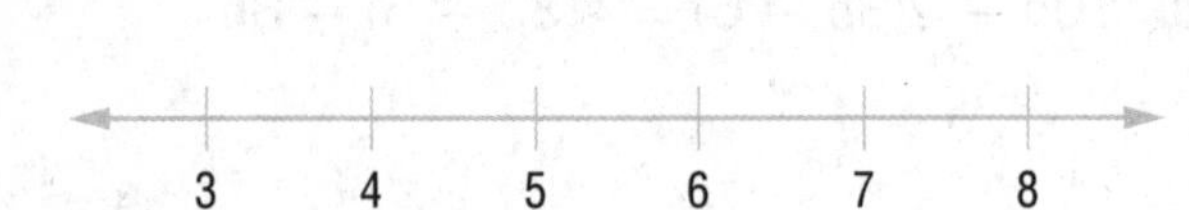

27. $x - 9 \geq -12$ ______

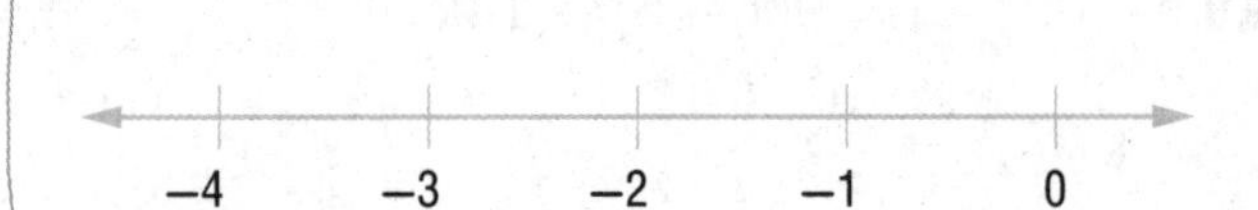

28. $5y \leq -30$ ______

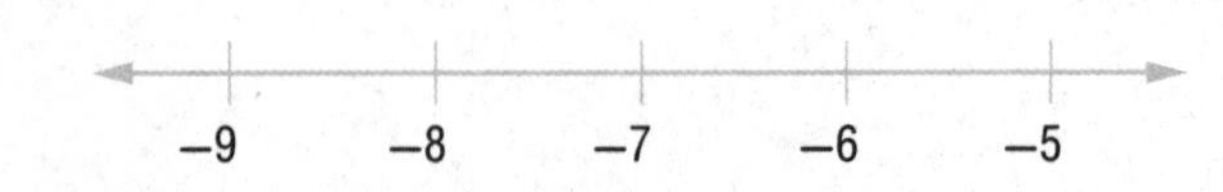

29. $-\frac{n}{4} > -2$ ______

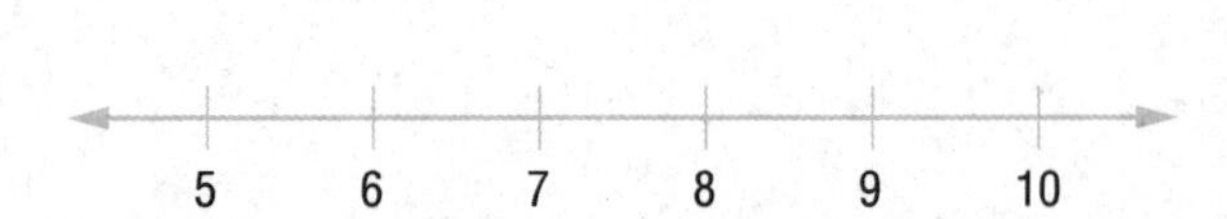

30. $4h - 7 \leq 13$ ______

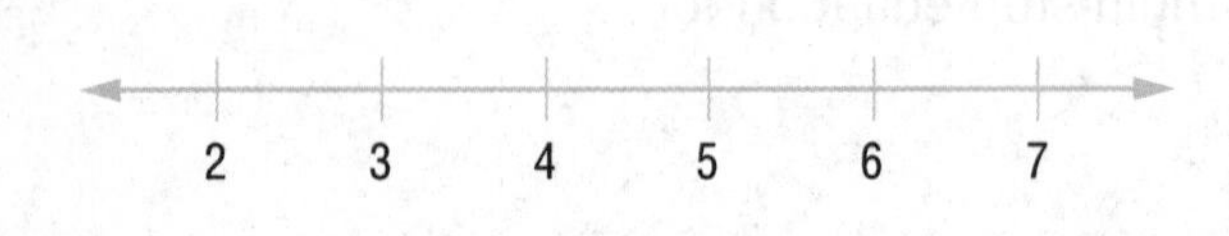

31. $-3m + 5 > 17$ ______

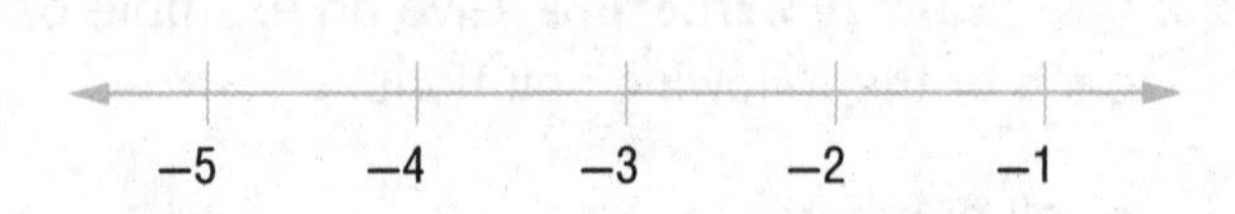

21ST CENTURY CAREER in Design

Skateboard Designer

If you love the sport of skateboarding, and you are creative and have strong math skills, you should think about a career designing skateboards. A skateboard designer applies engineering principles and artistic ability to design high-performance skateboards that are both strong and safe. To have a career in skateboard design, you should study physics and mathematics and have a good understanding of skateboarding.

Explore college and careers at ccr.mcgraw-hill.com

Is This the Career for You?

Are you interested in becoming a skateboard designer? Take some of the following courses in high school.

- Digital Design
- Geometry
- Physics
- Trigonometry

Turn the page to find out how math relates to a career in Design.

It's Great to Skate

Use the information in the table to solve each problem.

1. The total width of two standard shortboards and a technical shortboard is 23.5 inches. Write an equation to represent the situation.

2. Solve the equation from Exercise 1 to find the width of a standard shortboard.

3. The total length of two longboards and a standard shortboard is 113.4 inches. Write and solve an equation to find the length of a longboard.

4. The total width of three technical shortboards is 4.5 inches more than the total width of two longboards. Write and solve an equation to find the width of a longboard.

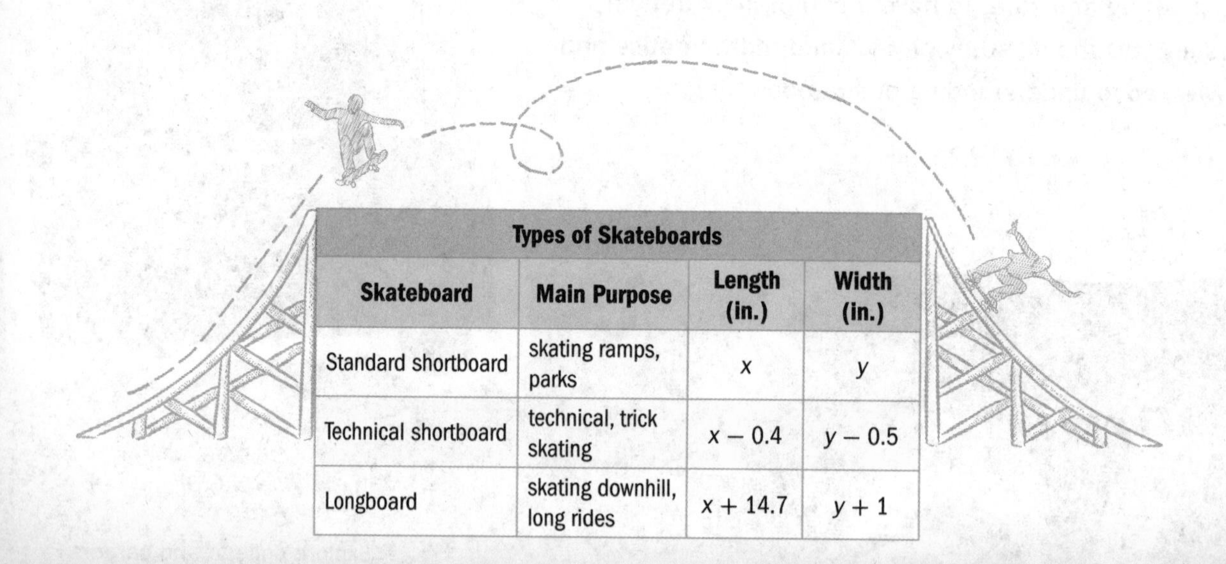

Types of Skateboards

Skateboard	Main Purpose	Length (in.)	Width (in.)
Standard shortboard	skating ramps, parks	x	y
Technical shortboard	technical, trick skating	$x - 0.4$	$y - 0.5$
Longboard	skating downhill, long rides	$x + 14.7$	$y + 1$

Career Project

It's time to update your career portfolio! Describe the skills that would be necessary for a skateboard designer to possess. Determine whether this type of career would be a good fit for you.

What problem-solving skill might you use as a skateboard designer?

Chapter Review

Vocabulary Check

Unscramble each of the clue words. After unscrambling all of the terms, use the numbered letters to find the name of a famous mathematician.

FOCICTIENEF

2 11

TECVALMIITUPLI SEVNEIR

6

13 3

DEYTINTI

4 9

LLUNETS

PERRITPEOS

1

REABIASVL

10 5

PEOCILRACR

12 8 7

1 2 3 4 5 6 7 8 9 10 W 11 12 13

Complete each sentence using one of the unscrambled words above.

1. The ____________ is the numerical factor of a term that contains a variable.
2. Another name for the reciprocal is the ____________.
3. In mathematics, statments that are true for all numbers are ____________.
4. When writing an equation from a real-world problem, it is important to define the ____________.

Key Concept Check

Use Your FOLDABLES

Use your Foldable to help review the chapter.

Tape here

Tab 1 Solving Equations	Write About It	Solve $6(x-3) + 10 = 2(4x-5)$	Tab 2

Got it?

Number the steps in the order needed to solve each equation. Then solve the equation.

1. $5(x + 3) = 170$

2 Subtract 15 from each side.

1 Multiply x and 3 by 5.

3 Divide each side by 5.

$x =$ 31

2. $2p - 9 = 6p + 7$

___ Divide each side by 4.

___ Subtract 7 from each side.

___ Subtract $2p$ from each side.

$p =$ ______

3. $-\frac{2}{3}(a + 3) = \frac{5}{3}a - 19$

___ Add $\frac{2}{3}a$ to each side.

___ Add 19 to each side.

___ Multiply a and 3 by $-\frac{2}{3}$.

___ Multiply each side by $\frac{3}{7}$.

$a =$ ______

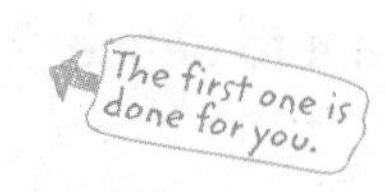

Problem Solving

1. The area of Arizona covered by desert is about 5,880 square miles. If 42% of the total area is desert, about how many square miles is Arizona's total area? (Lesson 1) ______

Flagstaff
Phoenix
Tucson
Nogales

2. Four adults spend \$37 for admission and \$3 for parking at the zoo. Solve the equation $4a + 3 = 40$ to find the cost of admission per person.

(Lesson 2) ______

3. CCSS **Reason Abstractly** Jerome completes 8 extra credit problems on the first day and then 4 problems each day until the worksheet is complete. There are 28 problems on the worksheet. Write and solve an equation to find how many days it will take Jerome to complete the worksheet after the first day. (Lesson 3) ______

4. Elin wants to fence in two different garden plots in her back yard with two rolls of fencing that are the same length. Write and solve an equation to find the value of x so that the figures below have the same perimeter. (Lesson 4) ______

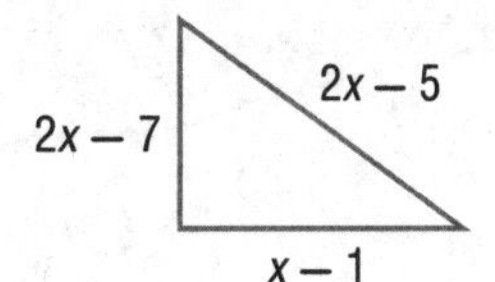

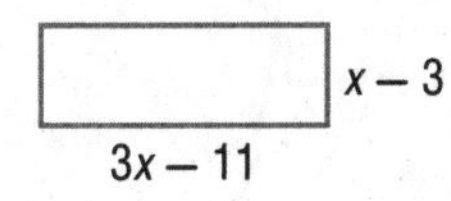

______

5. **Financial Literacy** Mr. and Mrs. Hawkins have budgeted \$500 for Marion's graduation party. The cost to rent the room is \$150. How much can they spend per person on food if each of the 30 guests receives a \$2.50 group photo? (Lesson 5) ______

Reflect

Answering the Essential Question

Use what you learned about equivalence to complete the graphic organizer. Draw or write an example for each category.

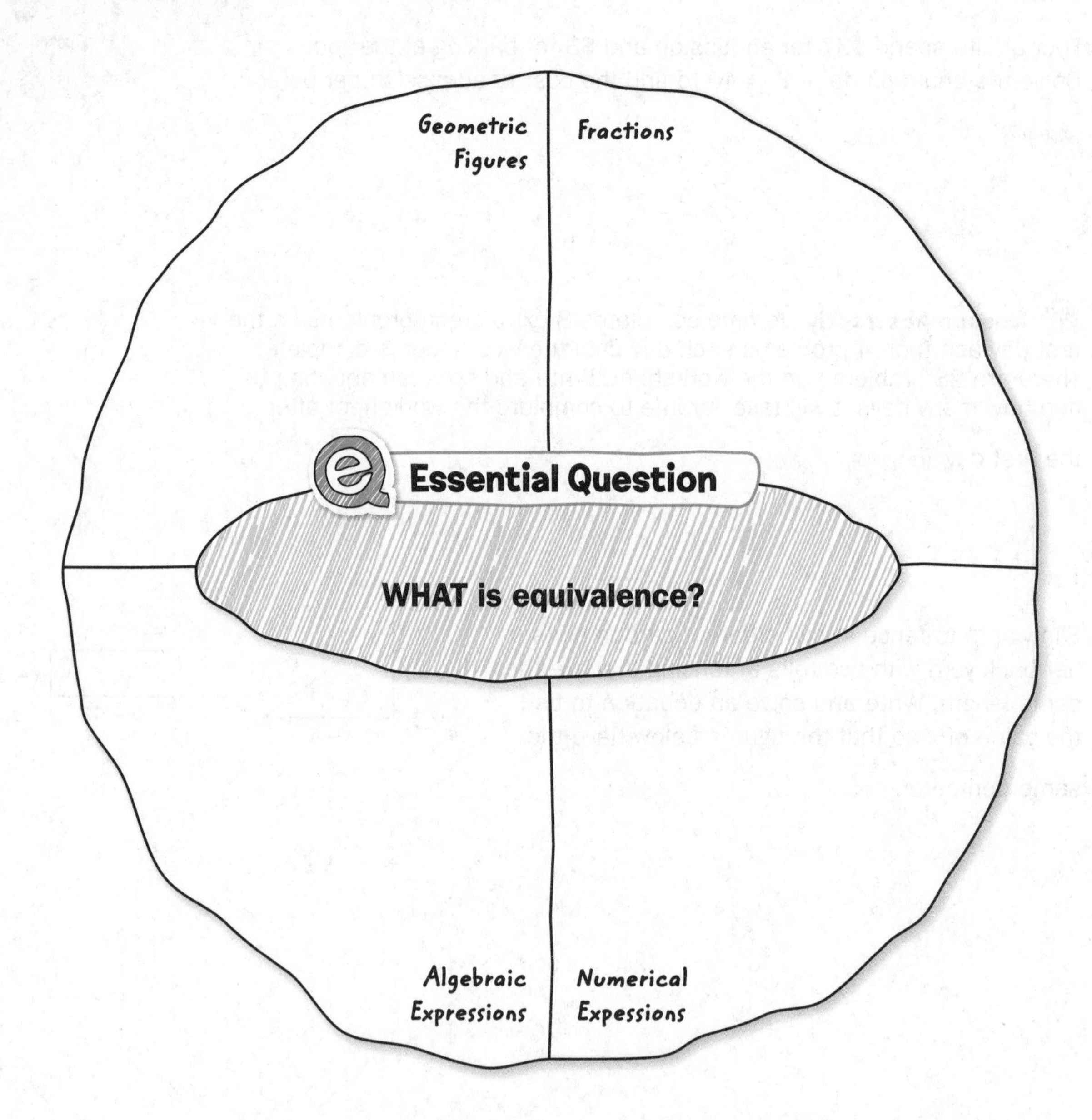

 **Answer the Essential Question.** WHAT is equivalence?

Chapter 3
Equations in Two Variables

Essential Question

WHY are graphs helpful?

Common Core State Standards

Content Standards
8.EE.5, 8.EE.6, 8.EE.8, 8.EE.8a, 8.EE.8b, 8.EE.8c, 8.F.2, 8.F.3, 8.F.4, 8.F.5

Mathematical Practices
1, 2, 3, 4, 5, 7

Math in the Real World

Games Bobbie is playing Attack of the Seas with Monique. Bobbie placed his ship between A5 and C5. To earn points, Monique guesses a point to locate one of Bobbie's ships.

Draw Bobbie's ship on the coordinate plane. If Monique guesses C8, will she locate one of Bobbie's ships? ______

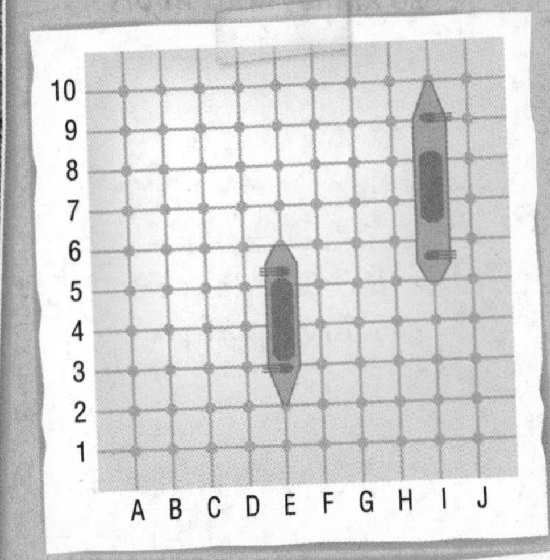

FOLDABLES® Study Organizer

1. Cut out the Foldable on page FL7 of this book.
2. Place your Foldable on page 256.
3. Use the Foldable throughout this chapter to help you learn about equations in two variables.

What Tools Do You Need?

Vocabulary

constant of proportionality	point-slope form	standard form
constant of variation	rise	substitution
constant rate of change	run	systems of equations
direct variation	slope	*x*-intercept
linear relationships	slope-intercept form	*y*-intercept

Review Vocabulary

Rates A rate is a ratio that compares two quantities with different kinds of units. Some common rates are $\frac{\text{miles}}{\text{hour}}$, $\frac{\text{price}}{\text{ounce}}$, and $\frac{\text{meters}}{\text{second}}$.

Unit Rates A rate is a unit rate when it has a denominator of 1 unit. You can find the unit rate by writing a ratio, then dividing the numerator by the denominator.

Circle the correct answer for each unit rate.

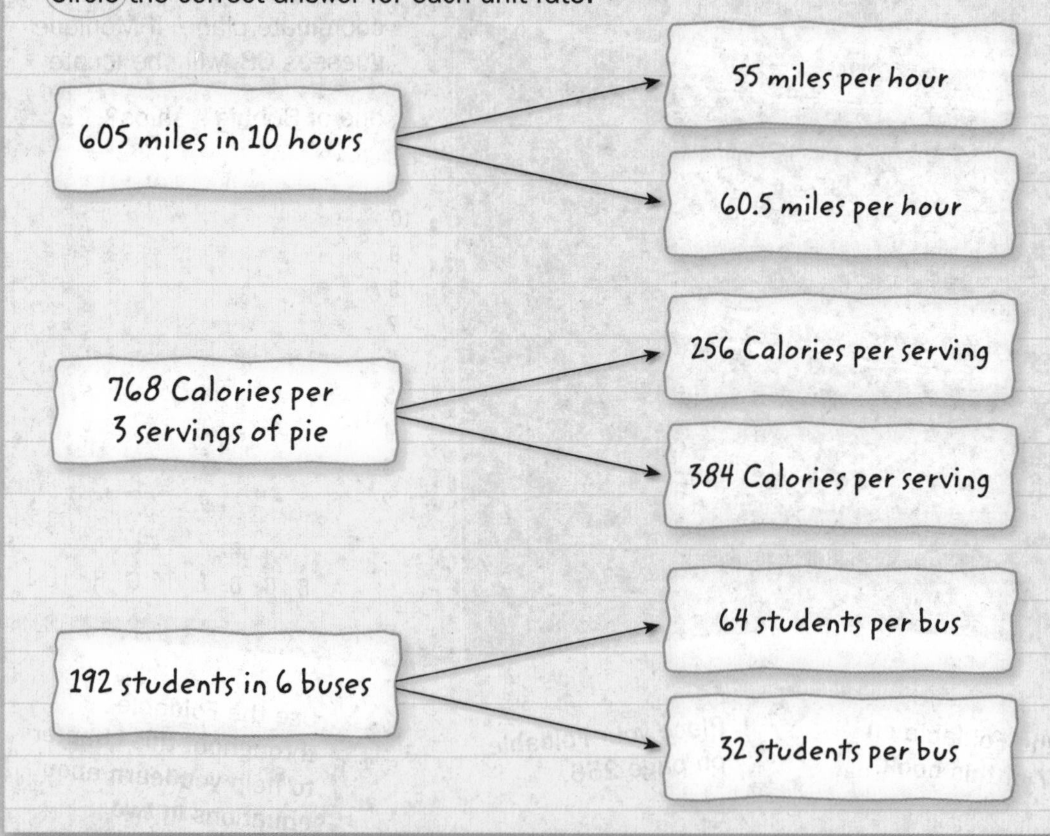

When Will You Use This?

Your Turn! You will solve this problem in the chapter.

Try the Quick Check below.
Or, take the Online Readiness Quiz.

Quick Review

Common Core Review 7.NS.1

Example 1

Find −15 − 8.

$-15 - 8 = -15 + (-8)$ To subtract 8, add −8.

$= -23$ Simplify.

Example 2

Evaluate $\frac{11 + 4}{9 - 4}$.

$\frac{11 + 4}{9 - 4} = \frac{15}{5}$ Simplify the numerator and denominator.

$= 3$ Simplify.

Quick Check

Subtract Integers Find each difference.

1. $5 - (-4) =$ ______
2. $10 - 8 =$ ______
3. $-4 - 3 =$ ______
4. $-6 - (-2) =$ ______
5. $12 - 6 =$ ______
6. $-5 - (-3) =$ ______

Numerical Expressions Evaluate each expression.

7. $\frac{6 - 2}{5 + 5} =$ ______
8. $\frac{7 - 4}{8 - 4} =$ ______
9. $\frac{3 - 1}{1 + 9} =$ ______
10. $\frac{5 + 7}{8 - 6} =$ ______
11. $\frac{2 - 4}{3 + 2} =$ ______
12. $\frac{1 - 5}{8 - 2} =$ ______

How Did You Do?

Which problems did you answer correctly in the Quick Check? Shade those exercise numbers below.

1 2 3 4 5 6 7 8 9 10 11 12

Lesson 1

Constant Rate of Change

What You'll Learn

Scan the lesson. Write the definitions of linear relationships and constant rate of change.

- ____________________
- ____________________

Essential Question

WHY are graphs helpful?

Vocabulary

linear relationship
constant rate of change

Common Core State Standards

Content Standards
Preparation for 8.EE.5

Mathematical Practices
1, 3, 4, 5

Real-World Link

Music Marcus can download two songs from the Internet each minute. This is shown in the table below.

Time (minutes), x	0	1	2	3	4
Number of Songs, y	0	2	4	6	8

1. Compare the change in the number of songs y to the change in time x. What is the rate of change?

2. Graph the ordered pairs from the table on the graph shown. Label the axes. Then describe the pattern shown on the graph.

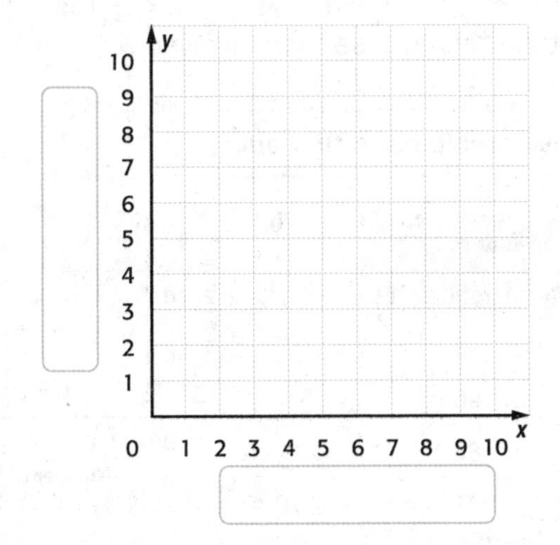

Work Zone

Linear Relationships

Relationships that have straight-line graphs, like the one on the previous page, are called **linear relationships**. Notice that as the number of songs increases by 2, the time in minutes increases by 1.

Number of Songs, y	0	2	4	6	8
Time (minutes), x	0	1	2	3	4

(Number of songs: +2, +2, +2, +2; Time: +1, +1, +1, +1)

Rate of Change
$\frac{2}{1}$ = 2 songs per minute

The rate of change between any two points in a linear relationship is the same or *constant.* A linear relationship has a **constant rate of change**.

Example

1. The balance in an account after several transactions is shown. Is the relationship between the balance and number of transactions linear? If so, find the constant rate of change. If not, explain your reasoning.

Number of Transactions	Balance ($)
3	170
6	140
9	110
12	80

(Number of transactions: +3, +3, +3; Balance: −30, −30, −30)

As the number of transactions increases by 3, the balance in the account decreases by $30.

Show your work.

Since the rate of change is constant, this is a linear relationship. The constant rate of change is $\frac{-30}{3}$ or −$10 per transaction. This means that each transaction involved a $10 *withdrawal.*

Got It? Do these problems to find out.

a.

Cooling Water	
Time (min)	Temperature (°F)
5	95
10	90
15	86
20	82

b.

Time (min) — y: 40, 60, 80

Number of Volunteers — x: 0, 8, 16, 24, 32

a. ____________

b. ____________

Proportional Linear Relationships

Key Concept

Words Two quantities *a* and *b* have a proportional linear relationship if they have a constant ratio and a constant rate of change.

Symbols $\frac{b}{a}$ is constant and $\frac{\text{change in } b}{\text{change in } a}$ is constant.

To determine if two quantities are proportional, compare the ratio $\frac{b}{a}$ for several pairs of points to determine if there is a constant ratio.

Example

2. Use the table to determine if there is a proportional linear relationship between a temperature in degrees Fahrenheit and a temperature in degrees Celsius. Explain your reasoning.

+5 +5 +5 +5

Degrees Celsius	0	5	10	15	20
Degrees Fahrenheit	32	41	50	59	68

+9 +9 +9 +9

Constant Rate of Change

$$\frac{\text{change in °F}}{\text{change in °C}} = \frac{9}{5}$$

Since the rate of change is constant, this is a linear relationship.

To determine if the two scales are proportional, express the relationship between the degrees for several columns as a ratio.

$$\frac{\text{degrees Fahrenheit}}{\text{degrees Celsius}} \longrightarrow \frac{41}{5} = 8.2 \quad \frac{50}{10} = 5 \quad \frac{59}{15} \approx 3.9$$

Since the ratios are not the same, the relationship between degrees Fahrenheit degrees Celsius is *not* proportional.

Check: Graph the points on the coordinate plane. Then connect them with a line.

The points appear to fall in a straight line so the relationship is linear. ✓

The line connecting the points does not pass through the origin so the relationship is not proportional. ✓

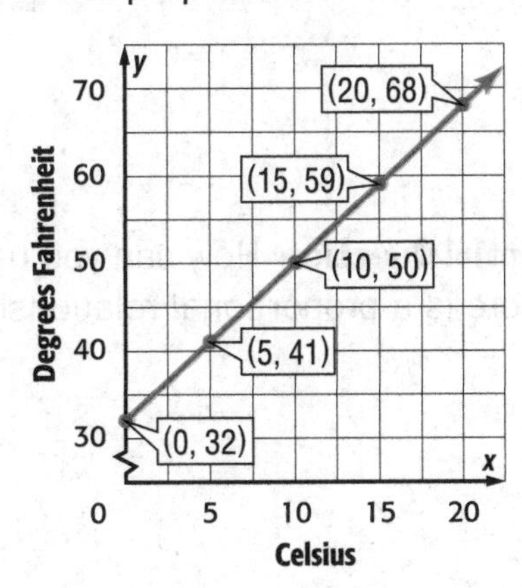

Proportional Relationships
Two quantities are proportional if they have a constant ratio.

Show your work.

Got It? Do this problem to find out.

c. ______________

c. Use the table to determine if there is a proportional linear relationship between mass of an object in kilograms and the weight of the object in pounds. Explain your reasoning.

Weight (lb)	20	40	60	80
Mass (kg)	9	18	27	36

Guided Practice

1. The amount of paint y needed to paint a certain amount of chairs x is shown in the table. Is the relationship between the two quantities linear? If so, find the constant rate of change. If not, explain your reasoning. (Example 1)

Paint Needed for Chairs	
Chairs, x	Cans of Paint, y
5	6
10	12
15	18

Show your work.

____

____

____

2. The altitude y of a certain airplane after a certain number of minutes x is shown in the graph. Is the relationship linear? If so, find the constant rate of change. If not, explain your reasoning. (Example 1)

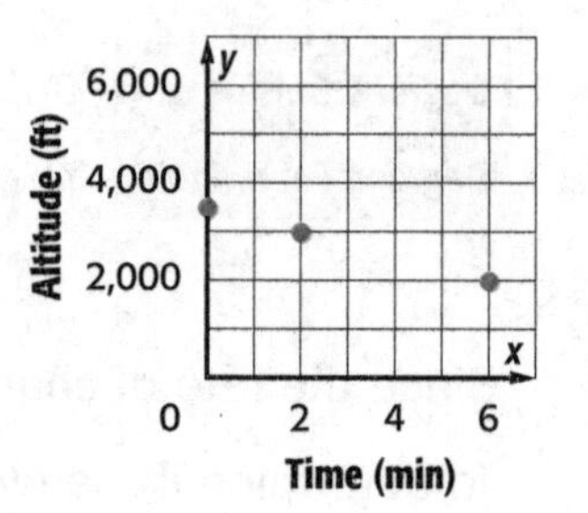

____

3. Determine whether a proportional relationship exists between the two quantities shown in Exercise 1. Explain your reasoning. (Example 2)

____

____

____

4. **Building on the Essential Question** How can you use a table to determine if there is a proportional relationship between two quantities? ______________________________

____

____

____

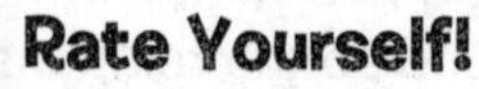

Rate Yourself!

Are you ready to move on? Shade the section that applies.

YES ? NO

For more help, go online to access a Personal Tutor.

Name ______ My Homework ______

Independent Practice

Go online for Step-by-Step Solutions

Determine whether the relationship between the two quantities shown in each table or graph is linear. If so, find the constant rate of change. If not, explain your reasoning. (Example 1)

1.

Cost of Electricity to Run Personal Computer	
Time (h)	Cost (¢)
5	15
8	24
12	36
24	72

2.

Distance Traveled by Falling Object				
Time (s)	1	2	3	4
Distance (m)	4.9	19.6	44.1	78.4

3.

Italian Dressing Recipe				
Oil (c)	2	4	6	8
Vinegar (c)	$\frac{3}{4}$	$1\frac{1}{2}$	$2\frac{1}{4}$	3

4.

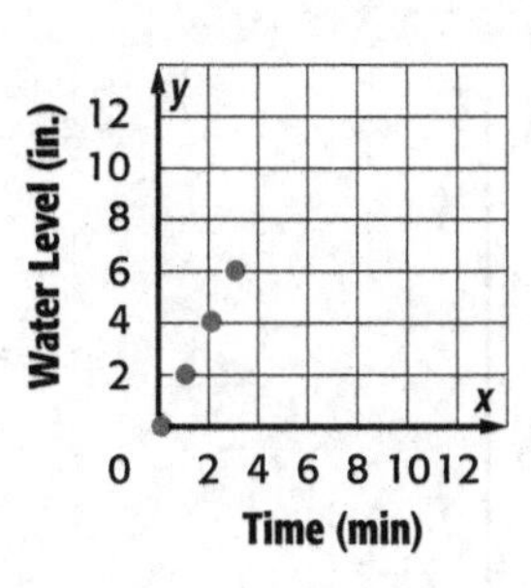

5.

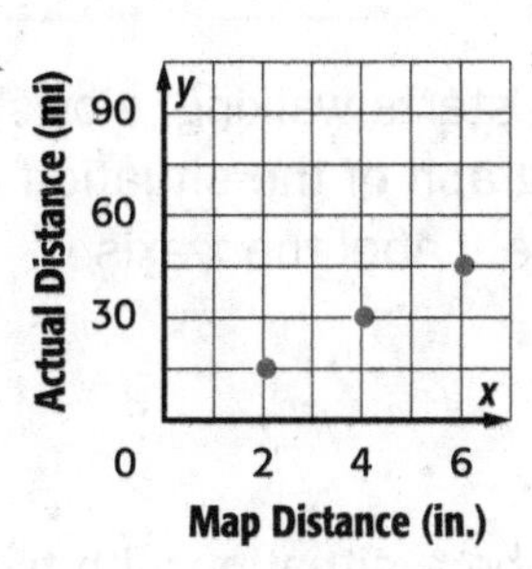

6.

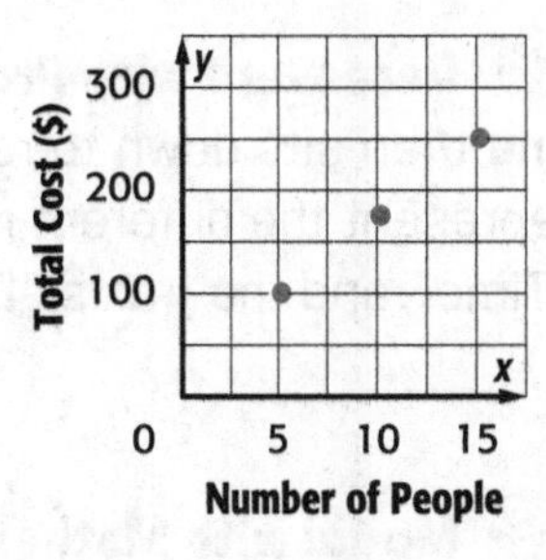

Determine whether a proportional relationship exists between the two quantities shown in the following Exercises. Explain your reasoning. (Example 2)

7. Exercise 1

8. Exercise 3

9. Exercise 5

10. **Use Math Tools** Match the table with its rate of change.

2.4 ft/min

10 ft/min

−0.8 ft/min

0.25 ft/min

Time (min)	20	30	40
Altitude (ft)	170	162	154

Time (min)	1	2	3
Distance (ft)	20	30	40

Time (min)	4	6	8
Height (ft)	1	1.5	2

Time (min)	5	10	15
Depth (ft)	12	24	36

H.O.T. Problems Higher Order Thinking

11. **Persevere with Problems** A dog starts walking, slows down, and then sits down to rest. Sketch a graph of the situation to represent the different rates of change. Label the *x*-axis "Time" and the *y*-axis "Distance".

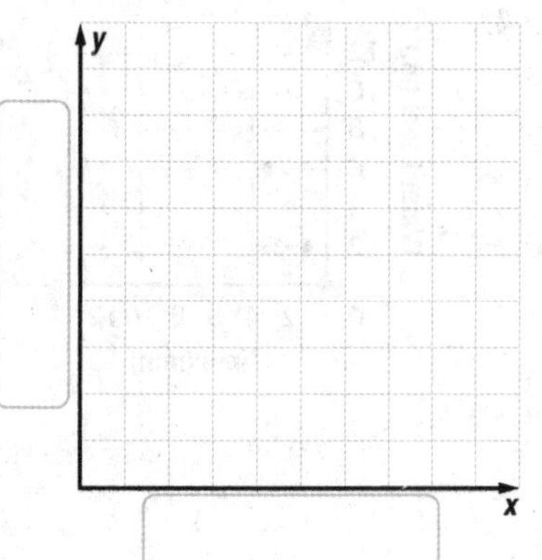

12. **Model with Mathematics** Describe a situation with two quantities that have a proportional linear relationship.

Standardized Test Practice

13. Tickets to the school play are $2.50 each. Which table contains values that fit this situation if *c* represents the total cost for *t* tickets?

Ⓐ

Cost of Play Tickets ($)				
t	1	2	3	4
c	2.50	3.25	4.00	4.75

Ⓑ

Cost of Play Tickets ($)				
t	1	2	3	4
c	3.50	6.00	8.50	11.00

Ⓒ

Cost of Play Tickets ($)				
t	1	2	3	4
c	3.50	4.00	4.50	5.00

Ⓓ

Cost of Play Tickets ($)				
t	1	2	3	4
c	2.50	5.00	7.50	10.00

Extra Practice

Determine whether the relationship between the two quantities shown in each table is linear. If so, find the constant rate of change. If not, explain your reasoning.

14.

Sale Price Comparison	
Retail Price ($)	**Sale Price ($)**
0	0
10	5
20	10
30	15
40	20
50	25
60	30

(Retail Price: +10, +10, +10, +10, +10, +10; Sale Price: +5, +5, +5, +5, +5, +5)

Yes; the rate of change between the sale price and retail price is a constant value of $\frac{1}{2}$.

15.

Total Number of Customers Helped at Jewelry Store	
Time (h)	**Total Helped**
1	12
2	24
3	36
4	60

16. Determine whether a proportional relationship exists between the two quantities in Exercise 14. Explain your reasoning. ______

CCSS Reason Abstractly **Find the constant rate of change for each graph and interpret its meaning.**

17.

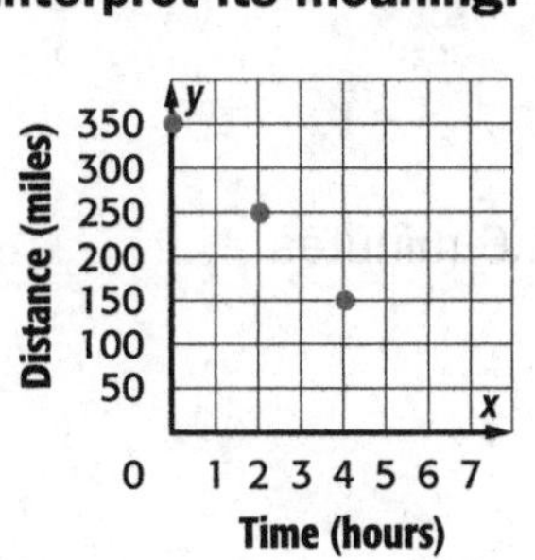

18.

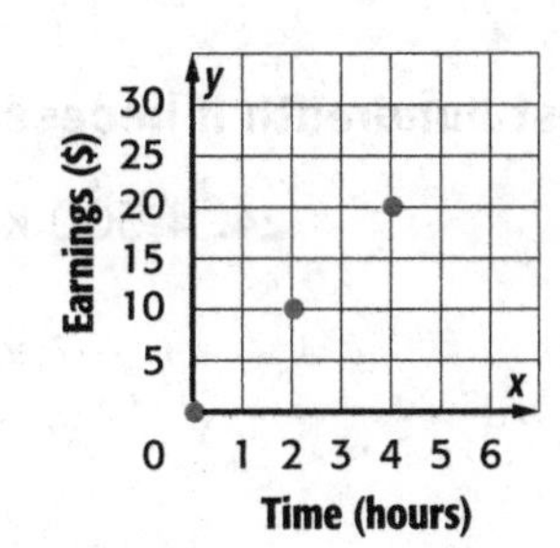

19.

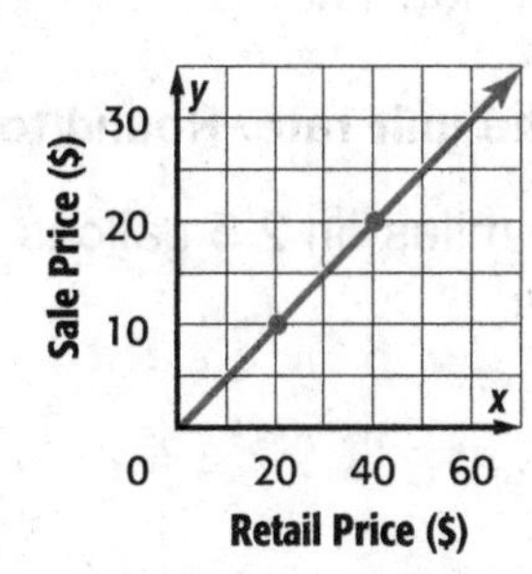

Standardized Test Practice

20. The graph shows the distance Bianca traveled on her 2-hour bike ride. Which of the following is true?

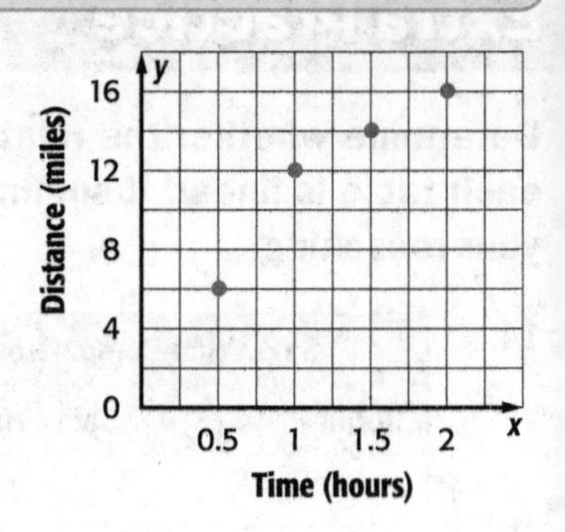

Ⓐ She traveled at a constant speed of 12 miles per hour for the entire ride.

Ⓑ She traveled at a constant speed of 8 miles per hour for the last hour.

Ⓒ She traveled at a constant speed of 4 miles per hour for the last hour.

Ⓓ She traveled at a constant speed of 8 miles per hour for the entire ride.

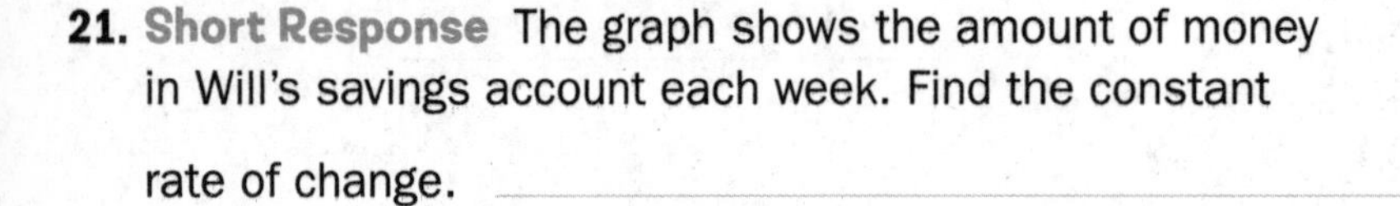

21. Short Response The graph shows the amount of money in Will's savings account each week. Find the constant rate of change. ______________________

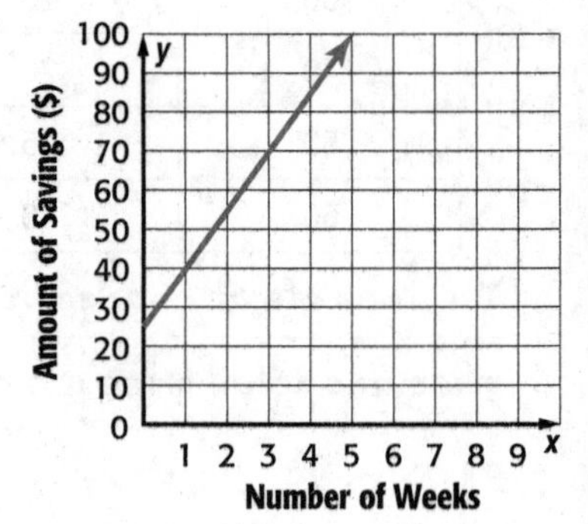

CCSS Common Core Review

22. Mickey charges $15, $30, $45, and $60 for babysitting 1, 2, 3, and 4 hours, respectively. Is the relationship between the number of hours and the amount charged proportional? If so, find the unit rate. If not, explain why not. 7.RP.2 ______________________

Find the unit rate. Round to the nearest hundredth if necessary. 7.RP.1

23. 60 miles on 2.5 gallons ____________

24. 4,500 kilobytes in 6 minutes ____________

25. 10 red peppers for $5.50 ____________

26. 72.6 meters in 11 seconds ____________

Inquiry Lab

Graphing Technology: Rate of Change

HOW can you use a graphing calculator to determine the rate of change?

CCSS Content Standards 8.EE.5

Mathematical Practices 1, 3

School At the school store, tickets to the football game are sold for $5 each. The equation $y = 5x$ can be used to find the total cost y of any number of tickets x. Find the rate of change.

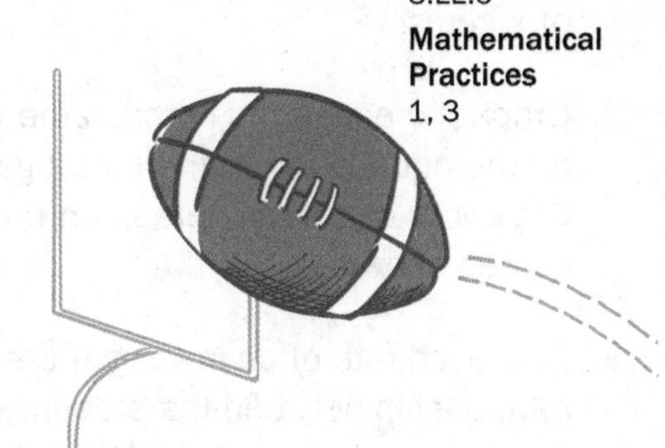

What do you know? ______

What do you need to find? ______

Investigation

Recall that a rate of change is a rate that describes how one quantity changes in relation to another.

Step 1 Enter the equation. Press [Y=] 5 [X,T,θ,n].

Step 2 Graph the equation in the standard viewing window. Press [Zoom] 6.

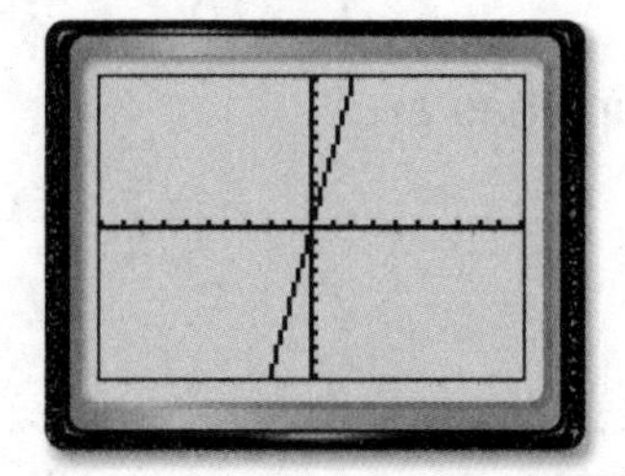

Step 3 Press [2nd] [TblSet] [▼] [▼] [ENTER] [▼] [ENTER] to generate the table automatically. Press [2nd] [Table] to access the table. Choose any two points on the line and find the rate of change.

$$\frac{\text{change in total cost}}{\text{change in number of tickets}} = \frac{\$(\square - \square)}{(\square - \square)\text{ tickets}}$$

$$= \frac{\square}{\square\text{ ticket}}$$

So, the rate of change is ______.

Collaborate

Work with a partner. School T-shirts are sold for $10 each and packages of markers are sold for $2.50 each.

1. For each item, write an equation that can be used to find the total cost y of x items. ______

2. Graph the equations in the same window as the equation from the Investigation. Copy your calculator screen on the blank screen shown.

3. Find each rate of change. Is there a relationship between the steepness of the lines on the graph and the rates of change? Explain.

Analyze

4. **CCSS Reason Inductively** Without graphing, predict which graph has a steeper line: $y = 3x$ or $y = \frac{1}{3}x$. Explain.

Reflect

5. **Inquiry** HOW can you use a graphing calculator to determine the rate of change?

Lesson 2
Slope

What You'll Learn

Scan the lesson. Predict two things you will learn about slope.

-

-

Essential Question

WHY are graphs helpful?

Vocabulary

slope
rise
run

Common Core State Standards

Content Standards
Preparation for 8.EE.5

Mathematical Practices
1, 3, 4

Vocabulary Start-Up

The term *slope* is used to describe the steepness of a straight line. **Slope** is the ratio of the **rise**, or vertical change, to the **run** or horizontal change.

Complete the graphic organizer.

I think this word means...	How is this concept related to other math concepts?
slope	
Where have I heard this word in my life?	What makes this an important word for me to know?

Real-World Link

A ride at an amusement park rises 8 feet every horizontal change of 2 feet. How could you determine the slope of the ride?

Work Zone

Find Slope Using a Graph or Table

Slope is a rate of change. It can be positive (slanting upward) or negative (slanting downward).

$$\text{slope} = \frac{\text{rise}}{\text{run}}$$

rise ← vertical change between any two points

run ← horizontal change between the same two points

Example

1. Find the slope of the treadmill.

$\text{slope} = \frac{\text{rise}}{\text{run}}$ Definition of slope

$= \frac{10 \text{ in.}}{48 \text{ in.}}$ rise = 10 in., run = 48 in.

$= \frac{5}{24}$ Simplify.

10 in.

48 in.

The slope of the treadmill is $\frac{5}{24}$.

Got It? Do this problem to find out.

a. A hiking trail rises 6 feet for every horizontal change of 100 feet. What is the slope of the hiking trail?

a. ____________

Examples

Tutor

2. The graph shows the cost of muffins at a bake sale. Find the slope of the line.

Choose two points on the line. The vertical change is 2 units and the horizontal change is 1 unit.

$\text{slope} = \frac{\text{rise}}{\text{run}}$ Definition of slope

$= \frac{2}{1}$ rise = 2, run = 1

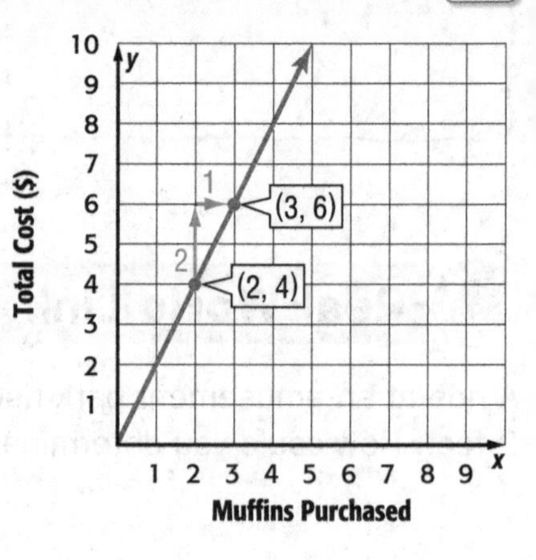

The slope of the line is $\frac{2}{1}$ or 2.

Translating Rise and Run

up → positive
down → negative
right → positive
left → negative

3. The table shows the number of pages Garrett has left to read after a certain number of minutes. The points lie on a line. Find the slope of the line.

Time (min), x	Pages left, y
1	12
3	9
5	6
7	3

Choose any two points from the table to find the changes in the x- and y-values.

$$\text{slope} = \frac{\text{change in } y}{\text{change in } x}$$ Definition of slope

$$= \frac{9 - 12}{3 - 1}$$ Use the points (1, 12) and (3, 9).

$$= \frac{-3}{2} \text{ or } -\frac{3}{2}$$ Simplify.

To check, choose two different points from the table and find the slope.

Check $$\text{slope} = \frac{\text{change in } y}{\text{change in } x}$$

$$= \frac{3 - 6}{7 - 5}$$

$$= \frac{-3}{2} \text{ or } -\frac{3}{2} \checkmark$$

Slope

In linear relationships, no matter which two points you choose, the slope, or rate of change, of the line is always constant.

Got It? **Do these problems to find out.**

Find the slope of each line.

b.

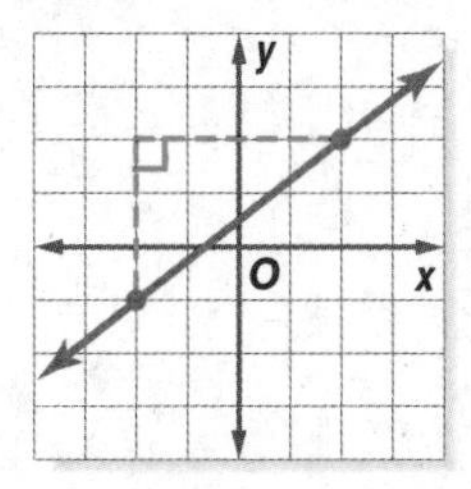

c.

x	−6	−2	2	6
y	−2	−1	0	1

b. ______

c. ______

Slope Formula

Key Concept

Words The slope m of a line passing through points (x_1, y_1) and (x_2, y_2) is the ratio of the difference in the y-coordinates to the corresponding difference in the x-coordinates.

Symbols $m = \frac{y_2 - y_1}{x_2 - x_1}$, where $x_2 \neq x_1$

Model

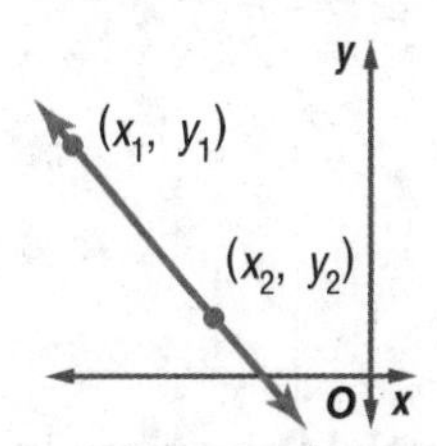

It does not matter which point you define as (x_1, y_1) and (x_2, y_2). However the coordinates of both points must be used in the same order.

Using the Slope Formula

To check Example 4, let $(x_1, y_1) = (-4, 3)$ and $(x_2, y_2) = (1, 2)$. Then find the slope.

d. ______

e. ______

Example

4. Find the slope of the line that passes through $R(1, 2)$, $S(-4, 3)$.

$m = \frac{y_2 - y_1}{x_2 - x_1}$ — Slope formula

$m = \frac{3 - 2}{-4 - 1}$ — $(x_1, y_1) = (1, 2)$; $(x_2, y_2) = (-4, 3)$

$m = \frac{1}{-5}$ or $-\frac{1}{5}$ — Simplify.

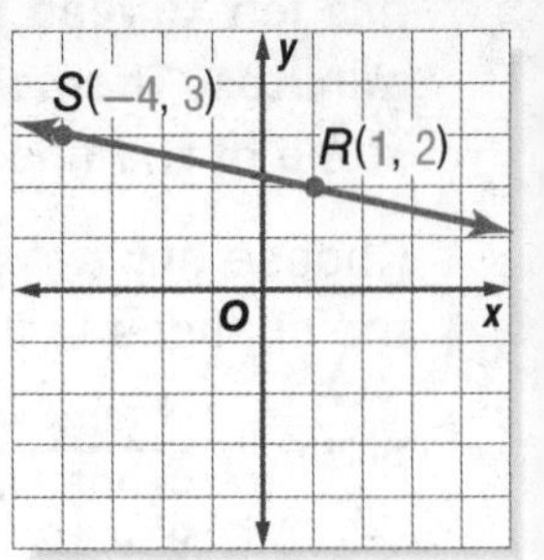

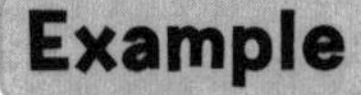

Got It? Do these problems to find out.

d. $A(2, 2)$, $B(5, 3)$ **e.** $J(-7, -4)$, $K(-3, -2)$

Guided Practice

1. Find the slope of the storage shed's roof. (Example 1)

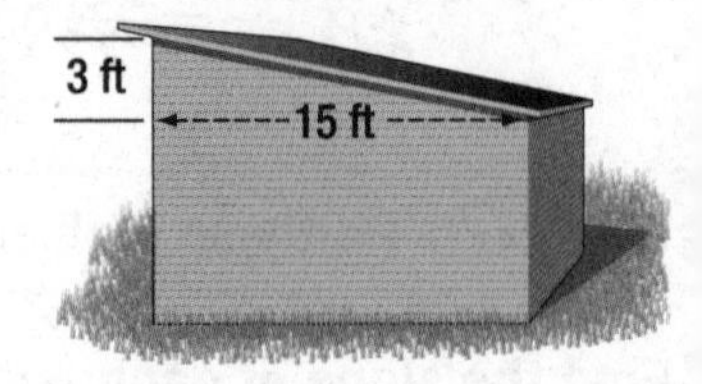

Find the slope of each line. (Examples 2 and 3)

2.

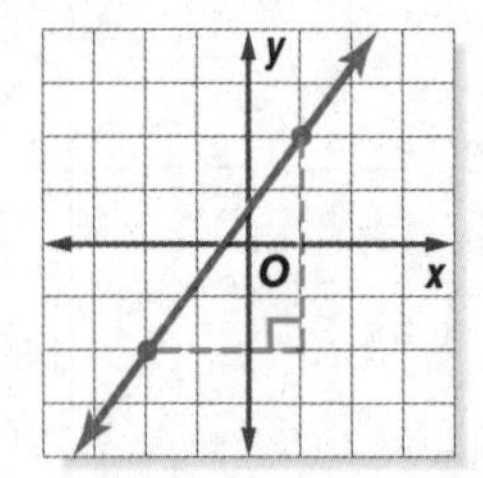

3.

x	0	1	2	3
y	1	3	5	7

Find the slope of the line that passes through each pair of points. (Example 4)

4. $A(-3, -2)$, $B(5, 4)$

5. $E(-6, 5)$, $F(3, -3)$

6.  **Building on the Essential Question** In any linear relationship, explain why the slope is always the same.

Rate Yourself!

How well do you understand slope? Circle the image that applies.

Clear — Somewhat Clear — Not So Clear

For more help, go online to access a Personal Tutor.

Independent Practice

Go online for Step-by-Step Solutions 

 1. Find the slope of a ski run that descends 15 feet for every horizontal change of 24 feet. (Example 1)

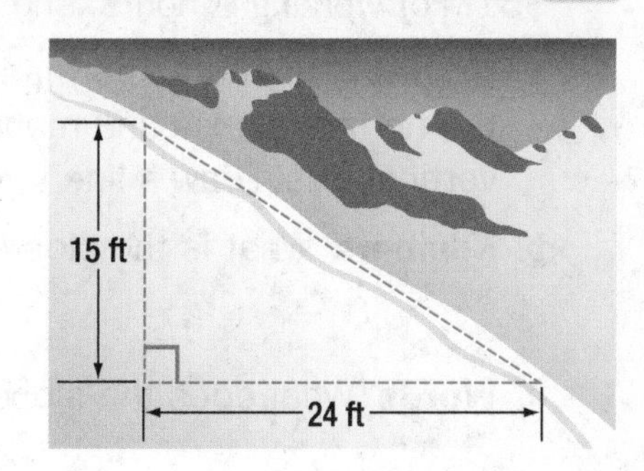

Find the slope of each line. (Example 2)

2. 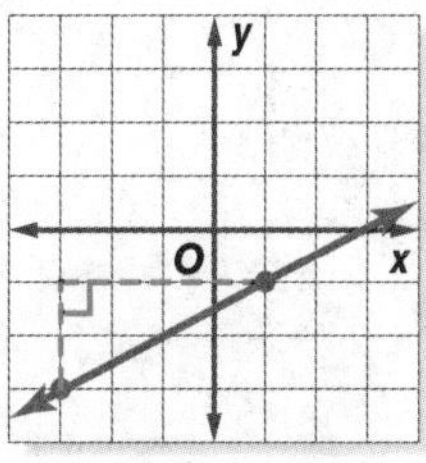

3. 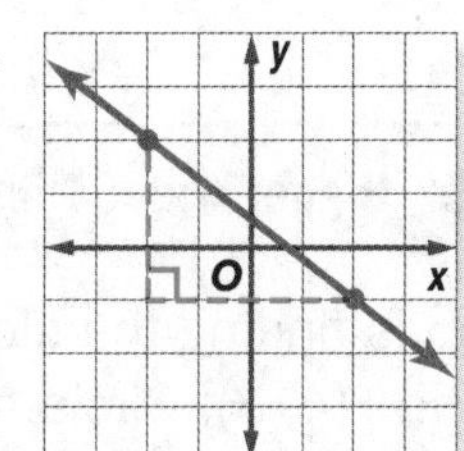

The points given in the table lie on a line. Find the slope of each line. (Example 3)

4.

x	0	2	4	6
y	9	4	−1	−6

5.

x	0	1	2	3
y	3	5	7	9

Find the slope of the line that passes through each pair of points. (Example 4)

6. $A(0, 1)$, $B(2, 7)$

 7. $C(2, 5)$, $D(3, 1)$

8. $E(1, 2)$, $F(4, 7)$

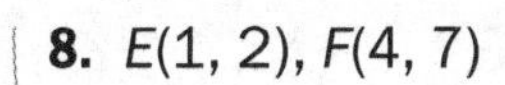

9. CCSS **Justify Conclusions** Wheelchair ramps for access to public buildings are allowed a maximum of one inch of vertical increase for every one foot of horizontal distance. Would a ramp that is 10 feet long and 8 inches tall meet this guideline? Explain your reasoning to a classmate.

10. **CCSS Multiple Representations** For working 3 hours, Sofia earns \$30.60. For working 5 hours, she earns \$51. For working 6 hours, she earns \$61.20.

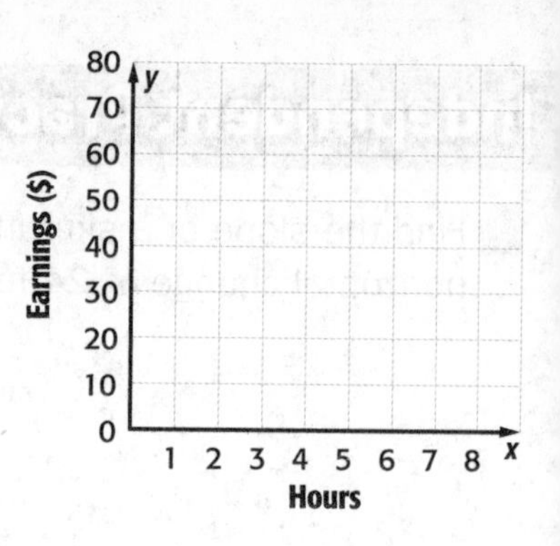

   a. **Graphs** Graph the information with hours on the horizontal axis and money earned on the vertical axis. Draw a line through the points.

   b. **Numbers** What is the slope of the line?

   c. **Words** What does the slope of the line represent?

H.O.T. Problems Higher Order Thinking

11. **CCSS Find the Error** Jacob is finding the slope of the line that passes through $X(0, 2)$ and $Y(4, 3)$. Circle his mistake and correct it.

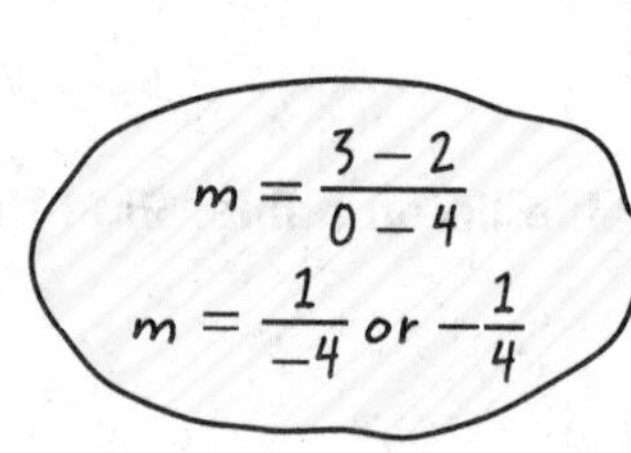

12. **CCSS Persevere with Problems** Two lines that are parallel have the same slope. Determine whether quadrilateral *ABCD* is a parallelogram. Justify your reasoning.

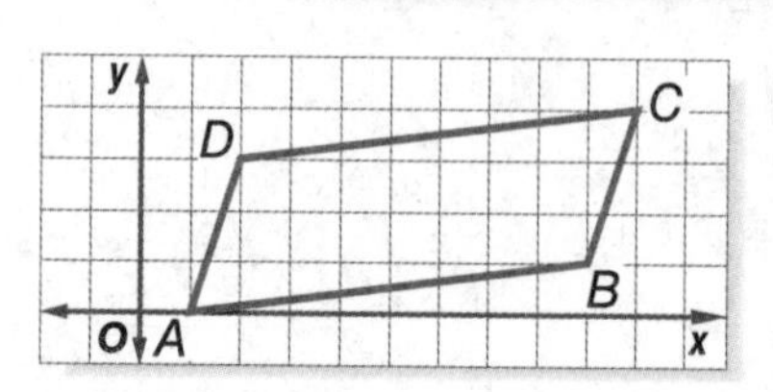

Standardized Test Practice

13. Line *AB* goes through points $A(-4, -3)$ and $B(-2, 0)$. What is the slope of the line?

   Ⓐ $-\frac{3}{2}$

   Ⓑ $-\frac{2}{3}$

   Ⓒ $\frac{2}{3}$

   Ⓓ $\frac{3}{2}$

Name ________________ My Homework ________________

Extra Practice

14. Find the slope of a road that rises 12 feet for every horizontal change of 100 feet. $\frac{3}{25}$

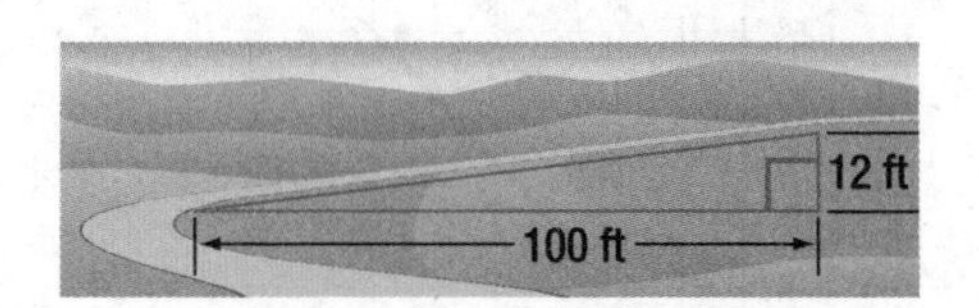

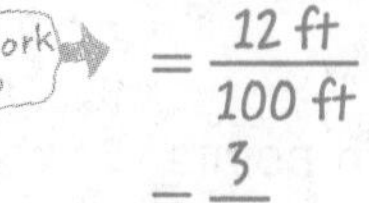

$$\text{slope} = \frac{\text{rise}}{\text{run}} \quad \text{Definition of slope}$$

$$= \frac{12 \text{ ft}}{100 \text{ ft}} \quad \text{rise} = 12 \text{ ft, run} = 100 \text{ ft}$$

$$= \frac{3}{25} \quad \text{Simplify.}$$

15. Wyatt is flying a kite in the park. The kite is a horizontal distance of 24 feet from Wyatt's position and a vertical distance of 72 feet.

Find the slope of the kite string. ________

Find the slope of each line.

16.

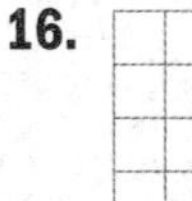

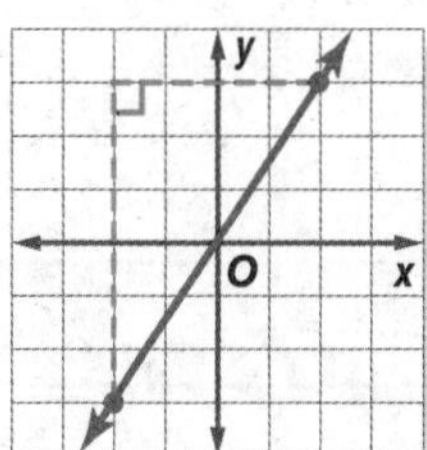

____

17.

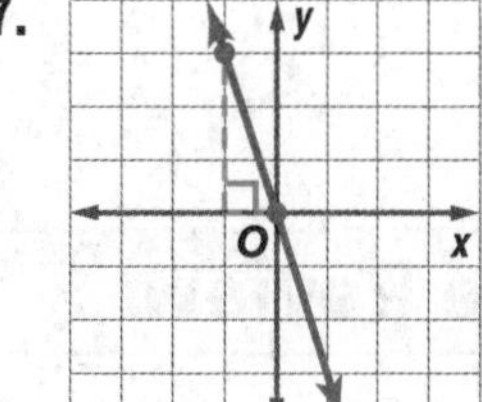

____

CCSS Use Math Tools **The points given in the table lie on a line. Find the slope of each line.**

18.

x	−3	3	9	15
y	−3	1	5	9

____

19.

x	−2	−1	1	2
y	−4	−2	2	4

____

Find the slope of the line that passes through each pair of points.

20. $M(-2, 3)$, $N(7, -4)$ ________

21. $G(-6, -1)$, $H(4, 1)$ ________

22. $J(-9, 3)$, $K(2, 1)$ ________

Standardized Test Practice

23. Line AB represents a steep hill.

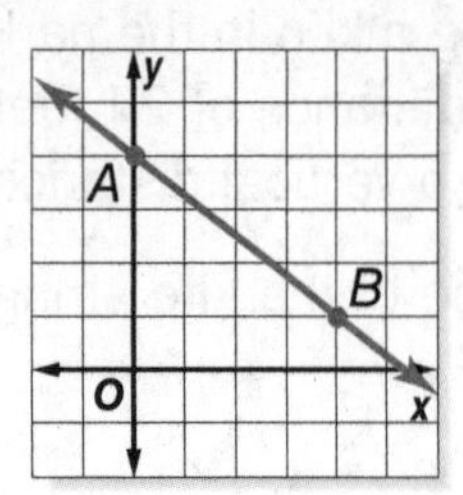

What is the slope of the hill?

Ⓐ $-\frac{4}{3}$

Ⓑ $-\frac{3}{4}$

Ⓒ $\frac{3}{4}$

Ⓓ $\frac{4}{3}$

24. **Short Response** Lionel charted the growth rate of his new puppy for several weeks and plotted the values on the graph below.

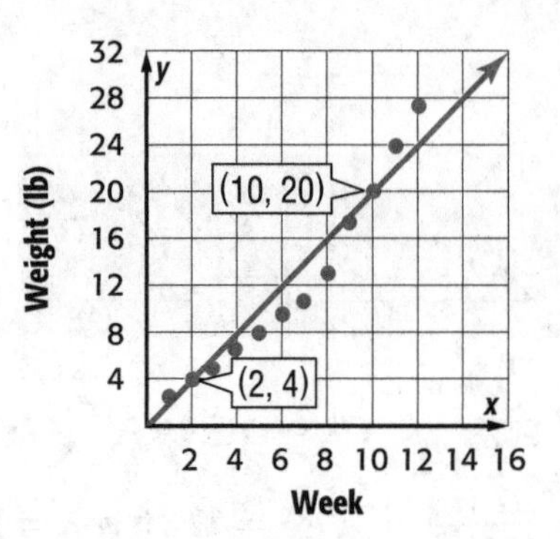

He drew a line passing through points (2, 4) and (10, 20) on the graph to estimate the puppy's weight during any week. What is the slope of the line he drew?

Common Core Review

25. The wait time to ride the Thunder boats is 30 minutes when 180 people are in line. Write and solve a proportion to find the wait time when 240 people are in line. 7.RP.2, 7.RP.3

Solve each proportion. 7.RP.2, 7.RP.3

26. $\frac{5}{7} = \frac{a}{35}$

27. $\frac{12}{p} = \frac{36}{45}$

28. $\frac{3}{9} = \frac{21}{k}$

29. $\frac{n}{15} = \frac{17}{34}$

30. $\frac{-7}{10} = \frac{3.5}{j}$

31. $\frac{12}{18} = \frac{-40}{x}$

Lesson 3

Equations in $y = mx$ Form

What You'll Learn

Scan the lesson. List two headings you would use to make an outline of the lesson.

- ______________________________
- ______________________________

Essential Question

WHY are graphs helpful?

Vocabulary

direct variation
constant of variation
constant of proportionality

Common Core State Standards

Content Standards
8.EE.5, 8.EE.6, 8.F.2, 8.F.4

Mathematical Practices
1, 3, 4

Real-World Link

Charity The amount of money David can raise for the Wish Upon A Rainbow Bike-a-thon is shown in the table.

Biking Time (h), x	Money Raised ($), y
2	20
4	40
6	60

Recall that when the ratio of two variable quantities is constant, a proportional relationship exists. This relationship is called a **direct variation**. The constant ratio is called the **constant of variation** or **constant of proportionality**.

Complete the steps below to derive the equation for a direct variation.

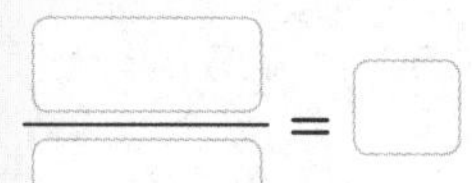
Slope formula

$\frac{y - 0}{x - 0} = m$ $(x_1, y_1) = (0, 0)$, $(x_2, y_2) = (x, y)$

y
(x, y)
O
(0, 0)
x

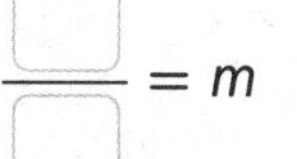
Simplify.

Multiplication Property of Equality

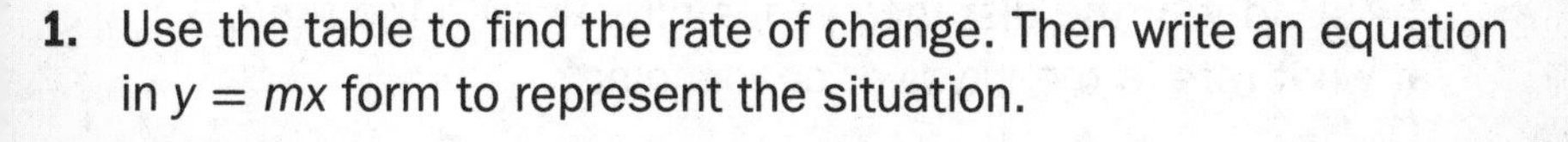

1. Use the table to find the rate of change. Then write an equation in $y = mx$ form to represent the situation.

Key Concept

Direct Variation

Work Zone

Words A linear relationship is a direct variation when the ratio of y to x is a constant, m. We say y varies directly with x.

Symbols $m = \frac{y}{x}$ or $y = mx$, where m is the constant of variation and $m \neq 0$

Example $y = 3x$

Graph

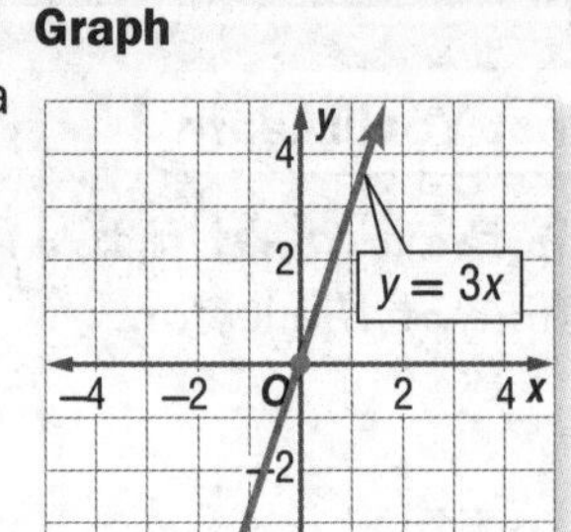

y* = *mx
In a direct variation equation $y = mx$, m represents the constant of variation, the constant of proportionality, the slope, and the unit rate.

The slope of the graph of $y = mx$ is m. Since (0, 0) is one solution of $y = mx$, the graph of a direct variation always passes through the origin.

Real World

Example

1. The amount of money Robin earns while babysitting varies directly with the time as shown in the graph. Determine the amount that Robin earns per hour.

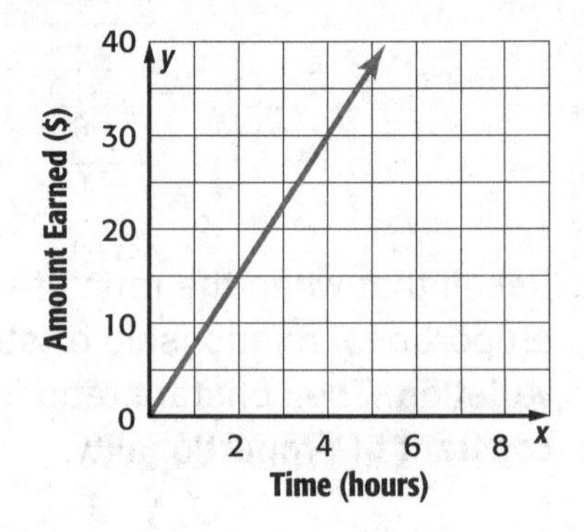

To determine the amount Robin earns per hour, or the unit rate, find the constant of variation.

Use the points (2, 15), (3, 22.5), and (4, 30).

$$\frac{\text{amount earned}}{\text{time}} \rightarrow \frac{15}{2} \text{ or } \frac{7.5}{1} \qquad \frac{22.5}{3} \text{ or } \frac{7.5}{1} \qquad \frac{30}{4} \text{ or } \frac{7.5}{1}$$

So, Robin earned $7.50 for each hour she babysits.

Got It? Do this problem to find out.

a. ____________

a. Two minutes after a skydiver opens his parachute, he has descended 1,900 feet. After 5 minutes, he descended 4,750 feet. If the distance varies directly with the time, at what rate is the skydiver descending?

Example

2. A cyclist can ride 3 miles in 0.25 hour. Assume that the distance biked in miles y varies directly with time in hours x. This situation can be represented by $y = 12x$. Graph the equation. How far can the cyclist ride per hour?

Make a table of values. Then graph the equation $y = 12x$. In a direct variation equation, m represents the slope. So, the slope of the line is $\frac{12}{1}$.

Hours, x	$y = 12x$	Miles, y
0	$y = 12(0)$	0
1	$y = 12(1)$	12
2	$y = 12(2)$	24

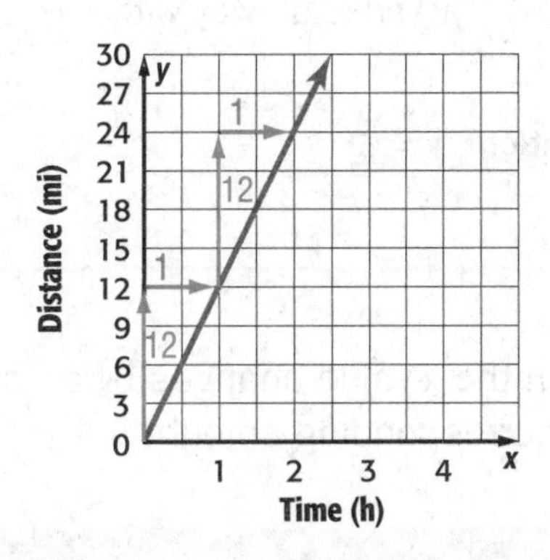

The unit rate is the slope of the line. So, the cyclist can ride 12 miles per hour.

Got It? **Do this problem to find out.**

b. A grocery store sells 6 oranges for \$2. Assume that the cost of the oranges varies directly with the number of oranges. This situation can be represented by $y = \frac{1}{3}x$. Graph the equation. What is the cost per orange?

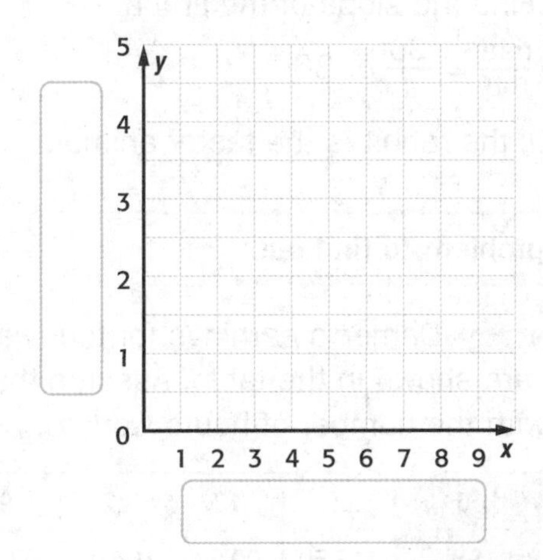

b. ____________

Key Concept: Compare Direct Variations

You can use tables, graphs, words, or equations to represent and compare proportional relationships.

Table

x	15	20	25	30
y	3	4	5	6

Graph

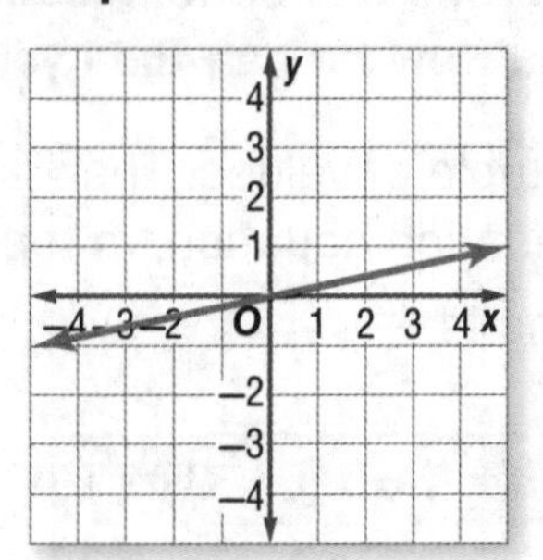

Words y varies directly with x

Equation $y = \frac{1}{5}x$

Work Zone

STOP and Reflect

In a proportional relationship, how is the unit rate represented on a graph? Explain below.

When the x-value changes by an amount A, the y-value will change by the corresponding amount mA.

Real World Example

3. The distance d in miles covered by a rabbit in t hours can be represented by the equation $d = 35t$. The distance covered by a grizzly bear is shown on the graph. Which animal is faster? Explain.

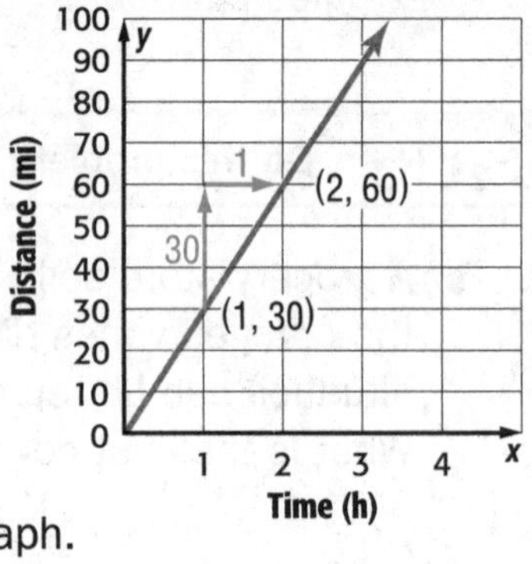

Distance (mi): 0, 10, 20, 30, 40, 50, 60, 70, 80, 90, 100; Time (h): 1, 2, 3, 4; points (1, 30), (2, 60); rise 30, run 1

Rabbit $d = 35t$

The slope or unit rate is 35 mph.

Grizzly Bear Find the slope of the graph.

$$\frac{\text{rise}}{\text{run}} = \frac{30}{1} \text{ or } 30$$

Since $35 > 30$, the rabbit is the faster animal.

Got It? Do this problem to find out.

c. **Financial Literacy** Damon's earnings for four weeks from a part time job are shown in the table. Assume that his earnings vary directly with the number of hours worked.

Time Worked (h)	15	12	22	9
Total Pay ($)	112.50	90.00	165.00	67.50

He can take a job that will pay him $7.35 per hour worked. Which job has the better pay? Explain.

c. ________

Example

4. A 3-year-old dog is often considered to be 21 in human years. Assume that the equivalent age in human years *y* varies directly with its age as a dog *x*. Write and solve a direct variation equation to find the human-year age of a dog that is 6 years old.

Let x represent the dog's actual age and let y represent the human-equivalent age.

$y = mx$	Direct variation
$21 = m(3)$	$y = 21, x = 3$
$7 = m$	Simplify.
$y = 7x$	Replace m with 7.

You want to know the human-year age or y-value when the dog is 6 years old.

$y = 7x$	Write the equation.
$y = 7 \cdot 6$	$x = 6$
$y = 42$	Simplify.

So, when a dog is 6 years old, the equivalent age in human years is 42.

Check
Graph the equation $y = 7x$.
The y-value when $x = 6$ is 42. ✓

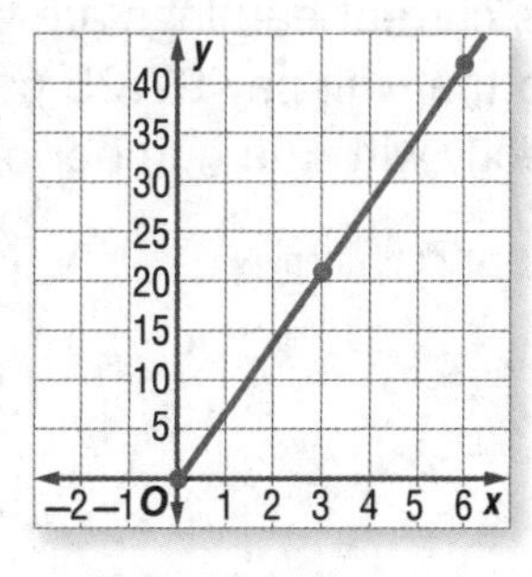

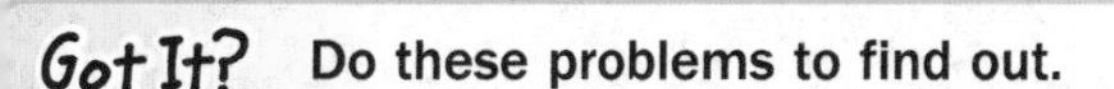

d. A charter bus travels 210 miles in $3\frac{1}{2}$ hours. Assume the distance traveled is directly proportional to the time traveled. Write and solve a direct variation equation to find how far the bus will travel in 6 hours.

e. A Monarch butterfly can fly 93 miles in 15 hours. Assume the distance traveled is directly proportional to the time traveled. Write and solve a direct variation equation to find how far the Monarch butterfly will travel in 24 hours.

d. __________

e. __________

Guided Practice

1. A color printer can print 36 pages in 3 minutes and 108 pages in 9 minutes. If the number of pages varies directly with the time, at what rate is the color printer printing? (Example 1)

2. A new compact car can travel 288 miles on nine gallons of gas. The distance driven in miles y varies directly with the number of gallons of gas x. This situation can be represented by the equation $y = 32x$. (Examples 2 and 3)

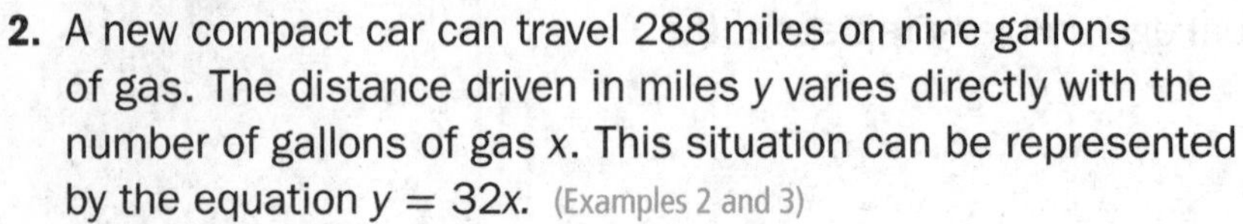
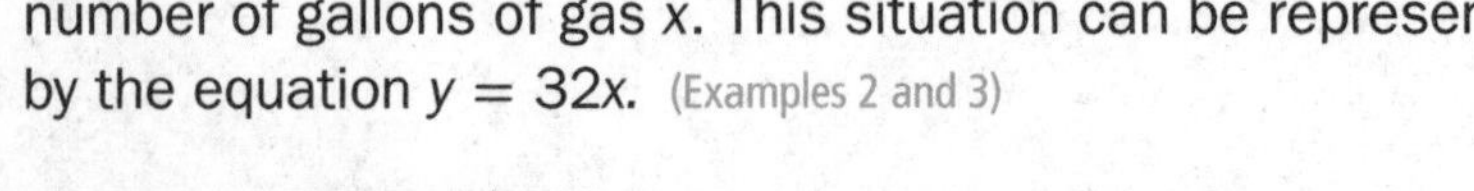
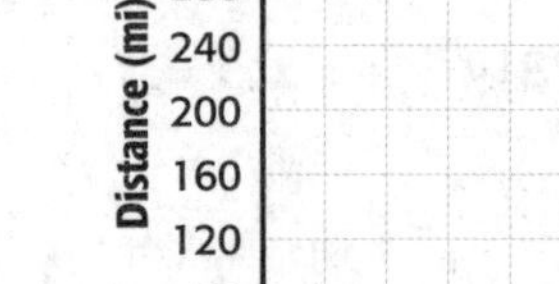

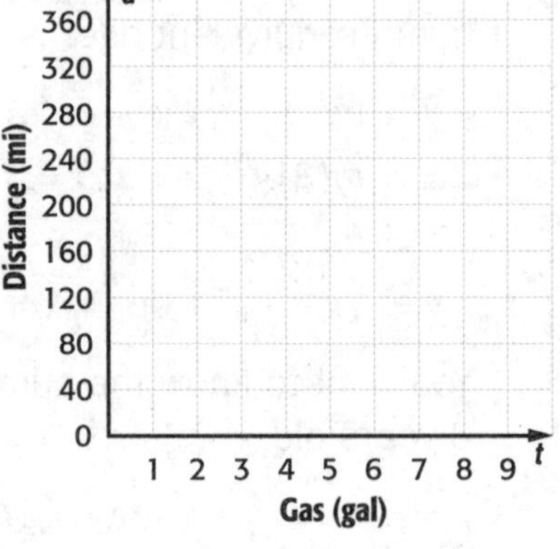

a. Graph the equation on the coordinate plane shown.

b. How many miles per gallon does the car get?

c. The distance y traveled by a hybrid car using x gallons of gas can be represented by $y = 42x$. Which car gets better gas mileage? Explain.

3. **Financial Literacy** Annie's current earnings are shown in the table. She was offered a new job that will pay $7.25 per hour. Assume that her earnings vary directly with the number of hours worked.

Which job pays more an hour? (Example 3)

Hours, x	Money Earned ($), y
2	13.00
3	19.50
4	26.00
5	32.50

4. The height of a wide-screen television screen varies directly with its width. A television screen that is 60 centimeters wide and 33.75 centimeters high. Write and solve a direct variation equation to find the height of a television screen that is 90 centimeters wide.

(Example 4)

5. **Building on the Essential Question** What is the relationship among the unit rate, slope, and constant rate of change of a proportional linear relationship?

Rate Yourself!

How well do you understand direct variation? Circle the image that applies.

Clear | Somewhat Clear | Not So Clear

For more help, go online to access a Personal Tutor.

Independent Practice

Go online for Step-by-Step Solutions

1. Dusty's earnings vary directly with the number of papers he delivers. The relationship is shown in the graph below. Determine the amount that Dusty earns for each paper he delivers. (Example 1)

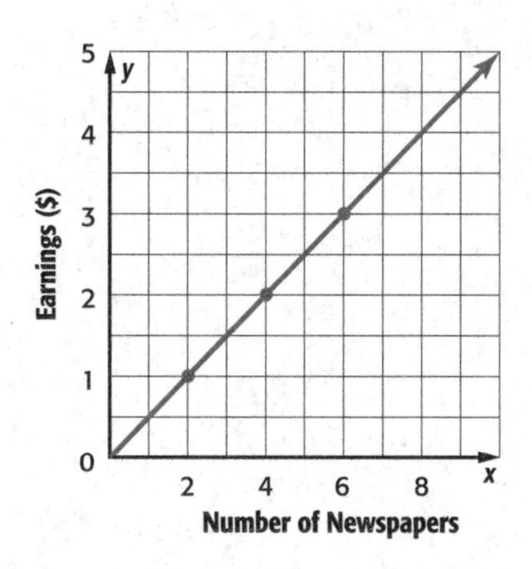

2. The Thompson family is buying a car that can travel 70 miles on two gallons of gas. Assume that the distance traveled in miles y varies directly with the amount of gas used x. This can be represented by $y = 35x$. Graph the equation on the coordinate plane. How many miles does the car get per gallon of gas? (Example 2) _______________

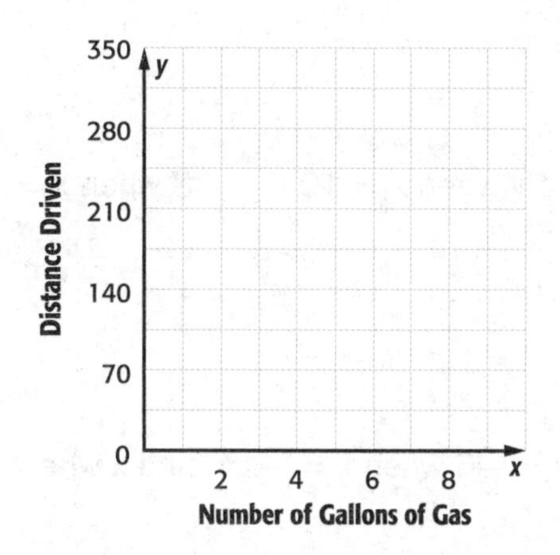

3 Tom was comparing computer repair companies. The cost y for Computer Access for x hours is shown in the graph. The cost for Computers R Us can be represented by the equation $y = 23.5x$. Which company's repair price is lower? Explain. (Example 3)

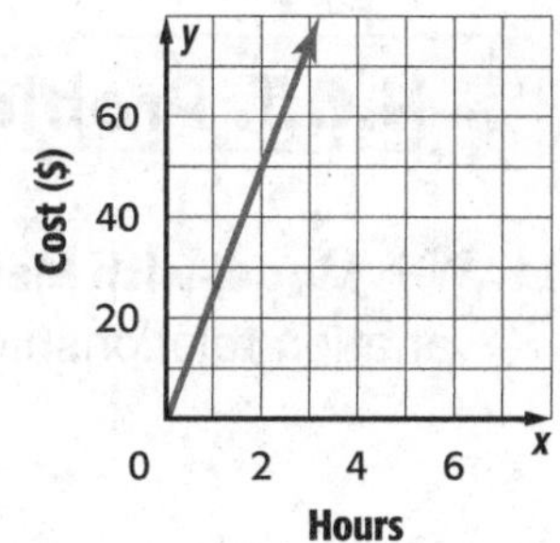

4. The weight of an object on Mars varies directly with its weight on Earth. An object that weighs 50 pounds on Mars weighs 150 pounds on Earth. If an object weighs 120 pounds on Earth, write and solve a direct variation equation to find how much an object would weigh on Mars. (Example 4)

Determine whether each linear function is a direct variation. If so, state the constant of variation. If not, explain why not.

5.

Pictures, x	5	6	7	8
Profit, y	20	24	28	32

6.

Age, x	10	11	12	13
Grade, y	5	6	7	8

7. The number of centimeters varies directly with the number of inches. Find the measure of an object in centimeters if it is 50 inches long. ______

Inches, x	6	9	12	15
Centimeters, y	15.24	22.86	30.48	38.10

CCSS **Persevere with Problems If y varies directly with x, write an equation for the direct variation. Then find each value.**

8. If $y = -12$ when $x = 9$, find y when $x = -4$. ______

9. Find y when $x = 10$ if $y = 8$ when $x = 20$. ______

10. If $y = -6$ when $x = -14$, find x when $y = -4$. ______

H.O.T. Problems Higher Order Thinking

11. CCSS **Model with Mathematics** Write three ordered pairs for a direct variation relationship where $y = 12$ when $x = 16$.

______

12. CCSS **Persevere with Problems** The amount of stain needed to cover a wood surface is directly proportional to the area of the surface. If 3 pints are required to cover a square deck with a side of 7 feet, how many pints of stain are needed to paint a square deck with a side of 10 feet 6 inches?

______

Standardized Test Practice

13. The Stratton family rented 3 DVDs for a total of $10.47. Which of the following represents the direct variation equation for this situation?

Ⓐ $y = 3x$

Ⓑ $y = 3.49x$

Ⓒ $y = 10.47x$

Ⓓ $y = 36.40x$

Extra Practice

Write and graph the direct variation equation that represents each situation.

14. Hector used 3 gallons of paint to cover 1,050 square feet and 5 gallons to paint an additional 1,750 square feet. The area covered varies directly with the amount of paint used. How many square feet will one can of paint cover?

$y = 350x$; 350 square feet per gallon

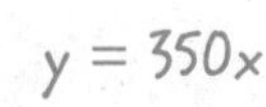

$y = mx$

$1{,}050 = m(3)$

$350 = m$

$y = 350x$

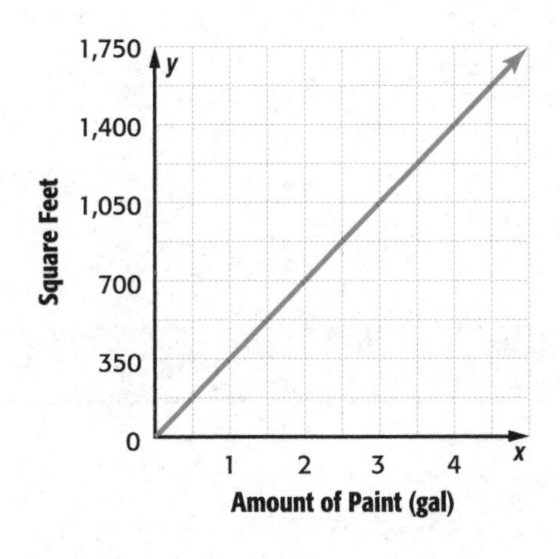

15. Nola purchased 2.5 pounds of cheese for \$10.50. Her mother purchased 3 pounds of the same cheese for \$12.60. The cost of cheese varies directly with the number of pounds purchased. How much does one pound of cheese cost?

____

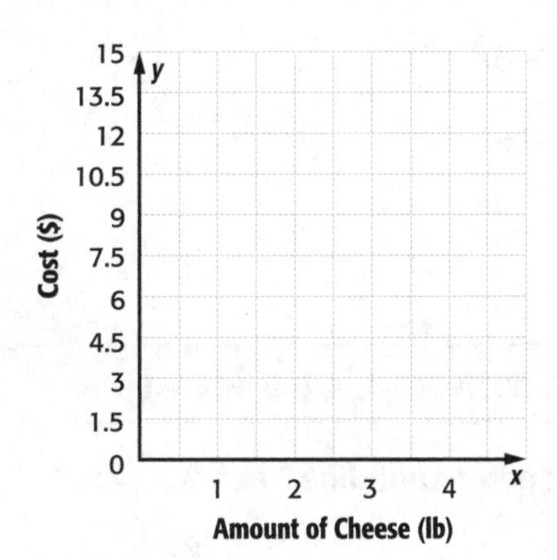

16. **STEM** When a 49 pound weight is attached to a spring, the spring stretches 7 inches. Assume that the length of the spring y varies directly with the weight attached x. Write and solve a direct variation equation to find the length of the spring when a 63 pound weight is attached.

____

17. CCSS **Justify Conclusions** The money raised by the Drama Club selling raffle tickets is shown in the table. They can also raise money by selling tickets to the play for \$6.25 per ticket. Assume that the money raised varies directly with the number of tickets sold. Which fundraiser has the potential to raise more money? Explain your reasoning to a classmate.

Raffle Tickets Sold	25	50	75	100
Money Raised ($)	125	250	375	500

____

____

Standardized Test Practice

18. Students in a science class recorded lengths of a stretched spring, as shown in the table below.

Length of Stretched Spring	
Distance Stretched, x (centimeters)	**Mass, y (grams)**
0	0
2	12
5	30
9	54
12	72

Which equation best represents the relationship between the distance stretched x and the mass of an object on the spring y?

Ⓐ $y = -6x$

Ⓑ $y = 6x$

Ⓒ $y = -\frac{x}{6}$

Ⓓ $y = \frac{x}{6}$

19. Short Response Nicole read 24 pages during a 30-minute independent reading period. At this rate, how many pages would she read in 45 minutes?

20. To make fruit punch, Kelli adds 8 ounces of pineapple juice for every 12 ounces of orange juice. Suppose she uses 32 ounces of orange juice. Which proportion can she use to find x, the number of ounces of pineapple juice needed to make the punch?

Ⓕ $\frac{8}{12} = \frac{32}{x}$

Ⓖ $\frac{8}{x} = \frac{32}{12}$

Ⓗ $\frac{8}{12} = \frac{x}{32}$

Ⓘ $\frac{x}{12} = \frac{8}{32}$

Common Core Review

Find the slope of each line. 8.EE.5

21. ______

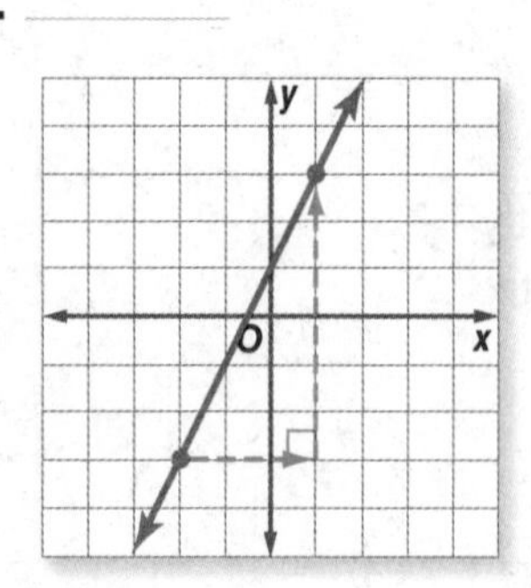

22. ______

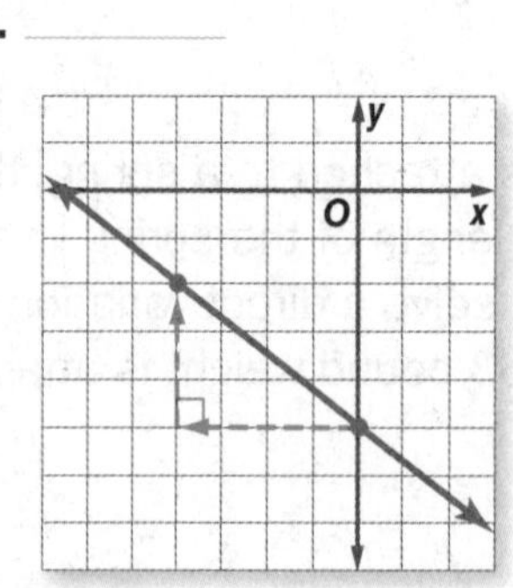

23. ______

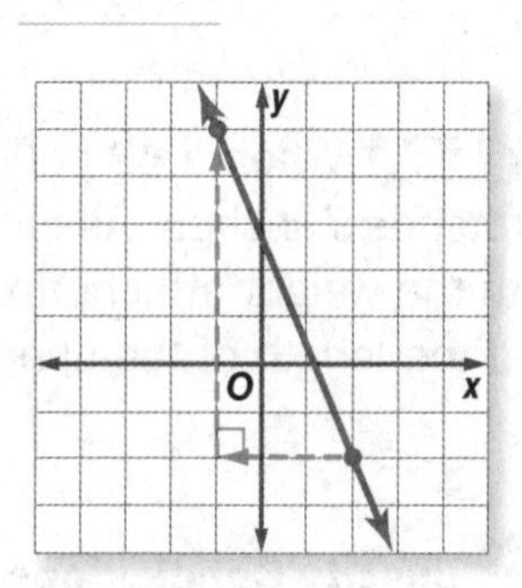

Find the slope of the line that passes through each pair of points. 8.EE.5

24. $(-1, 7)$ and $(5, 7)$ ______

25. $(1, 3)$ and $(1, 0)$ ______

26. $(1, 2)$ and $(5, 0)$ ______

Lesson 4

Slope-Intercept Form

What You'll Learn

Scan the lesson. Predict two things you will learn about the slope-intercept form of a linear equation.

- ______________________________

- ______________________________

Essential Question

WHY are graphs helpful?

Vocabulary

y-intercept
slope-intercept form

Common Core State Standards

Content Standards
8.EE.6, 8.F.3, 8.F.4

Mathematical Practices
1, 3, 4

Real-World Link

Football An interception in football is when a defensive player catches a pass made by an offensive player.

In a nonproportional linear relationship, the graph passes through the point $(0, b)$ or the y-intercept. The **y-intercept** of a line is the y-coordinate of the point where the line crosses the y-axis.

Complete the steps to derive the equation for a nonproportional linear relationship by using the slope formula.

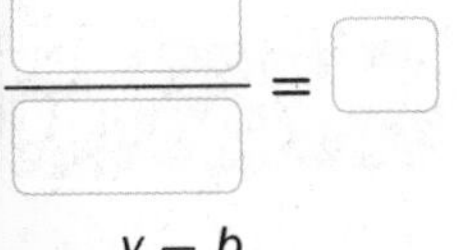

$\frac{\square}{\square} = \square$ Slope formula

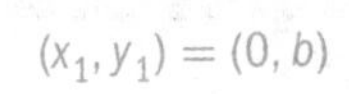

$(x_1, y_1) = (0, b)$

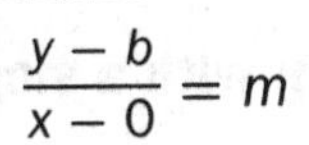

$\frac{y - b}{x - 0} = m$ $(x_2, y_2) = (x, y)$

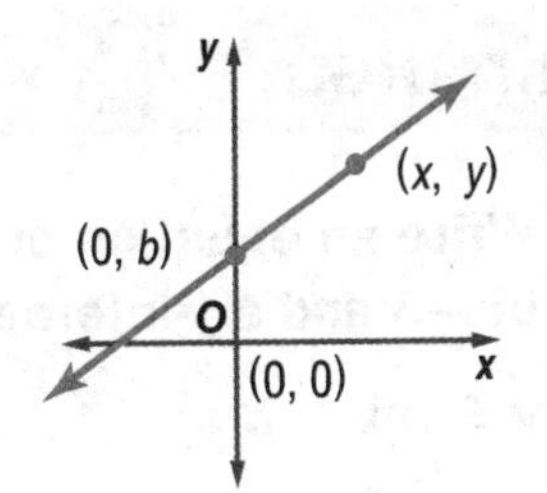

$\frac{\square}{\square} = m$ Simplify.

$y - b = \square \cdot \square$ Multiplication Property of Equality

$y = \square + \square$ Addition Property of Equality

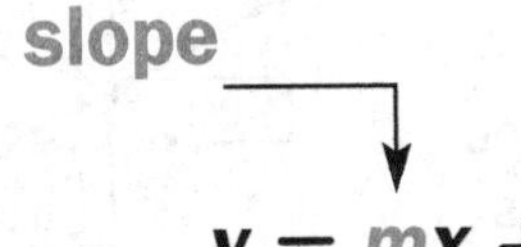

$y = mx + b$

How can knowing about an interception in football help you remember the definition of y-intercept?

____

____

Work Zone

Slope-Intercept Form of a Line

Nonproportional linear relationships can be written in the form $y = mx + b$. This is called the **slope-intercept form**. When an equation is written in this form, m is the slope and b is the y-intercept.

Examples

1. State the slope and the y-intercept of the graph of the equation $y = \frac{2}{3}x - 4$.

$y = \frac{2}{3}x + (-4)$ — Write the equation in the form $y = mx + b$.

$y = mx + b$ — $m = \frac{2}{3}, b = -4$

The slope of the graph is $\frac{2}{3}$, and the y-intercept is -4.

Got It? Do these problems to find out.

Show your work.

a. $y = -5x + 3$ **b.** $y = \frac{1}{4}x - 6$ **c.** $y = -x + 5$

a. ____________

b. ____________

c. ____________

Examples

2. Write an equation of a line in slope-intercept form with a slope of −3 and a y-intercept of −4.

$y = mx + b$ — Slope-intercept form

$y = -3x + (-4)$ — Replace m with -3 and b with -4.

$y = -3x - 4$ — Simplify.

3. Write an equation in slope-intercept form for the graph shown.

y-intercept: 4

1 unit

2 units

The y-intercept is 4. From (0, 4), you move down 1 unit and right 2 units to another point on the line.

So, the slope is $-\frac{1}{2}$.

$y = mx + b$ — Slope-intercept form

$y = -\frac{1}{2}x + 4$ — Replace m with $-\frac{1}{2}$ and b with 4.

Got It? Do these problems to find out.

d. Write an equation in slope-intercept form for the graph shown.

e. Write an equation of a line in slope-intercept form with a slope of $\frac{3}{4}$ and a y-intercept of -3.

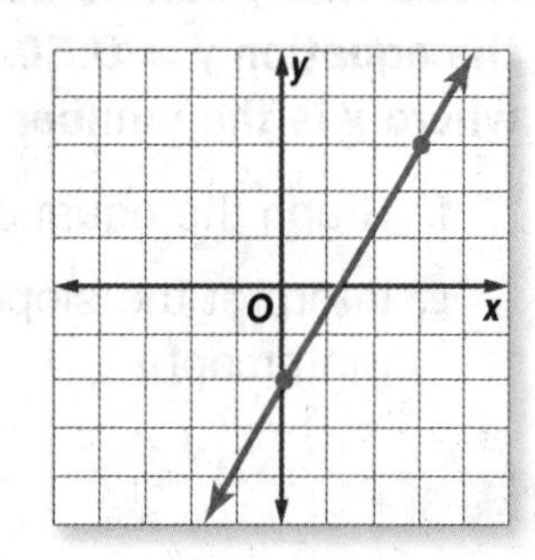

d. ______

e. ______

Interpret the y-intercept

When an equation in slope-intercept form applies to a real-world situation, the slope represents the rate of change and the y-intercept represents the initial value.

Examples

4. Student Council is selling T-shirts during spirit week. It costs \$20 for the design and \$5 to print each shirt. The cost y to print x shirts is given by $y = 5x + 20$. Graph $y = 5x + 20$ using the slope and y-intercept.

Step 1 Find the slope and y-intercept.

$y = 5x + 20$ — slope = 5, y-intercept = 20

Step 2 Graph the y-intercept (0, 20).

Step 3 Write the slope 5 as $\frac{5}{1}$. Use it to locate a second point on the line. Go up 5 units and right 1 unit. Then draw a line through the points.

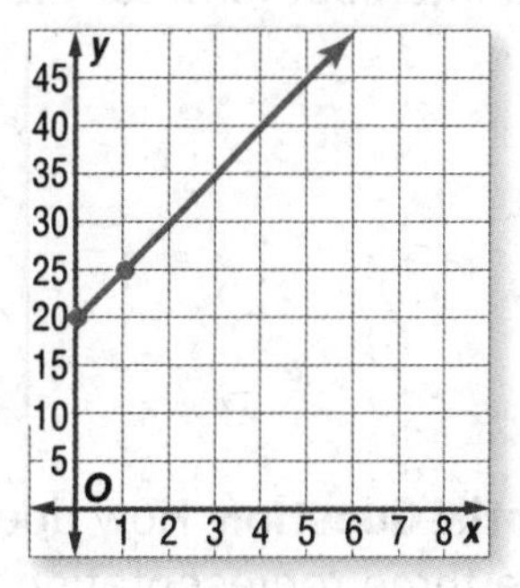

5. Interpret the slope and the y-intercept.

The slope 5 represents the cost in dollars per T-shirt. The y-intercept 20 is the one-time charge in dollars for the design.

Got It? Do these problems to find out.

A taxi fare y can be determined by the equation $y = 0.50x + 3.50$, where x is the number of miles traveled.

f. Graph the equation.

g. Interpret the slope and the y-intercept.

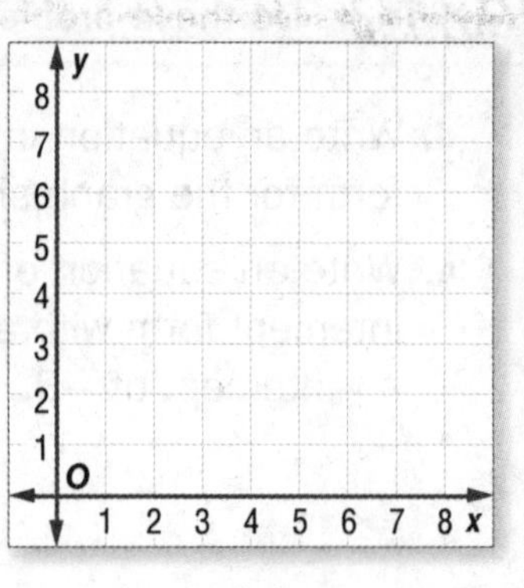

g. ______________

Guided Practice

1. Liam is reading a 254-page book for school. He can read 40 pages in one hour. The equation for the number of pages he has left to read is $y = 254 - 40x$, where x is the number of hours he reads. (Examples 1, 4, and 5)

   a. State the slope and the y-intercept of the graph of the equation. ______________

   b. Graph the equation.

   c. Interpret what the slope and the y-intercept represent.

   ____

   ____

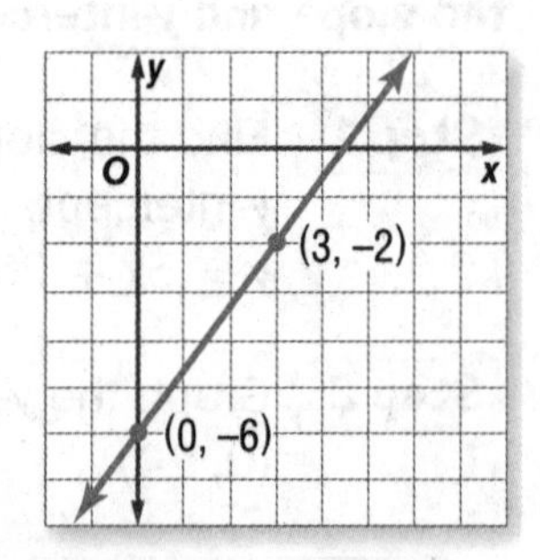

2. Write an equation in slope intercept form for the graph shown. (Examples 2 and 3) ______________

3. **Building on the Essential Question** How does the y-intercept appear in these three representations: table, equation, and graph? ______________

   ____

   ____

   ____

   ____

Rate Yourself!

How confident are you about equations in slope-intercept form? Check the box that applies.

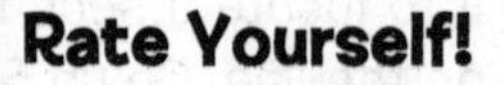

For more help, go online to access a Personal Tutor.

Name ______________________ My Homework ______________________

Independent Practice

Go online for Step-by-Step Solutions 

State the slope and the *y*-intercept for the graph of each equation. (Example 1)

1. $y = 3x + 4$ ________

2. $y = -\frac{3}{7}x - \frac{1}{7}$ ________

3. $3x + y = -4$ ________

Write an equation of a line in slope-intercept form with the given slope and *y*-intercept. (Example 2)

4. slope: $-\frac{3}{4}$, *y*-intercept: -2

____

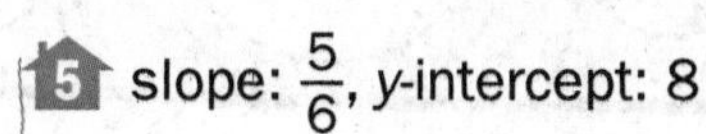

5. slope: $\frac{5}{6}$, *y*-intercept: 8

____

Write an equation in slope-intercept form for each graph shown. (Example 3)

6.

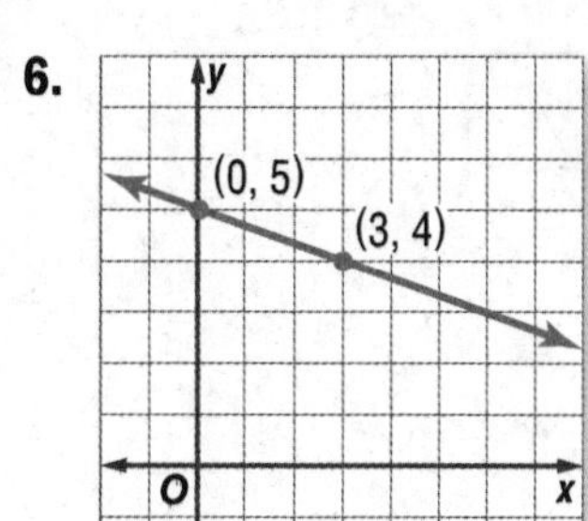

____

7.

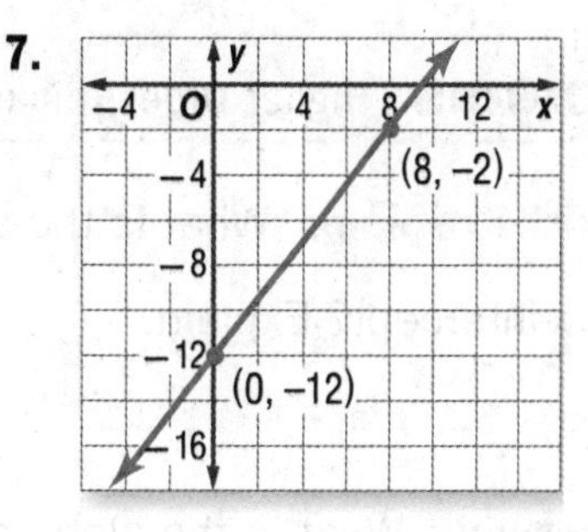

____

8. The Viera family is traveling from Philadelphia to Orlando for vacation. The equation $y = 1{,}000 - 65x$ represents the distance in miles remaining in their trip after *x* hours. (Examples 4 and 5)

a. Graph the equation.

b. Interpret the slope and the *y*-intercept. ________

____

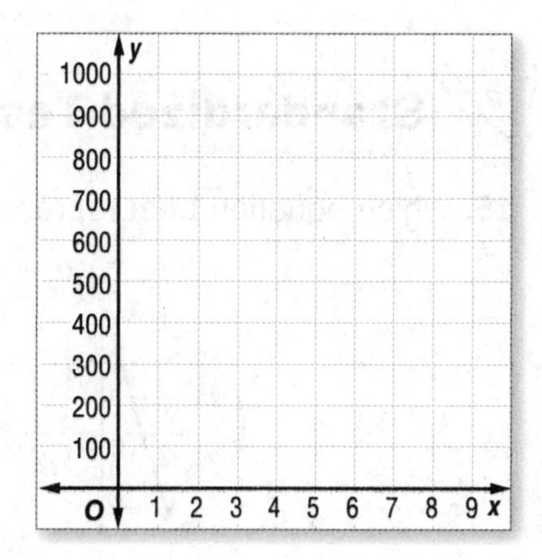

Copy and Solve. Graph each equation on a separate piece of grid paper.

9. $y = \frac{1}{3}x - 5$

10. $y = -x + \frac{3}{2}$

11. $y = -\frac{4}{3}x + 1$

12.  **Model with Mathematics** Refer to the graphic novel frame below for Exercises a–b.

a. Write an equation in slope-intercept form for the total cost of any number of tickets at 7 tickets for $5. __________

b. Write an equation in slope-intercept form for the total cost of a wristband for all you can ride. __________

H.O.T. Problems Higher Order Thinking

13. **Persevere with Problems** What is the slope of a line that has a *y*-intercept but no *x*-intercept? Explain. __________

14. **Reason Inductively** What is the slope and *y*-intercept of a vertical line? __________

Standardized Test Practice

15. Which equation best represents the graph shown?

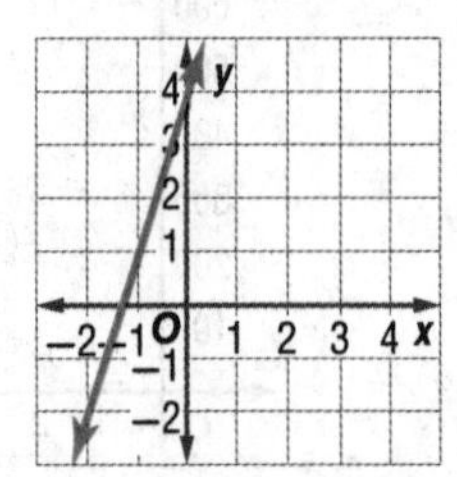

Ⓐ $y = -4x + 3$ Ⓒ $y = 3x + 4$

Ⓑ $y = -3x + 4$ Ⓓ $y = 4x + 3$

Name ______________________ My Homework ______________________

Extra Practice

State the slope and the y-intercept for the graph of each equation.

16. $y = -5x + 2$ $-5; 2$

In the equation, $m = -5$ and $b = 2$ so the slope is -5 and the y-intercept is 2.

17. $y = \frac{1}{2}x - 6$ ______

18. $y - 2x = 8$ ______

Write an equation of a line in slope-intercept form with the given slope and y-intercept.

19. slope: $\frac{1}{2}$; y-intercept: 6 ______

20. slope: -2; y-intercept: 3 ______

21. slope: $-\frac{3}{5}$; y-intercept: $-\frac{1}{5}$ ______

22. The Lakeside Marina charges a \$35 rental fee for a boat in addition to charging \$15 an hour for usage. The total cost y of renting a boat for x hours can be represented by the equation $y = 15x + 35$.

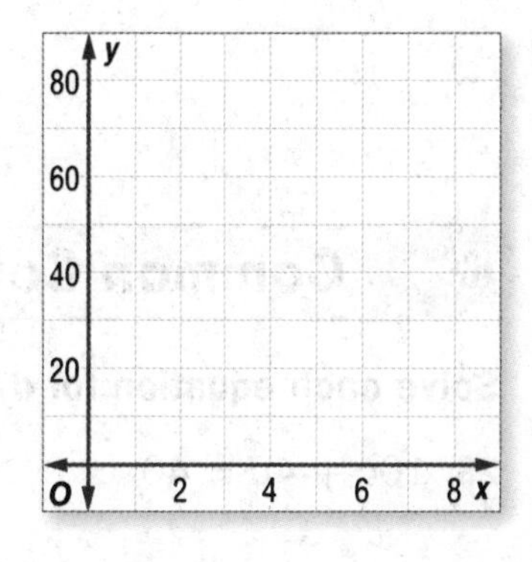

a. Graph the equation.

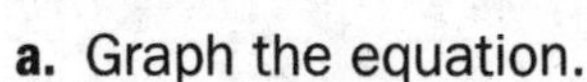

b. Interpret the slope and the y-intercept.

____

____

23. CCSS **Persevere with Problems** The equation $y = 15x + 37$ can be used to approximate the temperature y in degrees Fahrenheit based on the number of chirps x a cricket makes in 15 seconds. Graph the equation to estimate the number of chirps a cricket will make in 15 seconds if the temperature is 80°F.

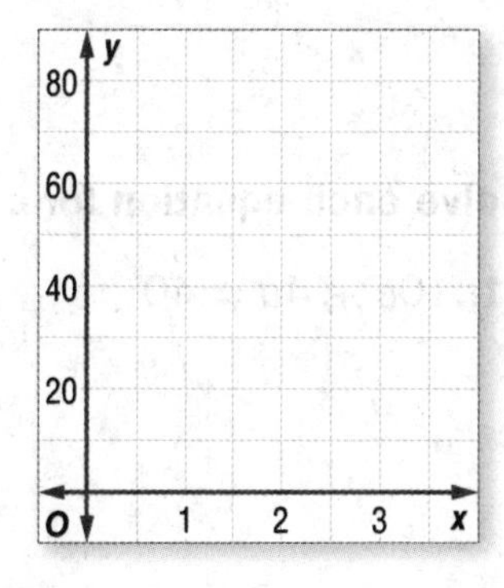

____

24. Write an equation in slope-intercept form for the table shown.

Number of Pizzas	0	1	2	3	4
Cost (\$)	5	13	21	29	37

____

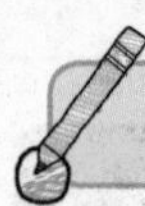

Standardized Test Practice

25. Which statement could be true for the graph below?

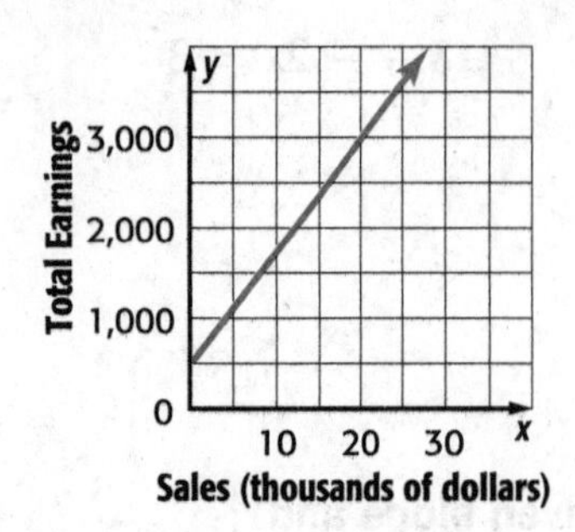

Ⓐ Mr. Blackwell will earn \$1,750 if his sales are \$10,000.

Ⓑ Ms. Chu will not earn any money if she has no sales.

Ⓒ Mr. Montoya earns \$250 for every \$1,000 he sells.

Ⓓ Ms. James earns \$2,500 if she sells \$2,500 worth of merchandise.

26. Jaquie has 20 postcards in her collection. She decides that from now on, every time she goes on vacation she will buy 8 postcards to add to the collection. The total number of postcards y can be represented by the equation $y = 8x + 20$. What does the slope represent?

Ⓕ the total number of postcards

Ⓖ the number of postcards when she began collecting

Ⓗ the number of vacations

Ⓘ the number of cards she buys on each vacation

27. Short Response A line has a slope of $-\frac{2}{3}$ and a y-intercept of -4. Write an equation in slope-intercept form for the line.

Common Core Review

Solve each equation for d when $c = 0$. 7.EE.4

28. $10c + 4d = 40$

29. $-5d = 2c + 10$

30. $-4c - 6d = 24$

Solve each equation for c when $d = 0$. 7.EE.4

31. $10c + 4d = 40$

32. $-5d = 2c + 10$

33. $-4c - 6d = 24$

Determine whether each linear relationship is proportional. If so, state the constant of proportionality. 7.RP.2

34.

Pictures, x	5	6	7	8
Profit, y	20	24	28	32

35.

Price, x	10	15	20	25
Tax, y	0.70	1.05	1.40	1.75

Inquiry Lab
Slope Triangles

HOW does graphing slope triangles on the coordinate plane help you analyze them?

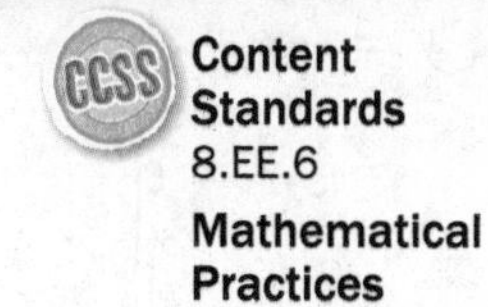

Content Standards
8.EE.6

Mathematical Practices
1, 3, 5

Skateboarding Donte ordered the plans shown to build a skateboard ramp. Each unit represents one foot. He wants to keep the same slope of the ramp and extend the base of the triangle three feet. How tall will the ramp be?

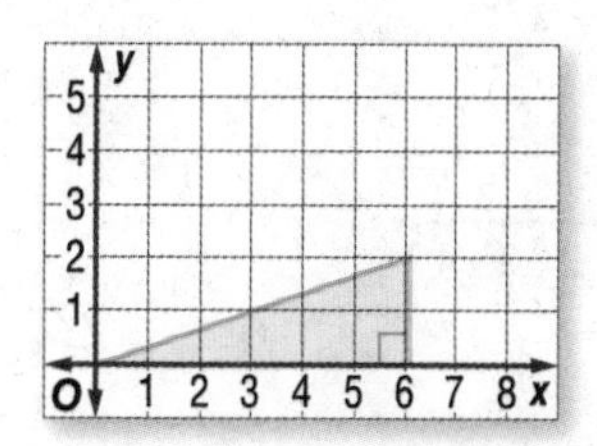

Investigation

Refer to the graph shown above. Triangle ABC is formed by the rise, run, and section of the line $y = \frac{1}{3}x$ between points A and B.

Step 1 Graph $y = \frac{1}{3}x$ on the grid paper. Draw a right triangle using the points $A(0, 0)$ and $B(6, 2)$. Label the third point C.

What is the slope of $\overline{AB}$? $\frac{\square}{\square}$

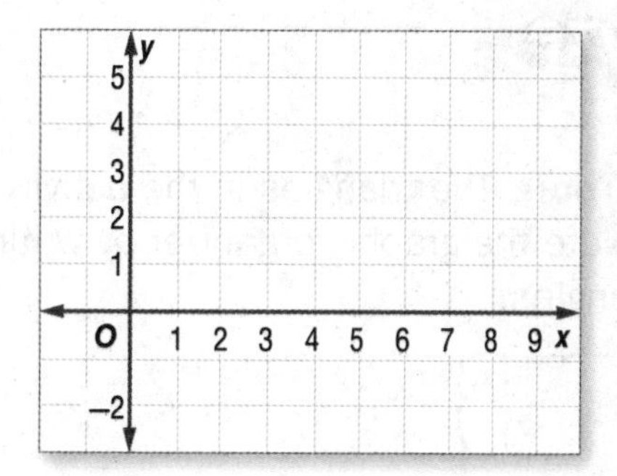

Step 2 Select any two different points on the line. Label them D and E. Draw another triangle from these two points.

Is the slope of $\overline{DE}$ the same as the slope of $\overline{AB}$? Explain.

____

____

Step 3 Donte wants to expand the base of the ramp 3 feet. Graph and give the coordinates of the point that will represent the extended base of the ramp. ____________

Create a right triangle using the line and that point. What will be the height of the new ramp? ____________

Work with a partner. Draw two right triangles for each exercise using the rise, run, and portions of the line.

1. $y = -x + 2$

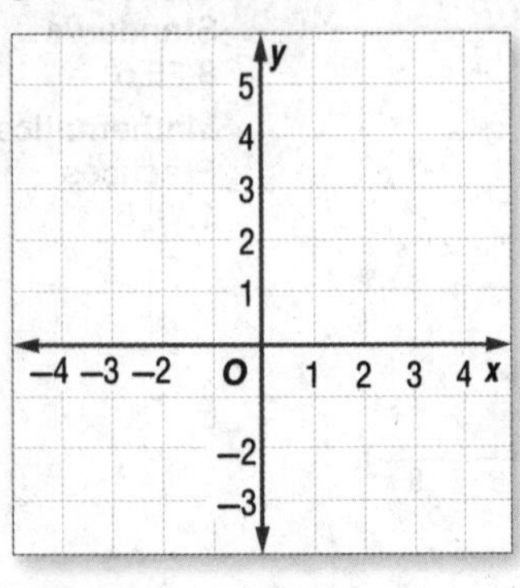

2. $y = x + 1$

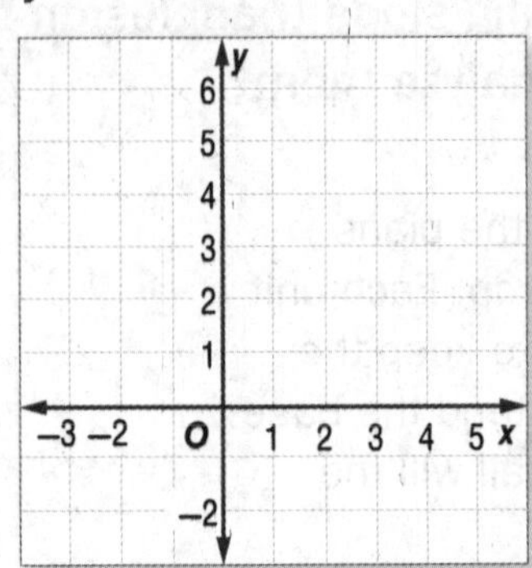

3. **Make a Conjecture** What do you notice about the shape and size of the pair of triangles in Exercises 1 and 2? ______________________________

Analyze

4. **Use Math Tools** The triangles in the activity are called *slope triangles*. Complete the graphic organizer by writing three observations about slope triangles.

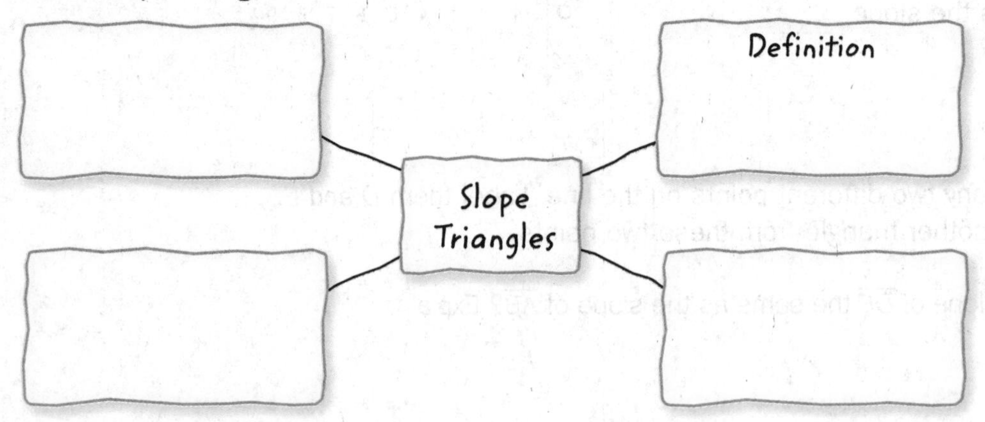

Reflect

5. Inquiry HOW does graphing slope triangles on the coordinate plane help you analyze them?

____

____

Lesson 5

Graph a Line Using Intercepts

What You'll Learn

Scan the lesson. Predict two things you will learn about intercepts.

- ______________________________

- ______________________________

Essential Question

WHY are graphs helpful?

Vocabulary

x-intercept
standard form

Common Core State Standards

Content Standards
Preparation for 8.EE.8c

Mathematical Practices
1, 3, 4

Real-World Link

Movies Mrs. Hodges spent $80 on movie tickets and drinks for her son and his friends. The total cost of x movie tickets and y drinks is represented by the equation $8x + 4y = 80$.

Item	Cost
ticket	$8
drink	$4

1. Complete the steps below to write the equation in slope-intercept form.

$8x + 4y = 80$

$\square \quad = \square$

$\frac{4y}{\square} = \frac{80 - 8x}{\square}$

$y = 20 - 2x$

$y = \square x + \square$

slope (first box) **y-intercept** (second box)

2. Graph the equation.

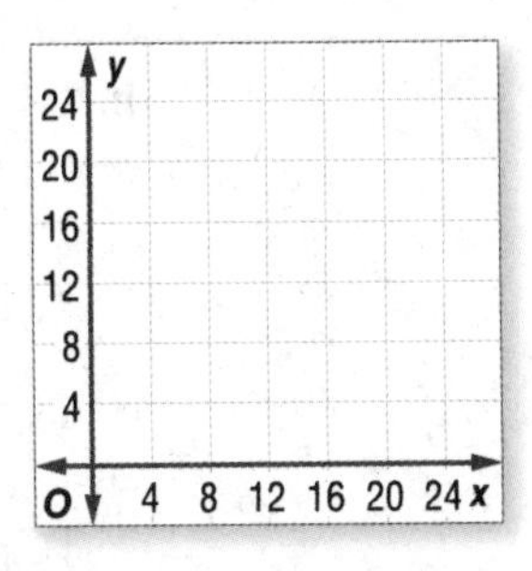

3. What does the point (0, 20) represent?

____

Slope-Intercept Form

The **x-intercept** of a line is the *x*-coordinate of the point where the graph crosses the *x*-axis. Since any linear equation can be graphed using two points, you can use the *x*- and *y*-intercepts to graph an equation.

Example

1. State the *x*- and *y*-intercepts of $y = 1.5x - 9$. Then use the intercepts to graph the equation.

Step 1 First find the *y*-intercept.

$y = 1.5x + (-9)$ Write the equation in the form $y = mx + b$.

$b = -9$

Step 2 To find the *x*-intercept, let $y = 0$.

$0 = 1.5x - 9$ Write the equation. Let $y = 0$.

$9 = 1.5x$ Addtion Property of Equality

$\frac{9}{1.5} = \frac{1.5x}{1.5}$ Division Property of Equality

$6 = x$ Simplify.

Step 3 Graph the points (6, 0) and (0, −9) on a coordinate plane. Then connect the points.

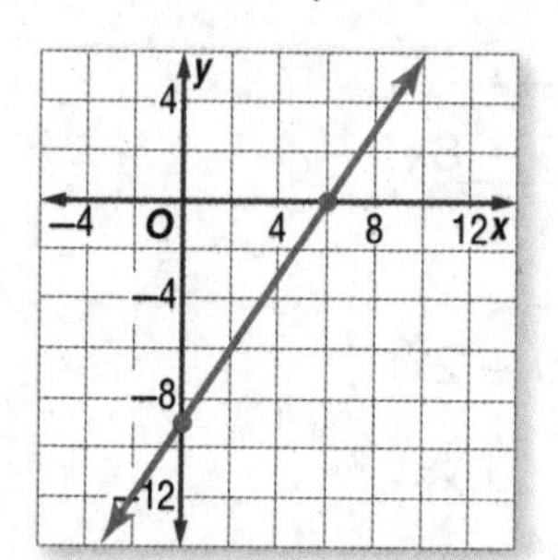

Show your work.

a. ____________

b. ____________

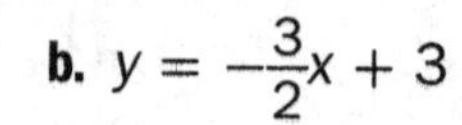

Got It? Do these problems to find out.

a. $y = -\frac{1}{3}x + 5$

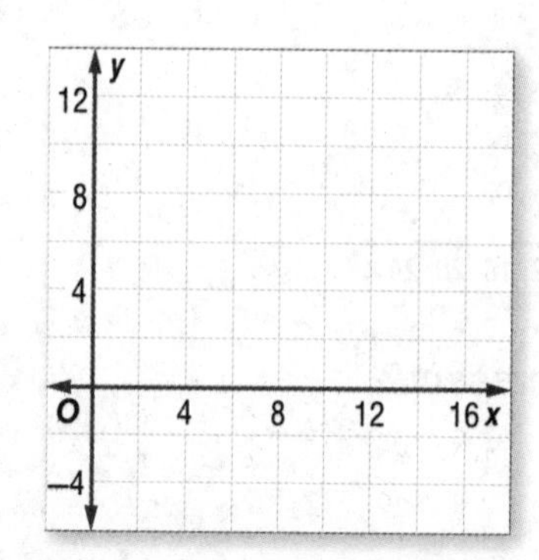

b. $y = -\frac{3}{2}x + 3$

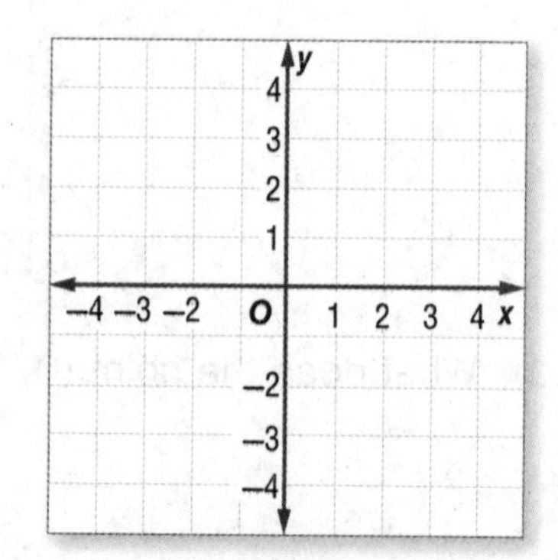

Work Zone

Standard Form

When an equation is written in the form $Ax + By = C$, where $A \geq 0$, and A, B, and C are integers, it is written in **standard form**.

Examples

Mauldin Middle School wants to make \$4,740 from yearbooks. Print yearbooks x cost \$60 and digital yearbooks y cost \$15. This can be represented by the equation $60x + 15y = 4{,}740$.

2. Use the x- and y-intercepts to graph the equation.

To find the x-intercept, let $y = 0$. To find the y-intercept, let $x = 0$.

$$60x + 15y = 4{,}740 \qquad 60x + 15y = 4{,}740$$
$$60x + 15(0) = 4{,}740 \qquad 60(0) + 15y = 4{,}740$$
$$60x = 4{,}740 \qquad 15y = 4{,}740$$
$$x = 79 \qquad y = 316$$

3. Interpret the x- and y-intercepts.

The x-intercept is at the point (79, 0). This means they can sell 79 print yearbooks and 0 digital yearbooks to earn \$4,740.

The y-intercept is at the point (0, 316). This means they can sell 0 print yearbooks and 316 digital yearbooks to earn \$4,740.

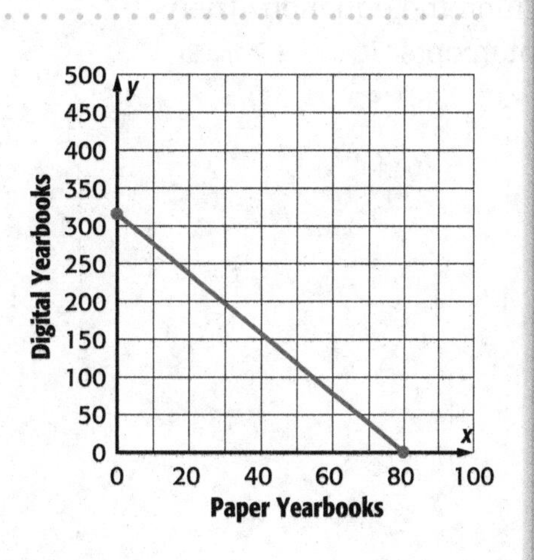

STOP and Reflect

Describe below two different methods for graphing a line.

y-intercept

When an equation is written in slope-intercept form, $y = mx + b$, the y-intercept is equal to b.

Got It? Do this problem to find out.

c. Mr. Davies spent \$230 on lunch for his class. Sandwiches x cost \$6 and drinks y cost \$2. This can be represented by the equation $6x + 2y = 230$. Use the x- and y-intercepts to graph the equation. Then interpret the intercepts.

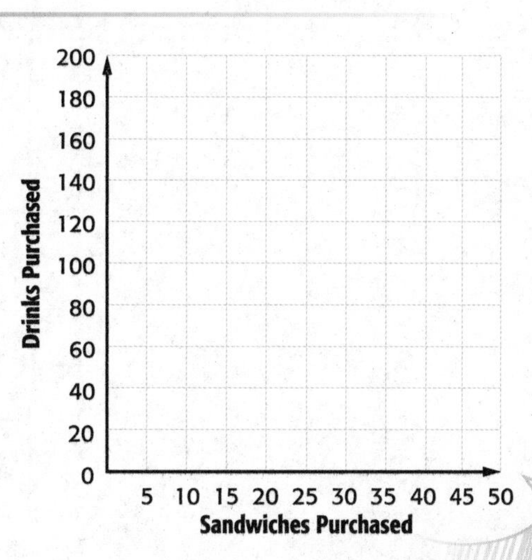

c. ______________

Guided Practice

State the *x*- and *y*-intercepts of each equation. Then use the intercepts to graph the equation. (Example 1)

1. $y = 3x - 9$

Show your work.

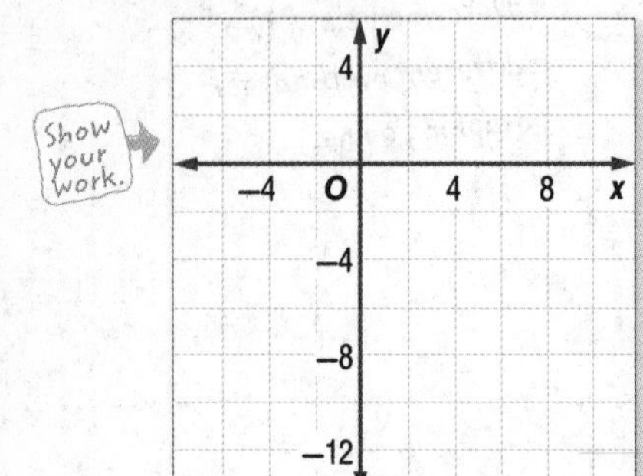

2. $y = \frac{1}{2}x + 2$

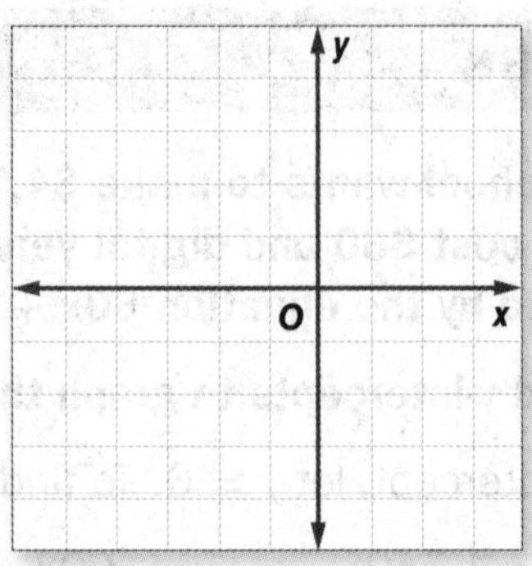

3. A store sells juice boxes in packages of 6 boxes and 8 boxes. They have 288 total juice boxes. This is represented by the function $6x + 8y = 288$. Use the *x*- and *y*-intercepts to graph the equation. Then interpret the *x*- and *y*-intercepts. (Examples 2 and 3)

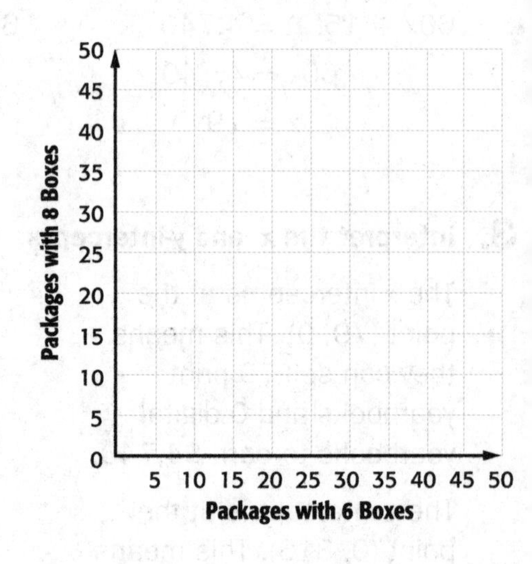

4. **Building on the Essential Question** How can the *x*-intercept and *y*-intercept be used to graph a linear equation?

Rate Yourself!

Are you ready to move on? Shade the section that applies.

I have a few questions.

I'm ready to move on.

I have a lot of questions.

For more help, go online to access a Personal Tutor.

Tutor

Independent Practice

Go online for Step-by-Step Solutions

State the *x*- and *y*-intercepts of each equation. Then use the intercepts to graph the equation. (Example 1)

1. $y = -2x + 7$

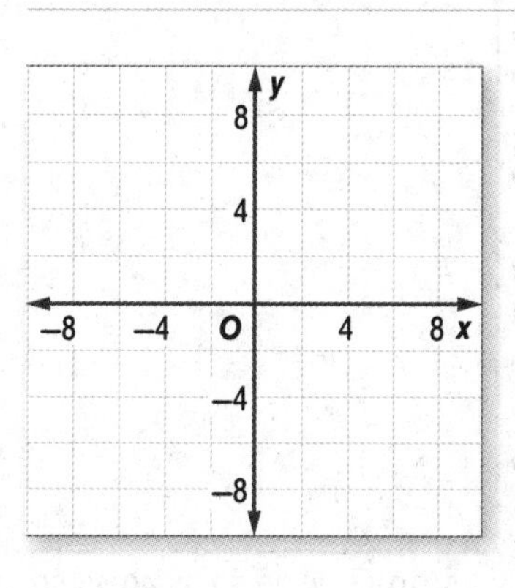

2. $y = \frac{3}{4}x + 3$

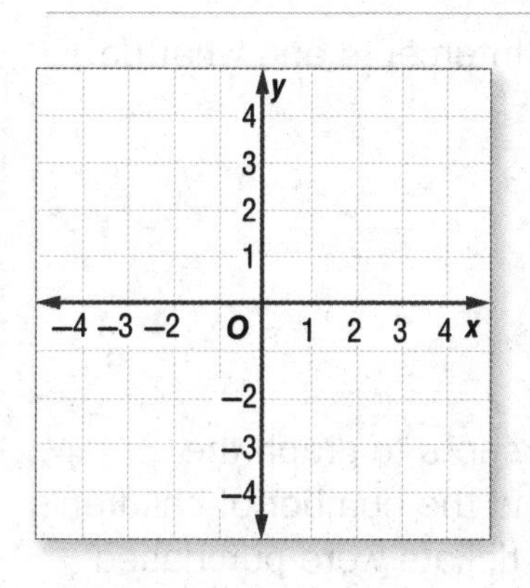

3. $12x + 9y = 15$

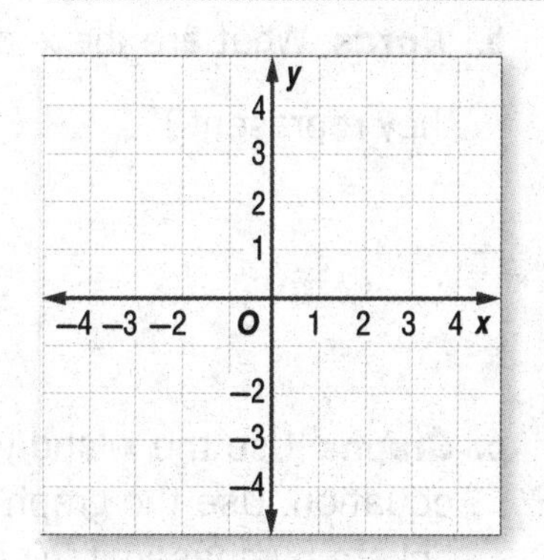

4. The table shows the cost for a clothing store to buy jeans and khakis. The total cost for Saturday's shipment, \$1,800, is represented by the equation $15x + 20y = 1{,}800$. Use the *x*- and *y*-intercepts to graph the equation. Then interpret the *x*- and *y*-intercepts. (Examples 2 and 3)

	Jeans	Khakis
Cost per Pair (\$)	15	20
Amount Shipped	*x*	*y*

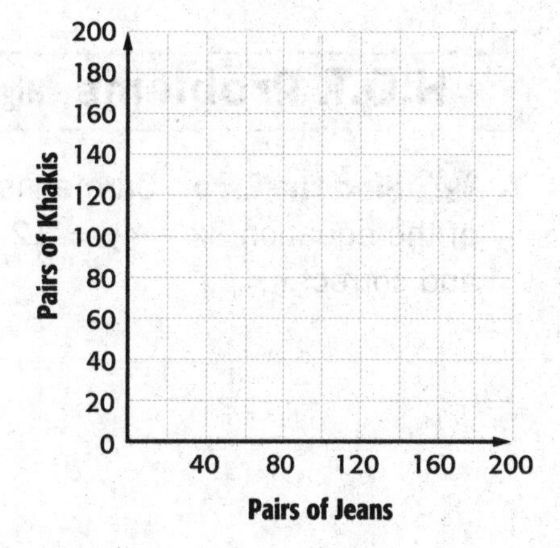

5. The total number of legs, 1,500, on four-legged and two-legged animals in a zoo can be represented by the equation $4x + 2y = 1{,}500$. Use the *x*- and *y*-intercepts to graph the equation. Then interpret the *x*- and *y*-intercepts. (Examples 2 and 3)

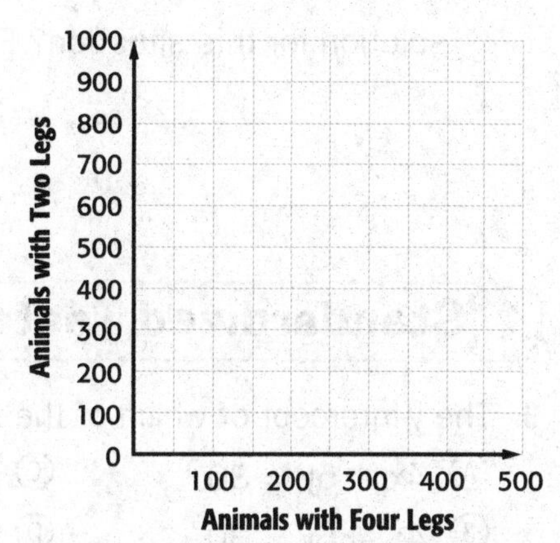

6. **Multiple Representations** The table shows the group rate for admission tickets for adults and children to an amusement park.

	Adult	Children
Ticket price ($)	45	30
Tickets purchased	x	y

a. **Symbols** The total cost of a group's tickets is $1,350. Write an equation to represent the number of adults' and children's tickets purchased.

b. **Words** What are the x- and y-intercepts and what do they represent?

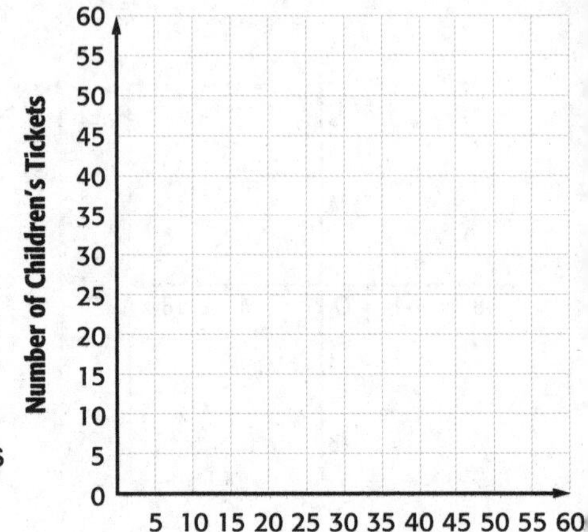

c. **Graphs** Use the x- and y-intercepts to graph the equation. Use the graph to find the number of children's tickets purchased if 20 adult tickets were purchased.

H.O.T. Problems Higher Order Thinking

7. **Find the Error** Carmen is finding the x-intercept of the equation $3x - 4y = 12$. Find her mistake and correct it.

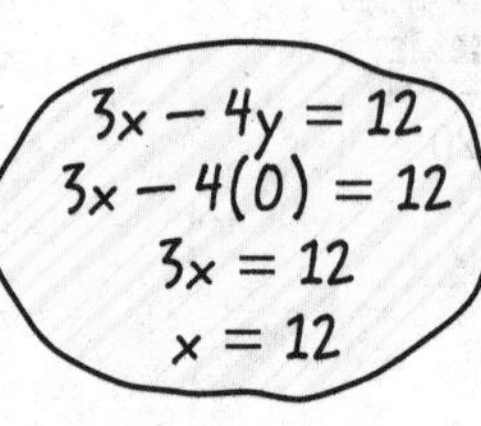

8. **Persevere with Problems** The perimeter of a rectangle that is x units wide and y units long is 24 centimeters.

a. Write an equation in standard form for the perimeter.

b. Find the x- and y-intercepts. Does either intercept make sense as a solution for this situation? Explain.

Standardized Test Practice

9. The y-intercept of which of the following equations is 5?

Ⓐ $4x - 5y = 30$

Ⓑ $4x - 6y = 30$

Ⓒ $4x + 5y = 30$

Ⓓ $4x + 6y = 30$

Name ____________________ My Homework ____________________

Extra Practice

10. State the x- and y-intercepts of the equation $y = \frac{2}{3}x - \frac{1}{3}$. Then use the intercepts to graph the equation.

Find the y-intercept.

$y = \frac{2}{3}x + (-\frac{1}{3})$

$b = -\frac{1}{3}$

Find the x- intercept.

$y = \frac{2}{3}x + (-\frac{1}{3})$

$0 = \frac{2}{3}x + (-\frac{1}{3})$

$\frac{1}{3} = \frac{2}{3}x$

$(\frac{3}{2})\frac{1}{3} = (\frac{3}{2})\frac{2}{3}x$

$\frac{1}{2} = x$

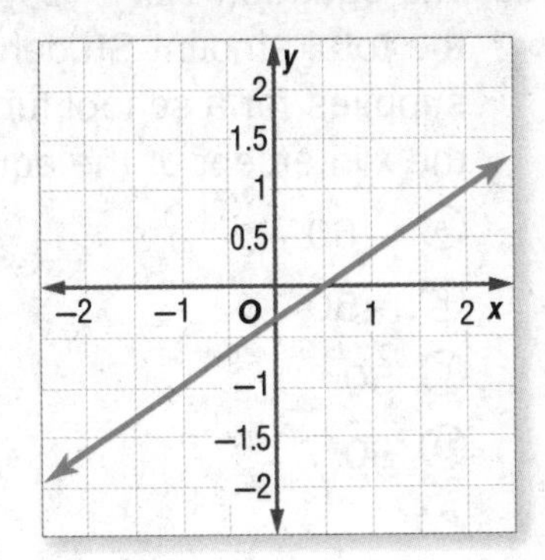

Copy and Solve **State the x- and y-intercepts of each equation. Then use the intercepts to graph each equation on a separate sheet of grid paper.**

11. $2x + 3y = 24$

12. $y = -\frac{8}{9}x - 16$

13. $5x + 3y = 30$

14. Tiffany has 15 teaspoons of chocolate chips. She uses $1\frac{1}{2}$ teaspoons for each muffin. The total amount of chocolate chips that she has left y after making x muffins can be given by $y = -\frac{3}{2}x + 15$. Graph the equation. Then interpret the x- and y-intercepts. ____________________

____

____

____

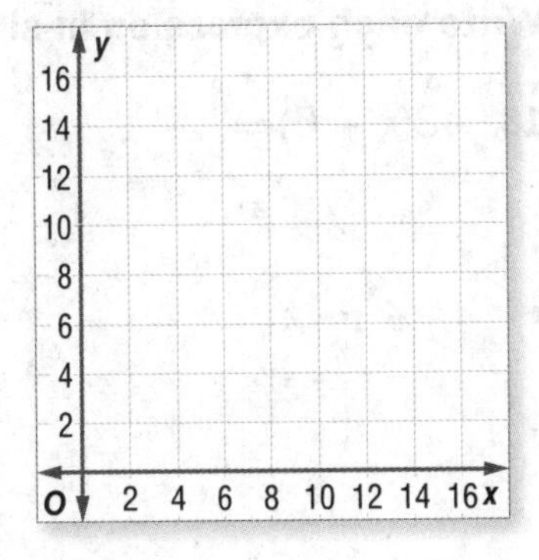

15. CCSS **Use Math Tools** Miriam has $440 to pay a painter to paint her basement. The painter charges $55 per hour. The equation $y = 440 - 55x$ represents the amount of money y she has after x number of hours worked by the painter. Graph the equation. Then interpret the x- and y-intercepts. ____________________

____

____

____

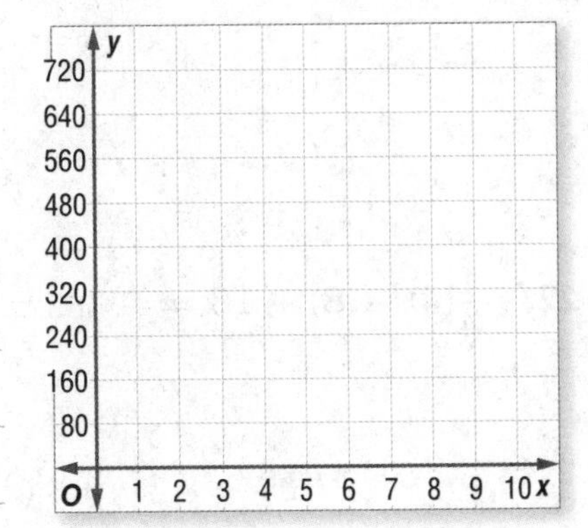

Standardized Test Practice

16. The equation $12x - 10y = 600$ represents the total amount Student Council spent on supplies for a school fundraiser. What is the x-intercept of the equation?

Ⓐ -60

Ⓑ -50

Ⓒ 50

Ⓓ 60

17. Which of the following is the graph of the equation $2x + 3y = 6$?

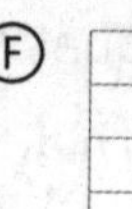
Ⓕ
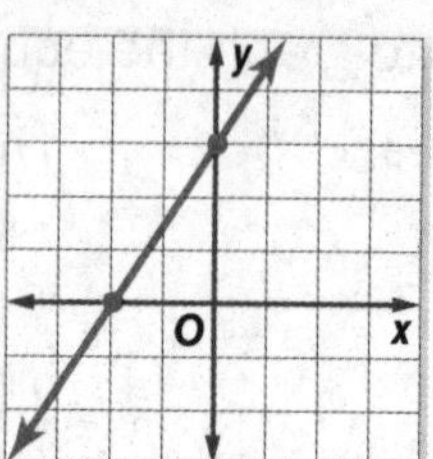

Ⓗ
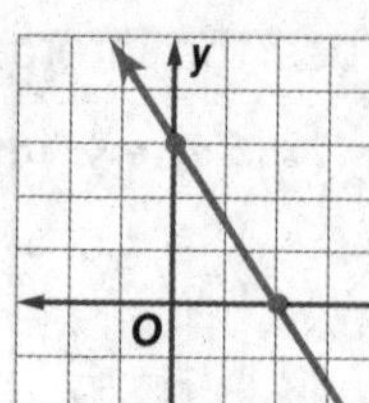

Ⓖ
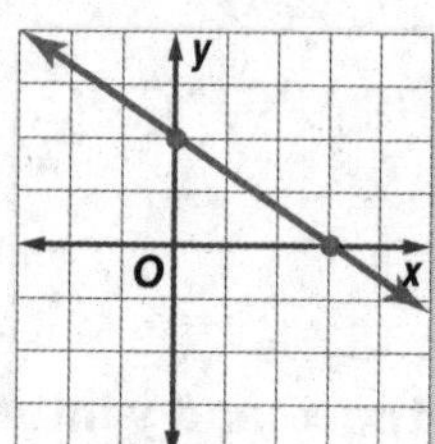

Ⓘ
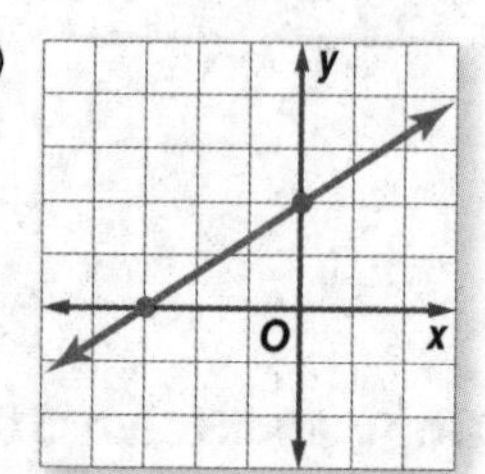

CCSS Common Core Review

Write each expression in simplest form. 7.EE.2

18. $-3(x + 6) =$ ________

19. $\frac{2}{3}(3x + 6) - 3 =$ ________

20. $4t + 10 - 5 - 3t =$ ________

21. $5x + 6 - x =$ ________

22. $-\frac{1}{4}(4x - 8) + 18 =$ ________

23. $2a + 4 - 8a - 10 =$ ________

Problem-Solving Investigation

Guess, Check, and Revise

Content Standards
8.EE.8

Mathematical Practices
1, 3, 4

Case #1 Polar Plunge

Adrienne's class is going to the zoo to see a polar bear exhibit. Student admission is $2 and adult admission is $4. They spent $66 on 30 tickets.

How many students and adults are going to the zoo?

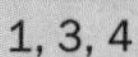

Understand What are the facts?

The student cost is $2 and the adult cost is $4. There are 30 people on the trip.

Plan What is your strategy to solve this problem?

Make a guess and check to see if your guess is correct.

Solve How can you apply the strategy?

Make a table.

Students → s, Adults → a

s	a	$2s + 4a$	Check
26	4	$2(26) + 4(4) = 68$	too high
29	1	$2(29) + 4(1) = 62$	too low
28	2	$2(28) + 4(2) =$ ☐	
27	3	$2(27) + 4(3) =$ ☐	

So, 27 students and 3 adults are going to the zoo.

Check Does the answer make sense?

$27 + 3 = 30$ and $2(27) + 4(3) = 66$; the guess is correct. ✓

Analyze the Strategy

Justify Conclusions Twenty-three students and 5 adults would also spend $66 to get into the zoo. Explain why this cannot be the correct solution.

Case #2 Coins

Gerardo has $2.50 in quarters, dimes, and nickels.

If he has 18 coins, how many of each coin does he have?

Understand

Read the problem. What are you being asked to find?

I need to find ______________________________.

Underline key words and values in the problem. What information do you know?

There are ☐ coins that have a sum of ☐.

The coins are a combination of ______________________________.

Is there any information that you do *not* need to know?

______________________________.

Plan

Choose a problem-solving strategy.

I will use the ______________________________ strategy.

Solve

Use your problem-solving strategy to solve the problem.

Q	D	N	Sum	Number of Coins	Check

So, ______________________________.

Check

Use information from the problem to check your answer.

(____ × 0.25) + (____ × 0.10) + (____ × 0.05) = $2.50; the answer is correct.

Collaborate **Work with a small group to solve the following cases. Show your work on a separate piece of paper.**

Case #3 Sport Trading Cards

Baseball cards come in packages of 8 and 12. Brighton bought some of each type for a total of 72 baseball cards.

How many of each package did he buy?

Case #4 Family

Three siblings have a combined age of 108 years. The oldest is 8 years older than the youngest.

What are the ages of the siblings?

Case #5 Birthday Surprise

Shyla was buying gifts for each of her 8 cousins. She bought everyone either a ring for \$6 or a toy for \$7.

If she spent a total of \$53, how many of each did she buy?

Circle a strategy below to solve the problem.

- Look for a pattern.
- Act it out.
- Work backward.
- Draw a diagram.

Case #6 Future Careers

One hundred fifteen students could sign up to hear three different speakers for career day. Seventy students heard the nurse speak, 37 heard the firefighter, and 63 heard the Webmaster. Some students heard more than one speaker. The results are shown in the table above.

Number of Students	Speaker
15	all three
20	nurse and firefighter
30	Webmaster and nurse
12	firefighter only

How many students signed up only for Webmaster?

Mid-Chapter Check

Vocabulary Check

1. **CCSS Be Precise** Define *linear relationship.* Give an example of a linear relationship. (Lesson 1)

Skills Check and Problem Solving

Find the slope of the line that passes through each pair of points. (Lesson 2)

2. $A(2, 5)$, $B(3, 1)$

3. $C(-1, 2)$, $D(-5, 2)$

4. $E(5, 2)$, $F(2, -3)$

5. $G(4, 3)$, $H(-2, -6)$

6. Ernesto baked 3 cakes in $2\frac{1}{2}$ hours. Assume that the number of cakes baked varies directly with the number of hours. Write and solve a direct variation equation to find how many cakes can he bake in $7\frac{1}{2}$ hours. (Lesson 3)

7. The total money y Aaron earned mowing x lawns is shown by the equation $y = 15x + 25$. What does the slope represent? (Lesson 4)

8. **Standardized Test Practice** What are the coordinates of the x- and y-intercepts of the graph shown? (Lesson 5)

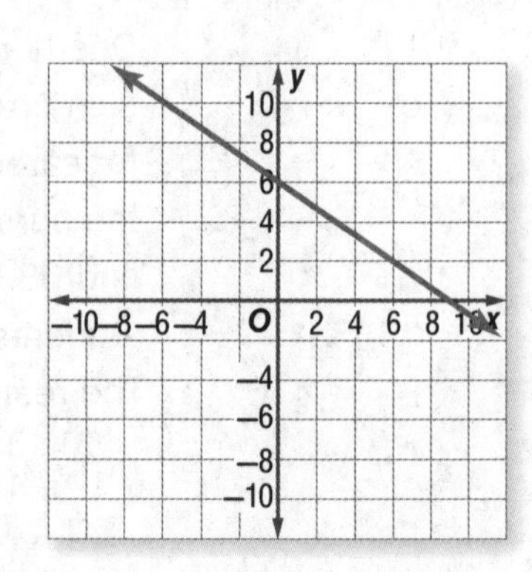

Ⓐ (0, 0) and (0, 0) Ⓒ (6, 0) and (9, 0)

Ⓑ (0, 9) and (6, 0) Ⓓ (0, 6) and (9, 0)

Lesson 6

Write Linear Equations

What You'll Learn

Scan the lesson. List two headings you would use to make an outline of the lesson.

- ______________________________
- ______________________________

Essential Question

WHY are graphs helpful?

Vocabulary

point-slope form

Common Core State Standards

Content Standards
Preparation for 8.EE.8c

Mathematical Practices
1, 2, 3, 4, 5, 7

Real-World Link

Zoo The cost for 1, 2, 3, and 4 people to go the zoo is shown in the table.

Number of People, x	1	2	3	4
Total Cost, y	\$13	\$22	\$31	\$40

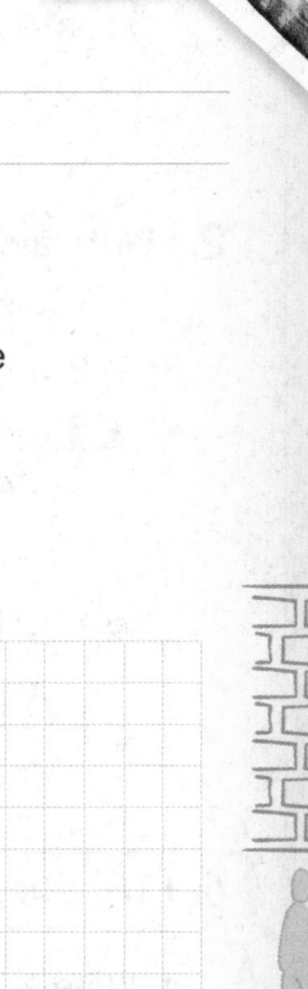

1. Is the relationship linear? Explain.

____

2. What is the slope of the related graph? ☐

3. Choose an ordered pair. (☐, ☐) Then substitute the values in the equation below.

$y = m \quad x + b$

$\square = \square \cdot \square + b$

4. Solve for b to find the y-intercept.

$b = \square$

5. Write an equation of the line in slope-intercept form.

____

6. Graph the data from the table on the coordinate plane.

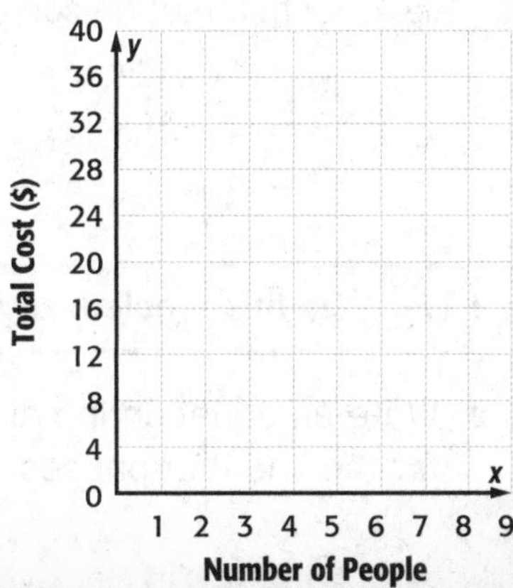

Key Concept

Point-Slope Form of a Linear Equation

Words The linear equation $y - y_1 = m(x - x_1)$ is written in point-slope form, where (x_1, y_1) is a given point on a nonvertical line and m is the slope of the line.

Symbols $y - y_1 = m(x - x_1)$

Graph

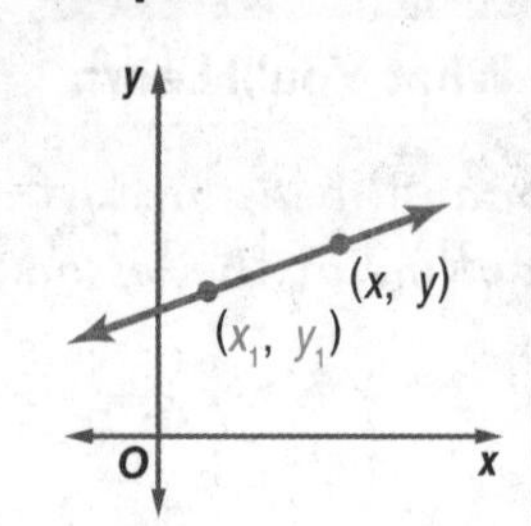

Work Zone

Slope

The point-slope form of a linear equation is tied directly to the definition of slope.

$$\frac{y - y_1}{x - x_1} = m$$

$$(y - y_1) = m(x - x_1)$$

You can write an equation of a line in slope-intercept form when you know the slope and the y-intercept. You can write an equation of a line in **point-slope form** when you are given the slope and the coordinates of a point on the line that is not the y-intercept.

Examples

1. Write an equation in point-slope form for the line that passes through (–2, 3) with a slope of 4.

$y - y_1 = m(x - x_1)$ Point-slope form

$y - 3 = 4[x - (-2)]$ $(x_1, y_1) = (-2, 3), m = 4$

$y - 3 = 4(x + 2)$ Simplify.

2. Write the slope-intercept form of the equation from Example 1.

$y - 3 = 4(x + 2)$ Write the equation.

$y - 3 = 4x + 8$ Distributive Property

$\underline{+3 = \quad +3}$ Addition Property of Equality

$y = 4x + 11$ Simplify.

Check: Substitute the coordinates of the given point in the equation.

$y = 4x + 11$

$3 \stackrel{?}{=} 4(-2) + 11$

$3 = 3$ ✓

Got It? **Do this problem to find out.**

Show your work.

a. Write an equation in point-slope form and slope-intercept form for the line that passes through $(-1, 2)$ and has a slope of $-\frac{1}{2}$.

a. ____________

Write a Linear Equation

Key Concept

From Slope and a Point • Substitute the slope m and the coordinates of the point in $y - y_1 = m(x - x_1)$.

From Slope and y-intercept • Substitute the slope m and y-intercept b in $y = mx + b$.

From a Graph • Find the y-intercept b and the slope m from the graph, then substitute the slope and y-intercept in $y = mx + b$.

From Two Points • Use the coordinates of the points to find the slope. Substitute the slope and coordinates of one of the points in $y - y_1 = m(x - x_1)$.

From a Table • Use the coordinates of the two points to find the slope, then substitute the slope and coordinates of one of the points in $y - y_1 = m(x - x_1)$.

The form you use to write a linear equation is based on the information you are given.

Example

3. Write an equation in point-slope form and slope-intercept form for the line that passes through (8, 1) and (−2, 9).

Step 1 Find the slope.

$m = \frac{y_2 - y_1}{x_2 - x_1}$ Slope formula

$m = \frac{9 - 1}{-2 - 8}$ $(x_1, y_1) = (8, 1), (x_2, y_2) = (-2, 9)$

$m = -\frac{8}{10}$ or $-\frac{4}{5}$ Simplify.

Step 2 Use the slope and the coordinates of either point to write the equation in point-slope form.

$y - y_1 = m(x - x_1)$ Point-slope form

$y - 1 = -\frac{4}{5}(x - 8)$ $(x_1, y_1) = (8, 1), m = -\frac{4}{5}$

So, the point-slope form of the equation is $y - 1 = -\frac{4}{5}(x - 8)$. In slope-intercept form, this is $y = -\frac{4}{5}x + \frac{37}{5}$.

Got It? Do these problems to find out.

c. (3, 0) and (6, −3)

d. (−1, 2) and (5, −10)

c. ________

d. ________

Example

4. The cost of assistance dog training sessions is shown in the table. Write an equation in point-slope form to represent the cost *y* of attending *x* dog training sessions.

Number of Sessions	Cost ($)
5	165
10	290

Find the slope of the line. Then use the slope and one of the points to write the equation of the line.

$m = \frac{290 - 165}{10 - 5}$ $(x_2, y_2) = (10, 290), (x_1, y_1) = (5, 165)$

$m = \frac{125}{5}$ or 25 Simplify.

$y - 165 = 25(x - 5)$ Replace (x_1, y_1) with (5, 165) and m with 25 in the point-slope form equation.

So, the equation of the line is $y - 165 = 25(x - 5)$.

Got It? **Do this problem to find out.**

e. The cost for making spirit buttons is shown in the table. Write an equation in point-slope form to represent the cost *y* of making *x* buttons.

Number of Buttons	Cost ($)
100	25
150	35

e. ____________

Guided Practice

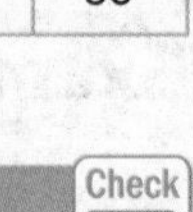

Write an equation in point-slope form and slope-intercept form for each line. (Examples 1–3)

1. passes through (2, 5), slope = 4

2. passes through (−3, 1) and (−2, −1)

3. Janelle is planning a party. The cost for 20 people is $290. The cost for 45 people is $590. Write an equation in point-slope form to represent the cost *y* of having a party for *x* people. (Example 4)

4. **Building on the Essential Question** How does using the point-slope form of a linear equation make it easier to write the equation of a line? ____________

Rate Yourself!

How confident are you about writing linear equations? Check the box that applies.

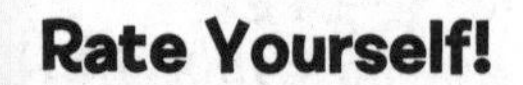

For more help, go online to access a Personal Tutor.

Name ______________________ My Homework ______________________

Independent Practice

Go online for Step-by-Step Solutions

Write an equation in point-slope form and slope-intercept form for each line.
(Examples 1–3)

1. passes through (1, 9), slope = 2

2. passes through (4, –1), slope = –3

3. passes through (–4, –5), slope = $\frac{3}{4}$

4. passes through (3, –6) and (–1, 2)

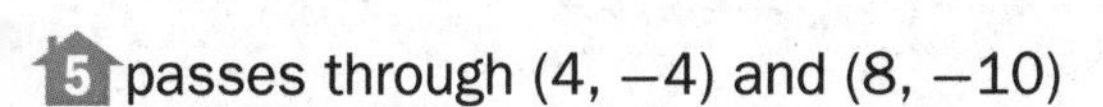

5 passes through (4, –4) and (8, –10)

6. passes through (3, 4) and (5, –4)

7. **STEM** For a science experiment, Mala measured the height of a plant every week. She recorded the information in the table. Assuming the growth is linear, write an equation in point-slope form to represent the height *y* of the plant after *x* weeks. (Example 4)

Weeks	Height (in.)
5	13
10	14

8. After 2 seconds on a penalty kick in soccer, the ball travels 160 feet. After 2.75 seconds on the same kick, the ball travels 220 feet. Write an equation in point-slope form to represent the distance *y* of the ball after *x* seconds.

(Example 4)

Write each equation in standard form.

9. $y - 4 = -3(x - 3)$

10. $y + 9 = 2(x + 5)$

11. **Identify Structure** Draw a line connecting the form of the equation to the correct equations.

	$5x + 3y = 12$
Slope-Intercept Form	$y = 2x - 8$
	$7x = y$
Standard Form	$y - 8 = \frac{1}{2}(x - 9)$
Point-Slope Form	$4x - 6y = 24$
	$y = 10 - 3x$

H.O.T. Problems Higher Order Thinking

12. **Reason Abstractly** Write a linear equation that is in point-slope form. Identify the slope and name a point on the line.

13. **Persevere with Problems** The equation of a line is $y = -\frac{1}{2}x + 6$. Write an equation in point-slope form for the same line. Explain the steps that you used.

Standardized Test Practice

14. Which point does the line $y + 5 = \frac{2}{3}(x - 7)$ pass through?

Ⓐ (7, −5)

Ⓑ (−7, 5)

Ⓒ (7, 5)

Ⓓ (−7, −5)

Name ______________________ My Homework ______________________

Extra Practice

Write an equation in point-slope form and slope-intercept form for each line.

15. passes through (−7, 10), slope = −4

$y - 10 = -4(x + 7)$; $y = -4x - 18$

$y - y_1 = m(x - x_1)$
$y - 10 = -4(x + 7)$
$y - 10 = -4x - 28$
$+10 = \quad +10$
$y = -4x - 18$

Homework Help

16. passes through (1, 2) and (3, 4)

$y - 4 = 1(x - 3)$; $y = x + 1$

$m = \frac{y_2 - y_1}{x_2 - x_1} = \frac{4 - 2}{3 - 1} = \frac{2}{2}$ or 1

$y - y_1 = m(x - x_1)$
$y - 4 = 1(x - 3)$
$y - 4 = x - 3$
$+4 = \quad +4$
$y = x + 1$

17. passes through (6, 2), slope = $\frac{2}{3}$

18. passes through (2, −2) and (4, −1)

Write each equation in standard form.

19. $y + 1 = \frac{4}{5}(x - 3)$

20. $y - 8 = -\frac{1}{2}(x + 4)$

Use Math Tools Write the point-slope form of an equation for each line graphed.

21.

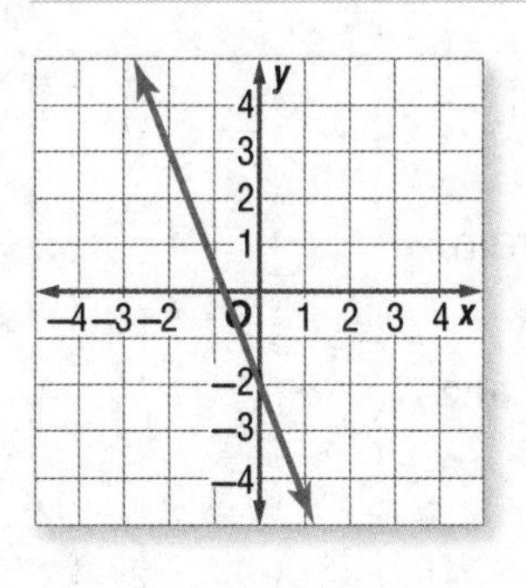

22.

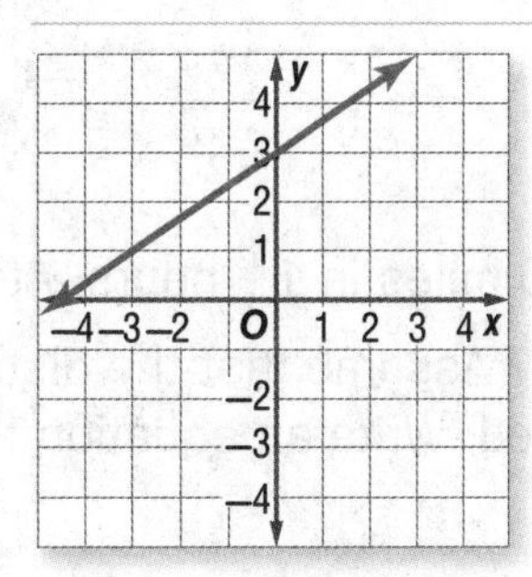

Standardized Test Practice

23. Which of the following equations in point-slope form represents the table of values shown?

x	−1	0	1	2
y	−6	−2	2	6

Ⓐ $y - 6 = -4(x - 1)$
Ⓑ $y - 2 = 4(x - 0)$
Ⓒ $y - 2 = 4(x - 1)$
Ⓓ $y + 6 = -4(x + 1)$

24. **Short Response** After 4 hours of driving, Jan is 248 miles away from home. After 6 hours of driving, she is 372 miles from home. Write an equation in point-slope form to determine her distance *y* from home after *x* hours.

25. Which of the following equations in point-slope form represents the graph?

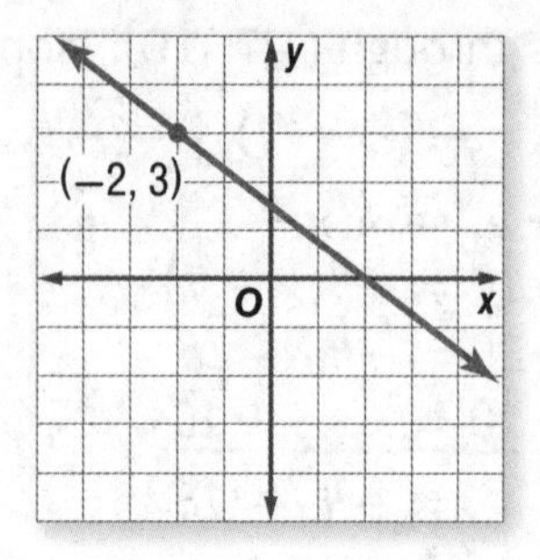

Ⓕ $y - 3 = -\frac{4}{3}(x - 2)$
Ⓖ $y - 3 = -\frac{3}{4}(x - 2)$
Ⓗ $y - 3 = -\frac{4}{3}(x + 2)$
Ⓘ $y - 3 = -\frac{3}{4}(x + 2)$

Common Core Review

26. Use the information in the table to find the constant rate of change in dollars per hour. 7.RP.2

Time (h)	0	1	2	3
Wage ($)	0	9	18	27

27. Graph $y = 4x$. 8.EE.5

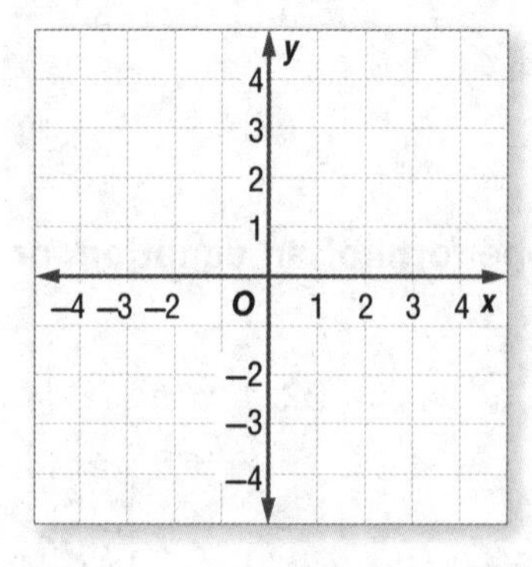

28. A train traveled 150 miles in $1\frac{1}{4}$ hours. At this rate, how far will the train travel after 5 hours? Assume that the distance traveled varies directly with the time traveled. Write an equation to represent the situation. 7.RP.2

Inquiry Lab

Graphing Technology: Model Linear Behavior

HOW does using technology help you to determine if situations display linear behavior?

Content Standards
8.F.4, 8.F.5

Mathematical Practices
1, 3, 5

Walking Simone and Lee walked to school at about 3 miles per hour. Use the Investigation to see if the relationship between time and distance is a linear relationship.

Investigation

Step 1 Connect a motion detector to your calculator. Start the data collection program by pressing APPS (CBL/CBR), ENTER, and then select Ranger, Applications, Meters, Dist Match.

Step 2 Place the detector on a desk or table so that it can read the motion of a walker.

Step 3 Mark the floor at a distance of 1 and 6 meters from the detector. Have a partner stand at the 1-meter mark.

Step 4 When you press the button to begin collecting data, have your partner begin to walk away from the detector at a slow but steady pace.

Step 5 Stop collecting data when your partner passes the 6-meter mark.

Step 6 Press ENTER to display a graph of the data. The *x*-values represent equal intervals of time in seconds. The *y*-values represent the distances from the detector in meters.

Describe the DISTANCE graph of the data. Does the relationship between time and distance appear to be linear? Explain.

Use Math Tools Refer to the Investigation. Work with a partner.

1. Use the TRACE feature on your calculator to find the *y*-intercept on the graph. Interpret its meaning. ______________________

____

2. Press STAT 1. The time data is in L1 and the distance data is in L2. Use these data to calculate the rate of change $\frac{\text{distance}}{\text{time}}$ for three pairs of points.

Point 1 (time, distance)	Point 2 (time, distance)	$\frac{\text{distance}_2 - \text{distance}_1}{\text{time}_2 - \text{time}_1}$	rate of change

What do you notice? ______________________

3. **Justify Conclusions** Does your answer to Exercise 2 support your conclusion about the graph in the Investigation? Explain. ______________________

____

4. Predict how the graph and answers to Exercise 2 would change if the person in the activity were to:

   a. move at a steady but *quick* pace *away* from the detector.

   ____

   b. move at a steady pace *toward* the detector.

   ____

5. **Reason Inductively** How could you change the situation to be one that does not display linear behavior? ______________________

6. Inquiry HOW does using technology help you to determine if a situation displays linear behavior?

Inquiry Lab

Graphing Technology: Systems of Equations

HOW can I use a graphing calculator to find one solution for a set of two equations?

CCSS Content Standards
8.EE.8, 8.EE.8a, 8.EE.8b, 8.EE.8c

Mathematical Practices
1, 3, 5, 7

Internet Shopping Web site A charges \$3 plus \$1 per pound to ship an item. Web site B charges \$1 plus \$2 per pound to ship the same item. For an object that weighs x pounds, the charges for Web site A are represented by $y = x + 3$. The charges for Web site B are represented by $y = 2x + 1$. At what point are the charges the same?

What do you know? ______________________________

____

What do you need to know? ______________________________

____

Investigation

Use a graphing calculator to generate a table of values for $y = x + 3$ and $y = 2x + 1$. Then use the table to find the total cost to ship objects that weigh 0, 1, 2, or 3 pounds.

Step 1 Press [Y=]. Then enter each equation.

Step 2 Set up the table. Press [2nd] [TblSet] to display the table setup screen. Press [▼] [▼] [▶] [ENTER] to highlight Indpnt: Ask. Then Press [▼] [ENTER] to highlight Depend: Auto.

Step 3 Access the table by pressing [2nd] [Table]. Now key in your input values, pressing [ENTER] after each one. Fill in the table. The first one is done for you.

So, the point when the charges are the same is ________.

Collaborate

Use Math Tools **Work with a partner. Refer to the Investigation.**

1. For what number of pounds are the charges for Web site A less than those for Web site B? ______

2. For what number of pounds are the charges for Web site A greater than the ones for Web site B? ______

3. Press GRAPH to graph both equations. Copy your calculator screen on the blank screen shown.

4. At what point do the two lines intersect? What does this ordered pair represent? ______

5. How does the point of intersection of the two lines compare to the answer to the Investigation? ______

Analyze

6. **Use Math Tools** Use a graphing calculator to graph each set of equations in the table. Find the point of intersection of the two lines.

Set of Equations	Point of Intersection
$y = 2x + 2$ $y = x + 2$	
$y = 2x + 6$ $y = 4x + 4$	
$y = -3x - 6$ $y = -5x - 2$	

7. **Identify Structure** Explain what the point of intersection represents.

______

Reflect

8. **Inquiry** HOW can I use a graphing calculator to find one solution for a set of two equations? ______

Lesson 7

Solve Systems of Equations by Graphing

What You'll Learn

Scan the lesson. List two headings you would use to make an outline of the lesson.

- ____________________

- ____________________

Essential Question

WHY are graphs helpful?

Vocabulary

systems of equations

Common Core State Standards

Content Standards
8.EE.8, 8.EE.8a, 8.EE.8b, 8.EE.8c

Mathematical Practices
1, 3, 4, 7

Real-World Link

Activities A campground offers tubing, kayak, and bicycle rentals as shown.

	Deposit ($)	Cost per Hour ($)
Tube	15	4.20
Kayak	25	6.50
Bicycle	20	7.50

1. Write an equation to represent the total cost y of renting a tube for any number of hours x. ____________

2. Write an equation to represent the total cost y of renting a kayak for any number of hours x. ____________

3. Write an equation to represent the total cost y of renting a bicycle for any number of hours x. ____________

4. Find the cost to rent each item for 1, 2, 3, 4, and 5 hours.

Hours	Cost of Tube ($)	Cost of Kayak ($)	Cost of Bicycle ($)
1			
2			
3			
4			
5			

Work Zone

Systems of Equations

Two or more equations with the same set of variables are called a **system of equations**. For example, $y = 4x$ and $y = 4x + 2$ together are a system of equations.

You can estimate the solution of a system of equations by graphing the equations on the same coordinate plane. The ordered pair for the point of intersection of the graphs is the solution of the system because the point of intersection simultaneously satisfies both equations.

Example

1. Solve the system $y = -2x - 3$ and $y = 2x + 5$ by graphing.

Graph each equation on the same coordinate plane.

The graphs appear to intersect at (−2, 1).

Check this estimate by replacing x with −2 and y with 1.

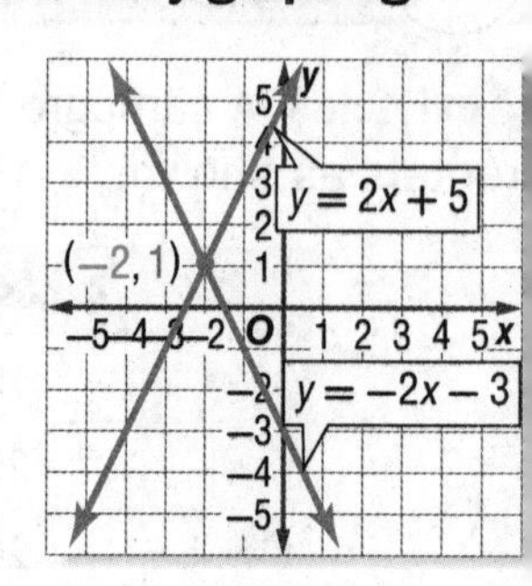

Check $y = -2x - 3$ $\qquad$ $y = 2x + 5$

$1 \stackrel{?}{=} -2(-2) - 3$ $\qquad$ $1 \stackrel{?}{=} 2(-2) + 5$

$1 = 1$ ✓ $\qquad$ $1 = 1$ ✓

The solution of the system is (−2, 1).

Got It? Do these problems to find out.

Show your work.

a. $y = x - 1$
$y = 2x - 2$

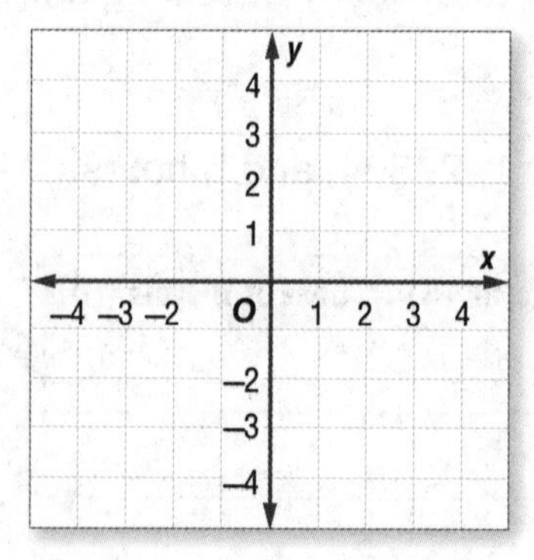

b. $y = 4x$
$y = x + 3$

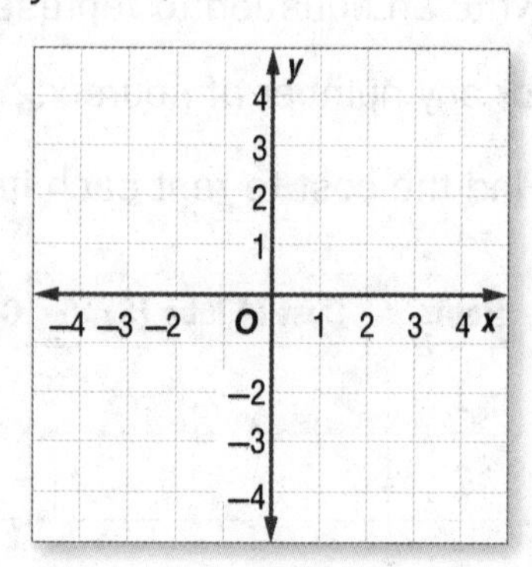

a. ________

b. ________

Examples

Gregory's Motorsports has motorcycles (two wheels) and ATVs (four wheels) in stock. The store has a total of 45 vehicles, that, together, have 130 wheels.

2. Write a system of equations that represents the situation.

Let y represent the motorcycles and x represent the ATVs.

$y + x = 45$ — The number of motorcycles and ATVs is 45.

$2y + 4x = 130$ — The number of wheels equals 130.

3. Solve the system of equations. Interpret the solution.

Write each equation in slope-intercept form.

$x + y = 45$	$2y + 4x = 130$
$y = -x + 45$	$2y = -4x + 130$
	$y = -2x + 65$

Graph both equations on the same coordinate plane. The equations intersect at (20, 25).

The solution is (20, 25). This means that the store has 20 ATVs and 25 motorcycles.

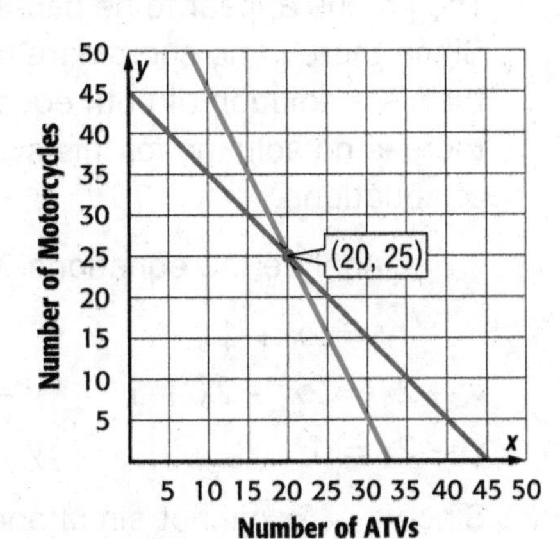

Check

$x + y = 45$	$2y + 4x = 130$
$20 + 25 \stackrel{?}{=} 45$	$2(25) + 4(20) \stackrel{?}{=} 130$
$45 = 45$ ✓	$130 = 130$ ✓

Got It? **Do this problem to find out.**

Show your work.

c. Creative Crafts gives scrapbooking lessons for \$15 per hour plus a \$10 supply charge. Scrapbooks Incorporated gives lessons for \$20 per hour with no additional charges. Write and solve a system of equations that represents the situation. Interpret the solution.

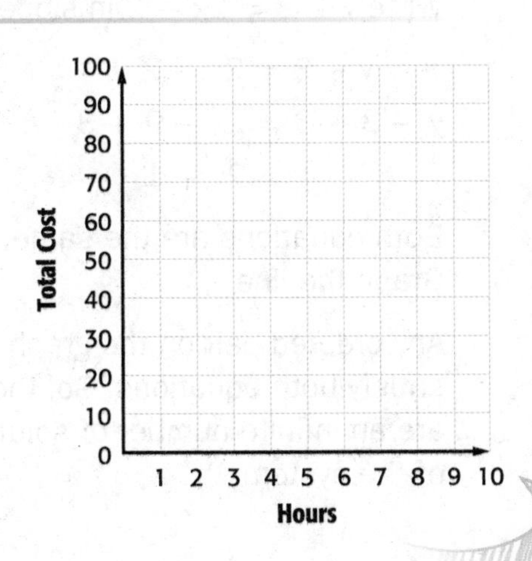

c. __________

Work Zone

Number of Solutions

The graph of a system of equations indicates the number of solutions.

- If the lines intersect, there is one solution.
- If the lines are parallel, there is no solution.
- If the lines are the same, there are an infinite number of solutions.

Examples

Solve each system of equations by graphing.

4. $y = 2x + 1$
$y = 2x - 3$

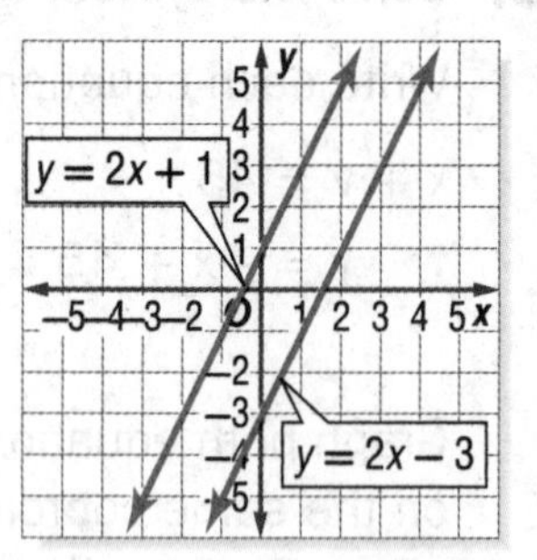

Graph each equation on the same coordinate plane.

The graphs appear to be parallel lines. Since there is no coordinate point that is a solution of both equations, there is no solution for this system of equations.

Check Analyze the equations. Write them in standard form.

$y = 2x + 1$ | $y = 2x - 3$

$y - 2x = 2x - 2x + 1$ | $y - 2x = 2x - 2x - 3$

$y - 2x = 1$ | $y - 2x = -3$

Since $y - 2x$ cannot simultaneously be 1 and -3, there is no solution. ✓

5. $y = 2x + 1$
$y - 3 = 2x - 2$

Write $y - 3 = 2x - 2$ in slope-intercept form.

$y - 3 = 2x - 2$ Write the equation.

$y - 3 + 3 = 2x - 2 + 3$ Add 3 to each side.

$y = 2x + 1$ Simplify.

Both equations are the same. Graph the line.

Any ordered pair on the graph will satisfy both equations. So, there are an infinite number of solutions of the system.

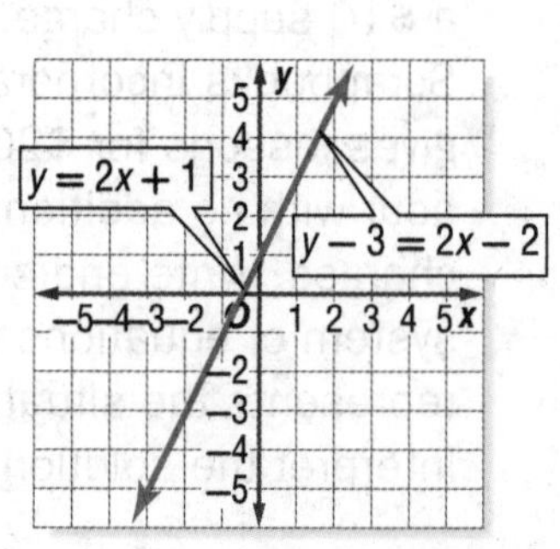

Got It? Do these problems to find out.

Solve each system of equations by graphing.

d. $y = \frac{2}{3}x + 3$
$3y = 2x + 15$

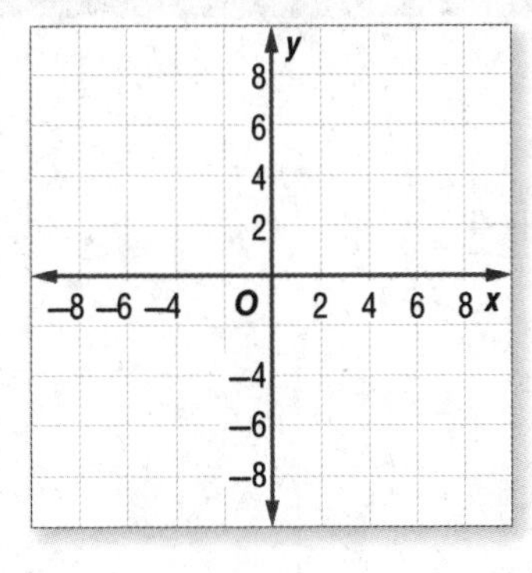

e. $y - x = 1$
$y = x - 2 + 3$

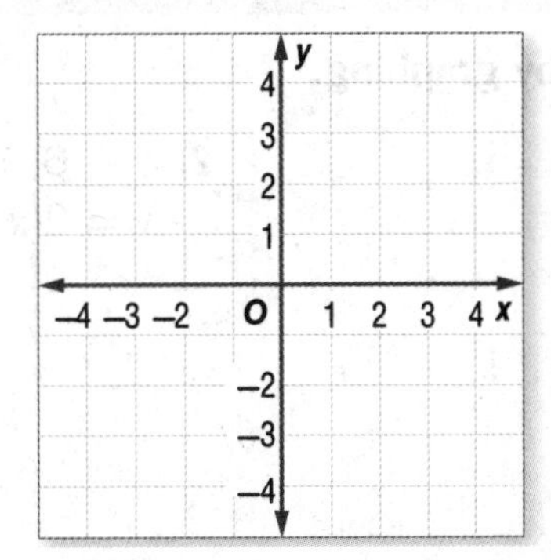

d. ______

e. ______

Example

6. A system of equations consists of two lines. One line passes through (2, 3) and (0, 5). The other line passes through (1, 1) and (0, −1). Determine if the system has *no solution*, *one solution*, or *an infinite number of solutions*.

To compare the two lines, write the equation of each line in slope-intercept form.

Find the slope of each line.

(2, 3) and (0, 5)

$\frac{y_2 - y_1}{x_2 - x_1} = \frac{5 - 3}{0 - 2}$ or −1

(1, 1) and (0, −1)

$\frac{y_2 - y_1}{x_2 - x_1} = \frac{-1 - 1}{0 - 1}$ or 2

Find the *y*-intercept for each line. Then write the equation.

Use the point (0, 5).
The *y*-intercept is 5.

$y = mx + b$
$y = -1x + 5$

Use the point (0, −1).
The *y*-intercept is −1.

$y = mx + b$
$y = 2x - 1$

Since the lines have different slopes and different *y*-intercepts, they intersect in exactly one point.

Check Graph each line on a coordinate plane.

The lines intersect at (2, 3) so there is exactly one solution. ✓

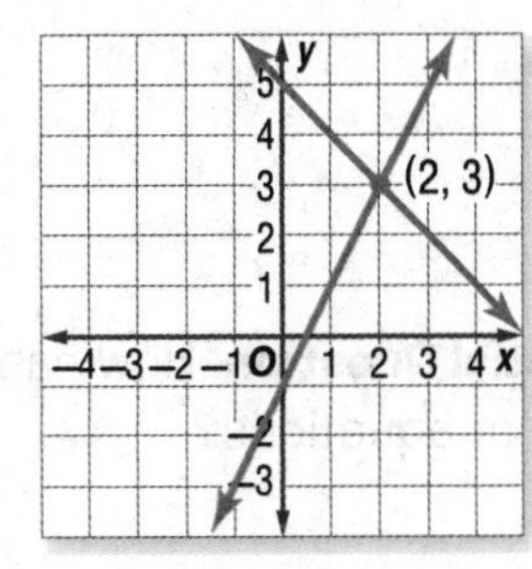

Slopes and Intercepts

When a linear system of equations has:

- different slopes and y-intercepts, there is one and only one solution.
- the same slope and different y-intercepts, there is no solution.
- the same slope and the same y-intercept, there is an infinite number of solutions.

Got It? Do this problem to find out.

f. __________

f. (0, 2), (1, 4) and (0, −1), (1, 1)

Guided Practice

Solve each system of equations by graphing. (Examples 1, 4, and 5)

1. $y = x + 3$
   $y = -2x - 3$ __________

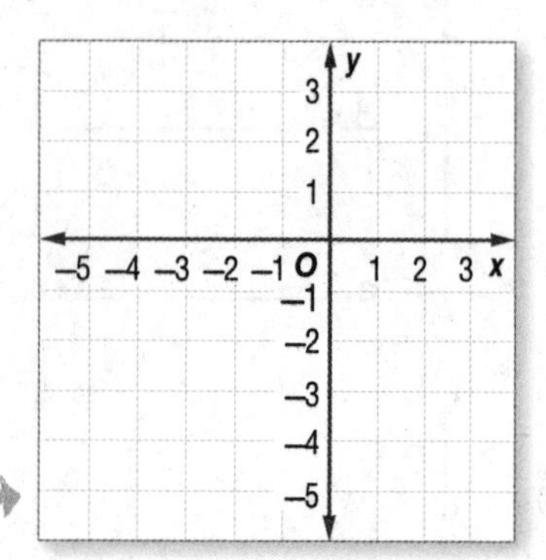

2. $y - 6 = 2x$
   $y = 2(x + 1) + 4$ __________

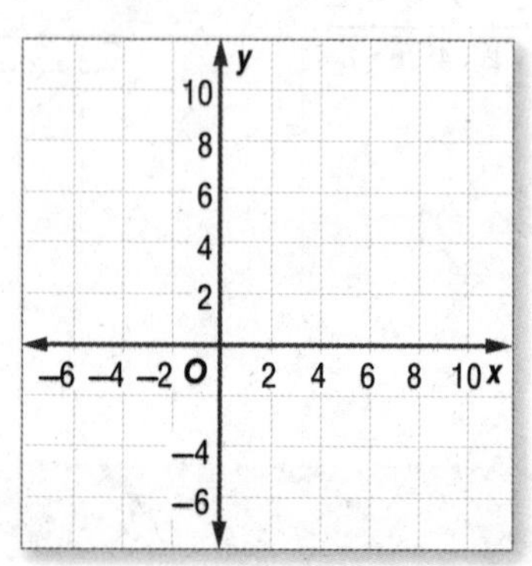

3. The sum of Sally's age plus twice Tomas' age is 12. The difference of Sally's age and Tomas' age is 3. Write and solve a system of equations to find their ages. Interpret the solution. (Examples 2 and 3)

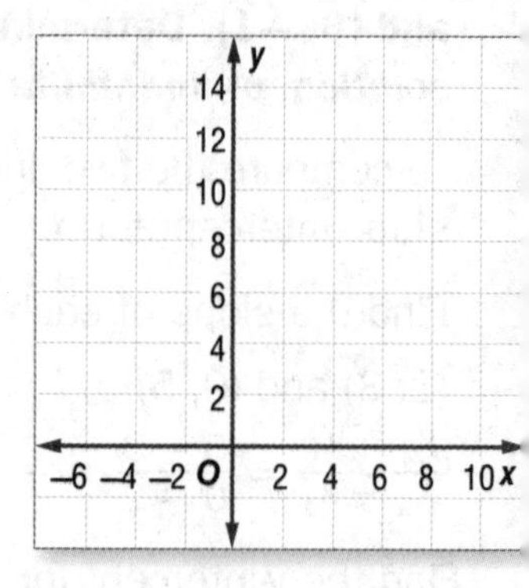

4. A system of equations consists of two lines. One line passes through (−1, 3) and (0, 1). The other line passes through (1, 4) and (0, 2). Determine if the system has *no solution, one solution,* or *an infinite number of solutions.* (Example 6) __________

5. **Building on the Essential Question** How can you use a graph to solve a system of equations?

Rate Yourself!

Are you ready to move on? Shade the section that applies.

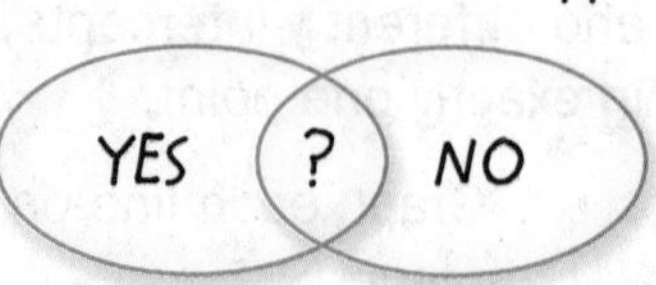

For more help, go online to access a Personal Tutor.

FOLDABLES Time to update your Foldable!

Name ______________________ My Homework ______________________

Independent Practice

Go online for Step-by-Step Solutions

Solve each system of equations by graphing. (Examples 1, 4, and 5)

1. $y = x$
$y = 2x - 4$ ________

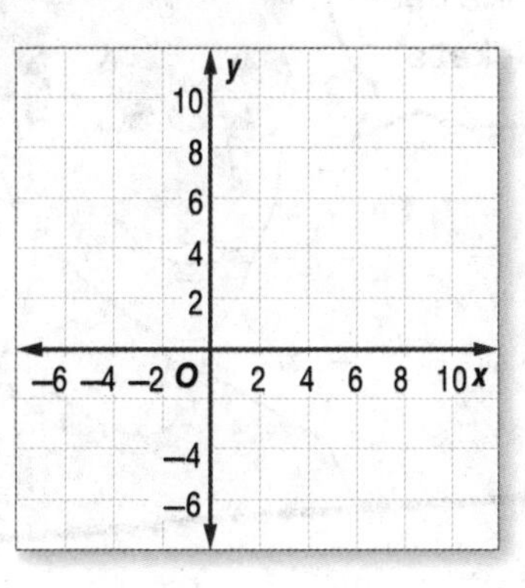

2. $y = -\frac{1}{2}x + 5$
$y = 3x - 2$ ________

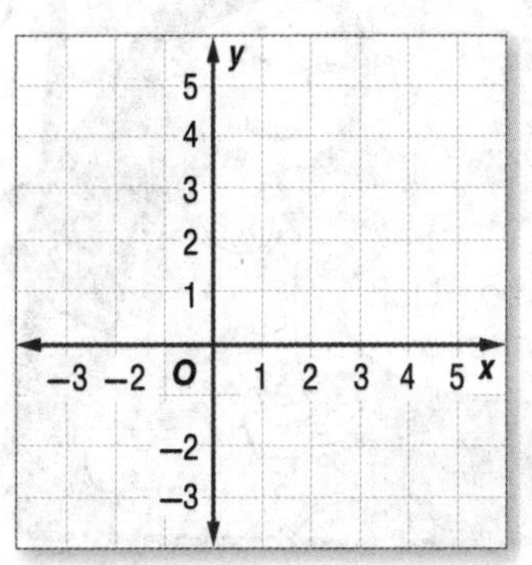

3 $y - 2x = 4$
$y = 2x$ ________

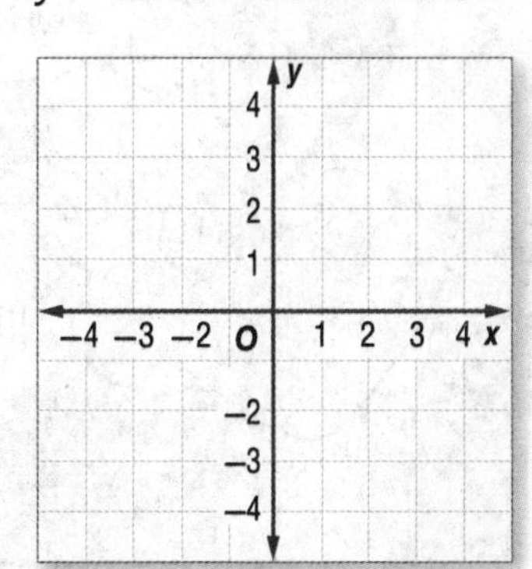

4. $y - 4x = 8$
$y = 2(2x + 4)$ ________

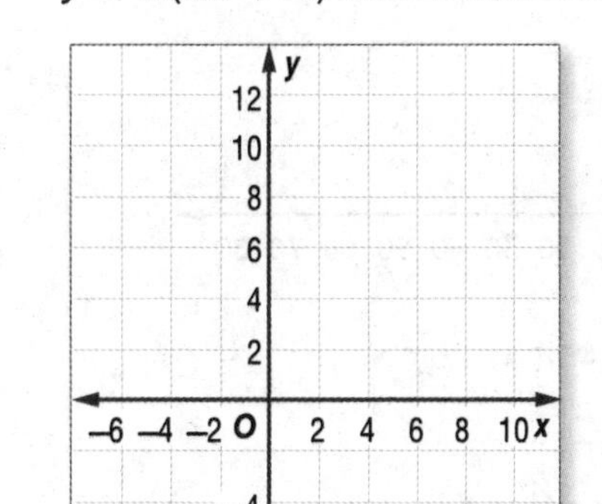

5. $x + y = 3$
$y = -3(2x - 1)$ ________

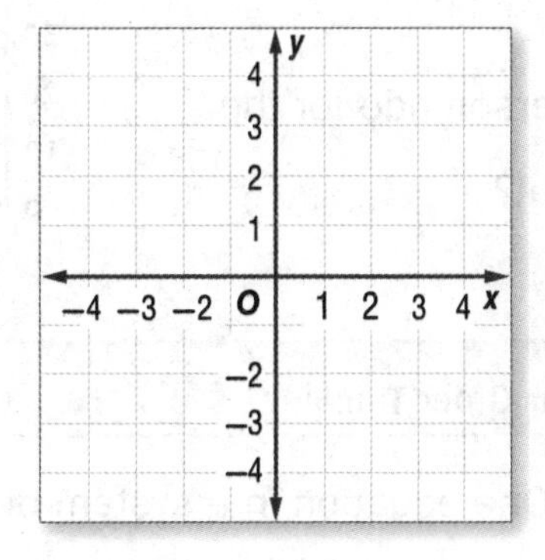

6. $-x + y = -2$
$y = 2$ ________

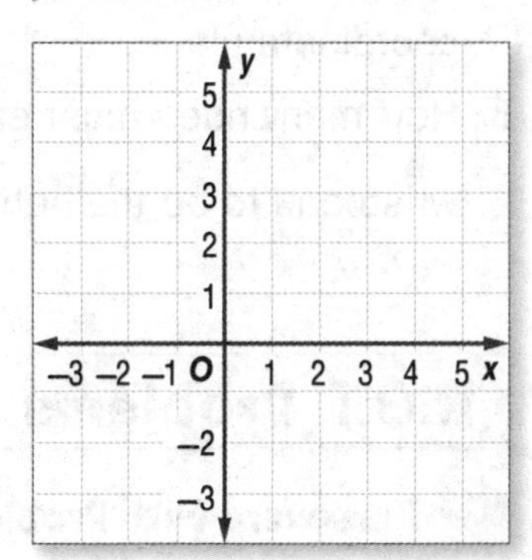

Copy and Solve

7. A pet store currently has a total of 45 cats and dogs. There are 7 more cats than dogs. Find the number of cats and dogs in the store. Write and solve a system of equations that represents the situation. Interpret the solution. (Examples 2 and 3)

____

____

Copy and Solve A line passes through each pair of points. Determine if the system has *no solution, one solution,* or *an infinite number of solutions.* Show your work on a separate piece of paper. (Example 6)

8. (0, 3) and (−2, 5); (5, −2) and (0, 3)

9. (4, 1) and (0, 1); (0, −4) and (4, 4)

10. (−2, −2) and (0, 2); (1, 1) and (0, −1)

11. **Model with Mathematics** Refer to the graphic novel frame below for Exercises a–b.

a. The equation $y = 0.71x$ represents the total cost y of x tickets at the rate of 7 tickets for $5. The equation $y = 25$ represents the cost of a wristband. Graph each equation on the same coordinate plane.

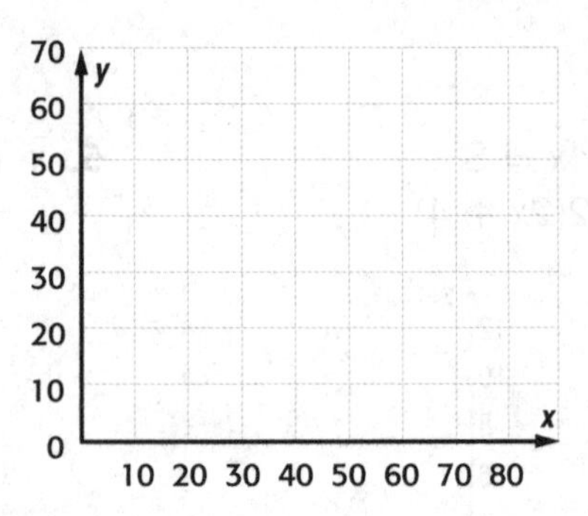

b. How many rides must each person ride for the wristband to be the better deal? ______

H.O.T. Problems Higher Order Thinking

12. **Persevere with Problems** One equation in a system of equations is $y = 2x + 1$.

a. Write a second equation so that the system has (1, 3) as its only solution. ______

b. Write an equation so that the system has no solution.

______

c. Write an equation so that the system has infinitely many solutions.

______

Standardized Test Practice

13. Katia baked 36 cookies. There are 8 more chocolate chip cookies than peanut butter. Which system can be used to find the number of each type of cookie?

Ⓐ $c + p = 36$, $p = c + 8$

Ⓑ $c + p = 36$, $c = p + 8$

Ⓒ $c + p = 8$, $p = c + 36$

Ⓓ $c + p = 8$, $c = p + 36$

Extra Practice

Solve each system of equations by graphing.

14. $y = 3x$
$y - 4 = 3x$ no solution

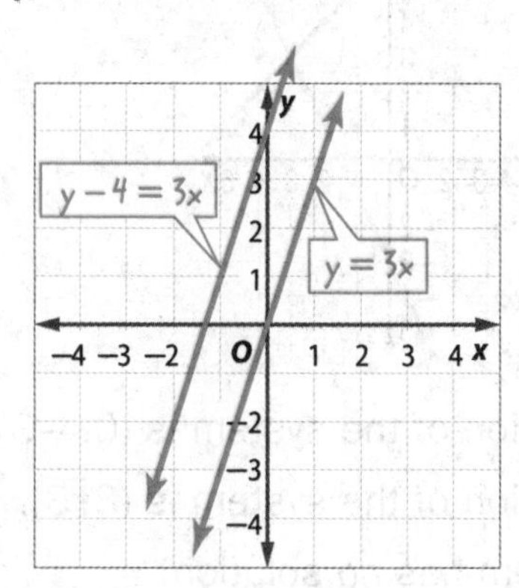

Homework Help

Write $y - 4 = 3x$ in slope-intercept form.

$$y - 4 = 3x$$
$$y - 4 + 4 = 3x + 4$$
$$y = 3x + 4$$

Graph the equations $y = 3x$ and $y = 3x + 4$ on the same coordinate plane.

The lines appear to be parallel, so there is no solution for this system of equations.

15. $y = 2x$
$y = x + 1$ ______

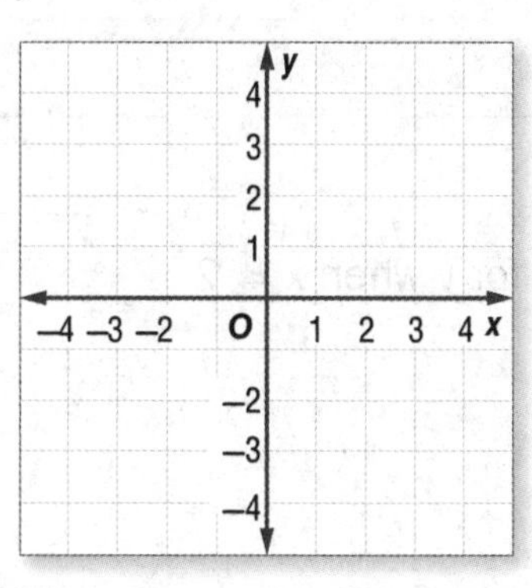

16. $y = \frac{3}{4}x$
$3x - 4y = 0$ ______

17. $y = \frac{1}{2}x + 1$
$y = \frac{1}{2}x - 2$ ______

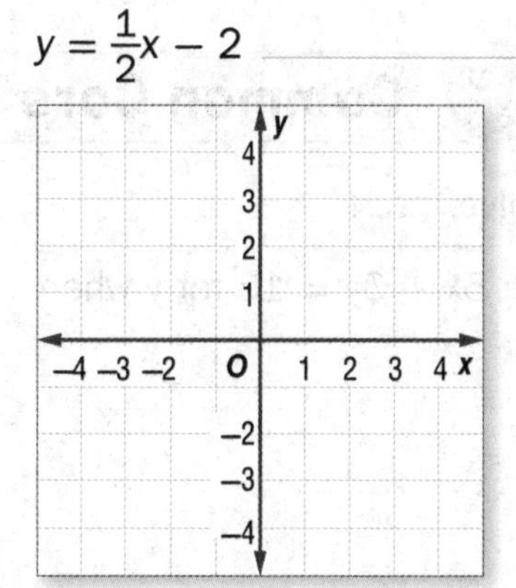

CCSS Identify Structure **Determine if each of the following systems of equations has *no solution, one solution,* or *an infinite number of solutions.* If there is a solution, find the solution. If not, explain why not.**

18. $2x + 3y = 6$
$2x + 3y = 7$ ______

19. $x + y = -2$
$y = x + 2$ ______

20. $x + y = -3$
$2x + y = 1$ ______

Standardized Test Practice

21. Aaron took three times as many pictures as Jennifer. Jennifer has 16 fewer pictures than Aaron. Which system of equations can be used to find the number of pictures each person took?

Ⓐ $a = 3j$
$a = j + 16$

Ⓑ $a = 3j$
$a = j - 16$

Ⓒ $j = 3a$
$j = a + 6$

Ⓓ $j = 3a$
$j = a - 16$

22. Two equations in a system are shown in the graph. Which of the following statements is true?

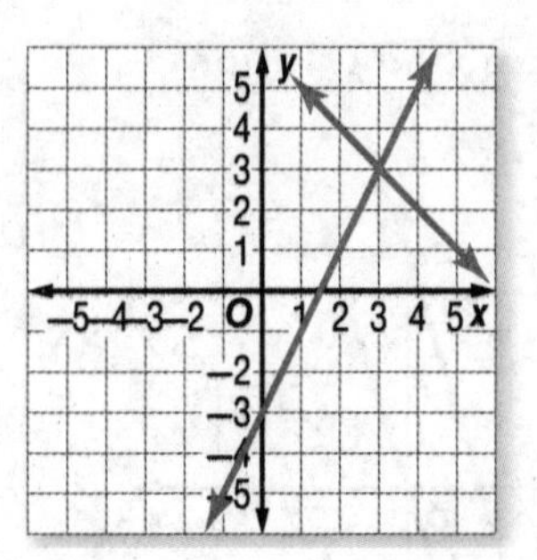

Ⓕ The solution of the system is (0, −3).

Ⓖ The solution of the system is (3, 3).

Ⓗ The system has no solution.

Ⓘ The system has infinitely many solutions.

Common Core Review

Solve. 7.EE.4

23. $5x + 3y = 15$ for y when $x = 0$.

24. $6x - 2y = 10$ for y when $x = 2$.

25. $\frac{1}{2}x + 3y = 4$ for x when $y = 6$.

26. $\frac{3}{4}x + 3y = 12$ for x when $y = 5$.

27. $7x - 4y = 20$ for y when $x = 3$.

28. $7x - 4y = 20$ for y when $x = 5$.

Lesson 8

Solve Systems of Equations Algebraically

What You'll Learn

Scan the lesson. List two headings you would use to make an outline of the lesson.

- ______________________________
- ______________________________

Essential Question

WHY are graphs helpful?

Vocabulary

substitution

Common Core State Standards

Content Standards
8.EE.8, 8.EE.8b, 8.EE.8c

Mathematical Practices
1, 3, 4, 7

Real-World Link

Jewelry Mary Anne sold 20 necklaces and bracelets at the craft fair. She sold 3 times as many necklaces as bracelets.

Step 1 The bar diagram below represents the situation

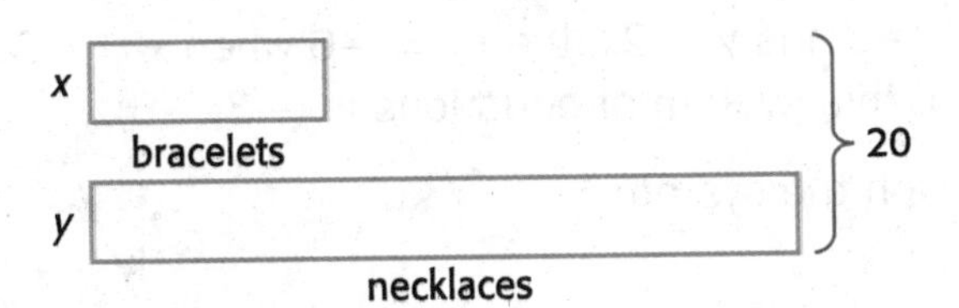

An equation to represent the bar diagram is $x + y = 20$.

Step 2 Mary Anne sold 3 times as many necklaces as bracelets. Divide the necklace bar into sections to represent this.

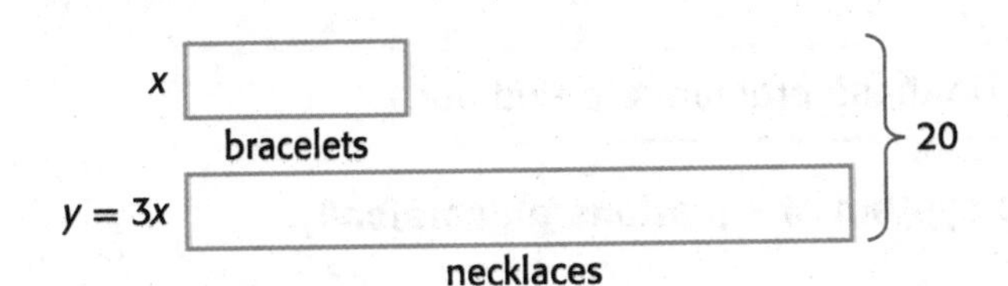

Write an equation using only x to represent the total number of necklaces and bracelets.

____

Step 3 Solve the equation from Step 2. What does the solution represent? ______________________________

1. How many bracelets and necklaces did Mary Anne sell?

☐ bracelets and ☐ necklaces

Work Zone

Solve a System Algebraically

In the previous lesson, you estimated the solution of a system of equations by graphing. **Substitution** is an algebraic model that can be used to find the exact solution of a system of equations.

Example

1. Solve the system of equations algebraically.
$y = x - 3$
$y = 2x$

Since y is equal to $2x$, you can replace y with $2x$ in the first equation.

$y = x - 3$	Write the equation.
$2x = x - 3$	Replace y with 2x.
$-x = -x$	Subtraction Property of Equality
$x = -3$	Simplify.

Since $x = -3$ and $y = 2x$, then $y = -6$ when $x = -3$. The solution of this system of equations is $(-3, -6)$.

Check Graph the system.

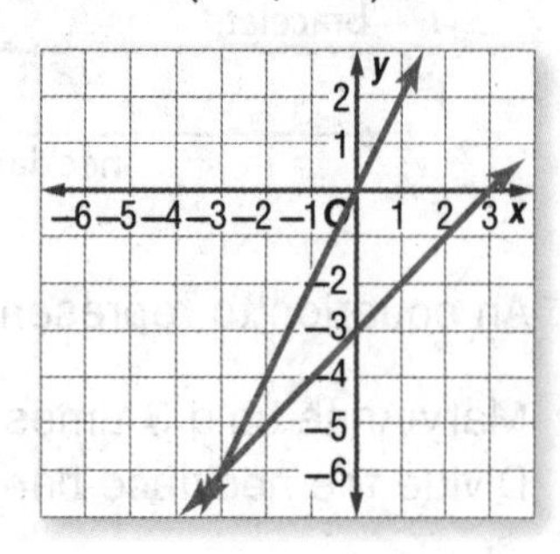

a. ____________

b. ____________

Got It? **Do these problems to find out.**

Solve each system of equations algebraically.

a. $y = x + 4$
$y = 2$

b. $y = x - 6$
$y = 3x$

Slope-Intercept and Standard Forms

Sometimes one or both equations are written in standard form. When solving a system by substitution, one of the equations should be solved for either x or y.

Example

2. Solve the system of equations algebraically.
$y = 3x + 8$
$8x + 4y = 12$

$8x + 4y = 12$	Write the equation.
$8x + 4(3x + 8) = 12$	Replace y with $3x + 8$.
$8x + 4 \cdot 3x + 4 \cdot 8 = 12$	Distributive Property
$8x + 12x + 32 = 12$	Simplify.
$20x + 32 = 12$	Collect like terms.
$20x + 32 = 12$ $-32 = -32$	Subtraction Property of Equality
$20x = -20$	Simplify.
$\frac{20x}{20} = \frac{-20}{20}$	Division Property of Equality
$x = -1$	Simplify.

Since $x = -1$, replace x with -1 in the equation $y = 3x + 8$ to find the value of y.

$y = 3x + 8$

$y = 3(-1) + 8$ or 5

The solution of this system is $(-1, 5)$.

Substitution
When you replace a variable with an expression, write the expression inside parentheses. This will help you apply the Distributive Property correctly.

Got It? Do these problems to find out.

c. $y = 2x + 1$
$3x + 4y = 26$

d. $2x + 5y = 44$
$y = 6x - 4$

c. ________

d. ________

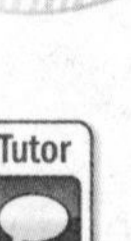

Examples

A total of 75 cookies and cakes were donated for a bake sale to raise money for the football team. There were four times as many cookies donated as cakes.

3. Write a system of equations to represent this situation.

Draw a bar diagram. Then write the system.

x	**cakes**			
y	**cookies**	**cookies**	**cookies**	**cookies**

(total: 75)

$y = 4x$ — There were 4 times as many cookies donated as cakes.

$x + y = 75$ — The total number of cakes and cookies is 75.

Write a system of equations that has the solution (1, 7).

4. Solve the system in Example 3 algebraically. Interpret the solution.

Since y is equal to $4x$, you can replace y with $4x$.

$x + y = 75$	Write the equation.
$x + 4x = 75$	Replace y with $4x$.
$5x = 75$	Simplify.
$\frac{5x}{5} = \frac{75}{5}$	Division Property of Equality
$x = 15$	Simplify.

Since $x = 15$ and $y = 4x$, then $y = 60$ when $x = 15$. The solution is (15, 60). This means that 15 cakes and 60 cookies were donated.

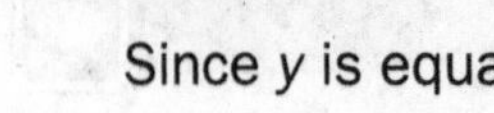

Do these problems to find out.

e. ____________

f. ____________

Mr. Thomas cooked 45 hamburgers and hot dogs at a cookout. He cooked twice as many hot dogs as hamburgers.

e. Write a system of equations to represent this situation.

f. Solve the system algebraically. Interpret the solution.

Guided Practice

Solve each system of equations algebraically. (Examples 1 and 2)

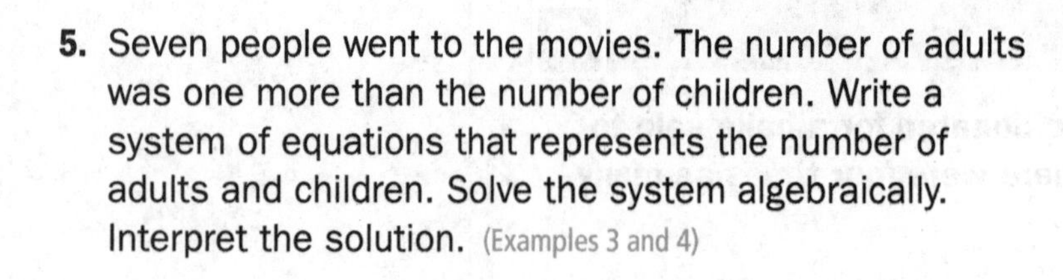

1. $y = x + 7$
   $y = 4$ ____________

2. $y = x + 5$
   $y = 3x$ ____________

3. $y = x - 9$
   $y = -4x$ ____________

4. $x + 3y = 1$
   $y = 2x + 5$ ____________

Show your work.

5. Seven people went to the movies. The number of adults was one more than the number of children. Write a system of equations that represents the number of adults and children. Solve the system algebraically. Interpret the solution. (Examples 3 and 4)

____

6. **Building on the Essential Question** How can you solve a system of equations? ____________

____

Rate Yourself!

How confident are you about solving systems of equations algebraically? Check the box that applies.

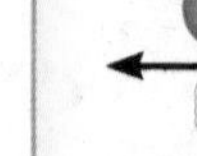

For more help, go online to access a Personal Tutor.

FOLDABLES Time to update your Foldable!

Independent Practice

Go online for Step-by-Step Solutions

Solve each system of equations algebraically. (Examples 1 and 2)

1. $y = x + 5$
   $y = 6$

2. $y = x + 12$
   $y = -18$

3. $y = x - 10$
   $y = -12$

4. $y = x + 15$
   $y = 2x$

5. $y = 2x - 3$
   $x + y = 18$

6. $y = \frac{1}{4}x$
   $x + 4y = 8$

7. $y = x + 12$
   $4x + 2y = 27$

8. $10x + 3y = 19$
   $y = 2x + 5$

Write and solve a system of equations that represents each situation. Use a bar diagram if needed. Interpret the solution. (Examples 3 and 4)

9. Elaine bought a total of 15 shirts and pairs of pants. She bought 7 more shirts than pants. How many of each did she buy?

10. Together, Preston and Horatio have 49 video games. Horatio has 11 more games than Preston. How many games does each person have?

11. The cost of 8 muffins and 2 quarts of milk is $18. The cost of 3 muffins and 1 quart of milk is $7.50. How much does 1 muffin and 1 quart of milk cost?

12. **Multiple Representations** The table shows the rates at which Ajay and Tory are biking along the same trail.

Person	Rate (m/min)
Ajay	200
Tory	250

a. **Algebra** Suppose Ajay began the trail 325 meters ahead of Tory. Write a system of equations to represent the distance y each person will travel after any number of minutes x. ______

b. **Words** Which person was farther along the trail after 5 minutes?

______

c. **Graphs** Graph the system. Use the graph to determine when Tory will catch up to Ajay.

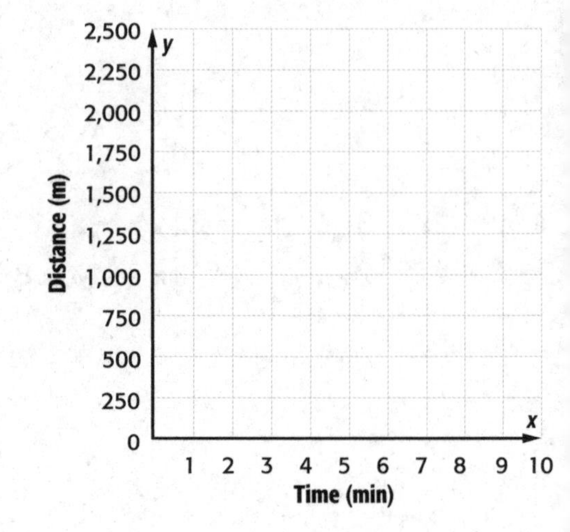

______

______

d. **Algebra** Solve the system of equations algebraically. Interpret your solution. How does your solution compare to your estimate in part **c**?

______

______

______

______

H.O.T. Problems Higher Order Thinking

13. **Persevere with Problems** What is the solution to the system $5x + y = 2$ and $y = -5x + 8$? Explain. ______

______

14. **Identify Structure** Describe when it is better to use substitution to solve a system of equations rather than graphing. ______

______

Standardized Test Practice

15. Leanne has three times as many points as Jay. Jay has 20 fewer points than Leanne. Which system of equations can be used to find each player's points?

Ⓐ $j = 3\ell$
$j = \ell + 20$

Ⓑ $j = 3\ell$
$j = \ell - 20$

Ⓒ $\ell = 3j$
$\ell = j + 20$

Ⓓ $\ell = 3j$
$\ell = j - 20$

Extra Practice

Solve each system of equations algebraically.

16. $y = 2x$

$y = x + 1$ (1, 2)

$y = x + 1$
$2x = x + 1$
$-x = -x$
$x = 1$

Since $x = 1$ and $y = 2x$, $y = 2$.

17. $y = 4x + 45$

$x = 4y$ ________

18. $y = -2x$

$x = 0$ ________

19. $x + y = -3$

$y = x + 3$ ________

20. $y = x + 4$

$y = 0$ ________

21. $x - y = 6$

$y = -1$ ________

22. The length of the rectangle is 3 meters more than the width. The perimeter is 26 meters. Write and solve a system of equations that represents this situation. What are the dimensions of the rectangle?

____

____

23. CCSS **Model with Mathematics** Ms. Corley wants to take her class on a trip to either the nature center or the zoo. The nature center charges \$4 per student plus \$95 for a 1-hour naturalist program. The zoo charges \$9 per student plus \$75 for a 1-hour guided tour.

a. Write a system of equations to represent this situation.

____

b. Solve the system of equations algebraically and by graphing. Interpret the solution.

____

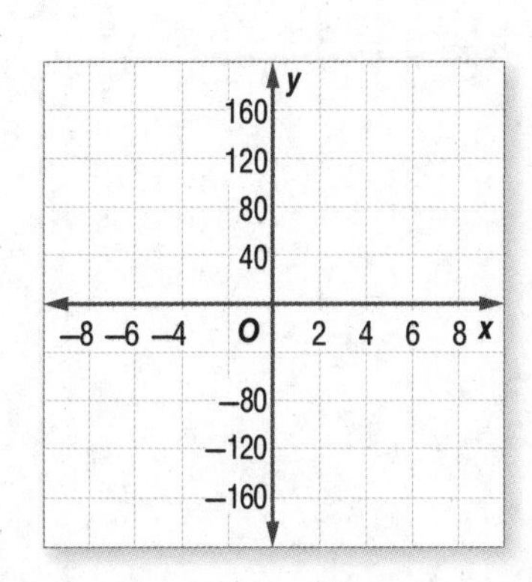

c. Ms. Corley has 22 students in her class. Determine which would cost less, the nature center or the zoo.

____

Standardized Test Practice

24. A veterinarian examined twice as many cats y as dogs x. She examined a total of 30 cats and dogs. Which system of equations represents this situation?

Ⓐ $x = 2y$
$y = 2x$

Ⓑ $y = 2x$
$x - y = 30$

Ⓒ $y = 2x$
$y = x + 30$

Ⓓ $2x = y$
$x + y = 30$

25. Short Response In one season, Naomi made 4 times as many goals as Kennedy. Together, they made 15 goals. How many goals did Naomi make?

______

CCSS Common Core Review

Solve. 6.EE.7

26. $p - 12 = 20$ ______

27. $31 = r - 36$ ______

28. $m + 1\frac{3}{8} = 5$ ______

29. $56.9 = 34 + p$ ______

30. $0.97 + a = 2.6$ ______

31. $x - 24 = 73$ ______

32. $t + 5 = 30$ ______

33. $r - 15 = 63$ ______

Inquiry Lab

Analyze Systems of Equations

HOW can you solve real-world mathematical problems using two linear equations in two variables?

Content Standards
8.EE.8, 8.EE.8a, 8.EE.8b, 8.EE.8c

Mathematical Practices
1, 3, 5

Maps A map uses a coordinate grid to show the locations of cities and towns. The map locations for four towns are shown in the table. Suppose Brent travels from Town A to Town B and Maria travels from Town C to Town D. Do Brent's and Maria's routes pass through a common location?

Town	Location
A	(0, 6)
B	(5, 1)
C	(0, 4)
D	(4, 8)

What do you know? ______________________________

____

What do you need to know? ______________________________

Investigation

Step 1 Plot and label the points of each town on the coordinate plane shown.

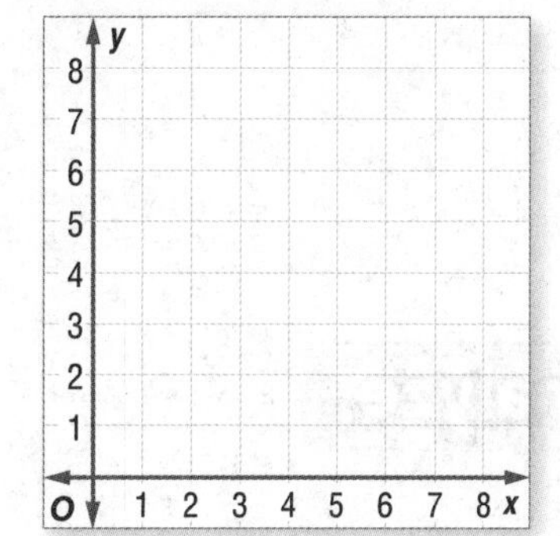

Step 2 Draw a red line segment to represent Brent's route and draw a blue line segment to represent Maria's route.

Step 3 Find the slope of the lines that represent Brent's route and Maria's route.

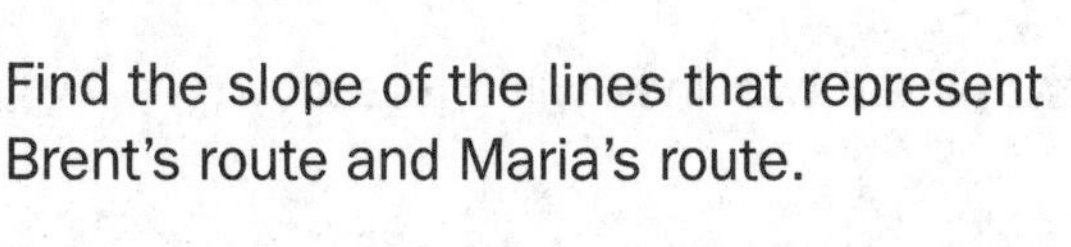

____

What do the slopes tell you about the lines? Explain.

____

____

Step 4 Where do the two lines intersect? ______________________________

So, Brent's and Maria's routes pass through the common location (☐, ☐).

Refer to the Investigation. Work with a partner.

1. Write an equation for the lines that represent Brent's routes and Maria's routes.

   Brent's route: ______________ Maria's route: ______________

2. Solve the system of equations from Exercise 1 algebraically. ______________

CCSS **Use Math Tools** **Write an equation for the line that passes through each pair of points. Use a graphing calculator to solve the system. Then copy your calculator screen on the blank screen shown and describe the slope of each pair of lines.**

3. $(0, -1)$ and $(4, 3)$; $(2, 1)$ and $(0, 3)$

   Equations: ______________

   Solution: ______________

4. $(0, 3)$ and $(3, 9)$; $(0, 2)$ and $(3, 8)$

   Equations: ______________

   Solution: ______________

Analyze

5. CCSS **Reason Inductively** How can you determine if two lines will intersect using the slope? ______________

6. Inquiry HOW can you solve real-world mathematical problems using two linear equations in two variables? ______________

21ST CENTURY CAREER in Music

Mastering Engineer

Do you love listening to music? Are you interested in the technical aspects of music-making? If so, a career creating digital masters might be something to think about! A mastering engineer produces digital masters and is responsible for making songs sound better, having the proper spacing between songs, removing extra noises, and assuring all the songs have consistent levels of tone and balance. Having a great-sounding master helps increase radio airplay and sales for recording artists.

Explore college and careers at ccr.mcgraw-hill.com

Is This the Career for You?

Are you interested in a career as a mastering engineer? Take some of the following courses in high school.

- Algebra
- Music Appreciation
- Recording Techniques
- Sound Engineering

Turn the page to find out how math relates to a career in Music.

Mastering the Music

Use the information in the tables to solve each problem.

1. At Engineering Hits, is the relationship between the number of songs and the cost linear? Explain your reasoning. ______

2. Is there a proportional linear relationship between number of songs and cost at Dynamic Mastering? Explain your reasoning.

3. Find the slope of the line represented in the Mastering Mix table. What does the slope represent? ______

4. Is the linear relationship represented in the Mastering Mix table a direct variation? Explain.

5. Write a direct variation equation to represent number of songs x and cost y at Dynamic Mastering. How much does it cost to master 11 songs? ______

6. For 4 or more songs at Engineering Hits, the cost varies directly as the number of songs. How much does it cost to master 6 songs?

Engineering Hits	
Number of Songs	Cost ($)
1	100
2	160
3	210
4	250

Dynamic Mastering	
Number of Songs	Cost ($)
2	120
4	240
6	360
8	480

Mastering Mix	
Number of Songs	Cost ($)
1	125
3	275
5	425
7	575

Career Project

It's time to update your career portfolio! Find the name of the mastering engineer on one of your CDs. Use the Internet or another source to write a short biography of this engineer. Include a list of other artists whose songs he or she has mastered.

Do you think you would enjoy a career as a mastering engineer? Why or why not?

Chapter Review

Check

Vocabulary Check

Complete the crossword puzzle using the vocabulary list at the beginning of the chapter.

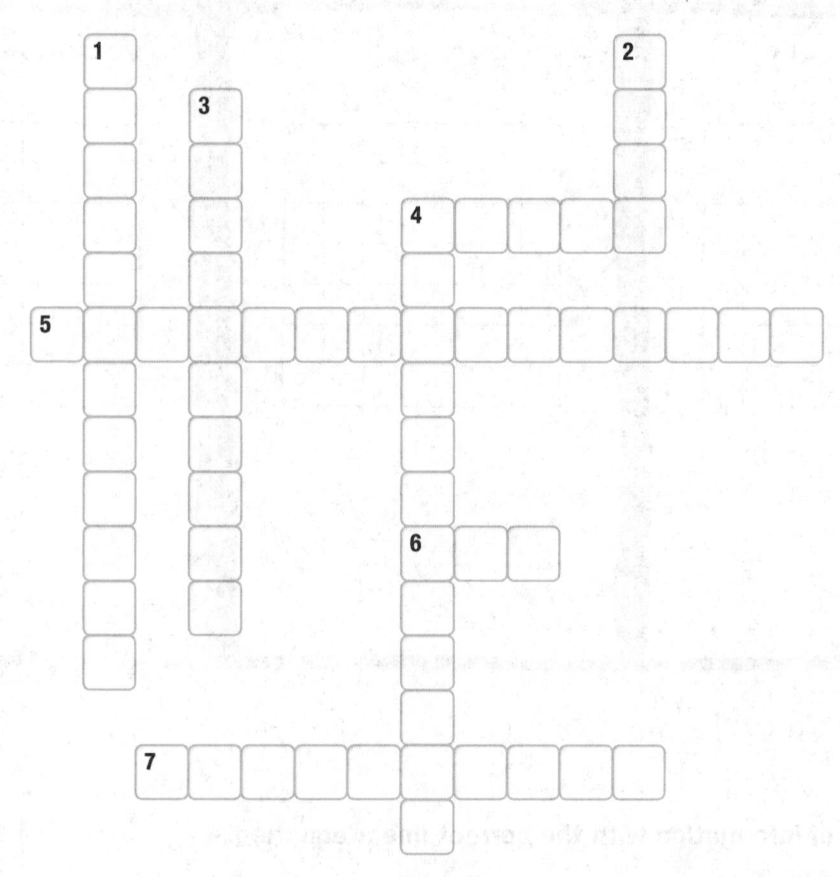

Across

4. to describe the steepness of a straight line
5. a relationship in which the ratio of two variables quantities is constant
6. the horizontal change between the same two points
7. the *x*-coordinate of the point where the graph crosses the *x*-axis

Down

1. an algebraic model used to find the exact solution of a system of equations
2. the vertical change between any two points
3. the *y*-coordinate of the point where the line crosses the *y*-axis
4. when an equation is written in the form $Ax + By = C$

Key Concept Check

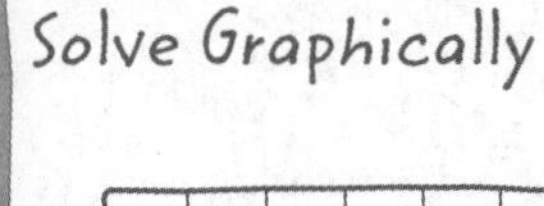

Use your Foldable to help review the chapter.

Tape here

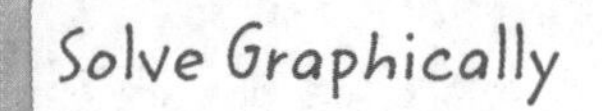

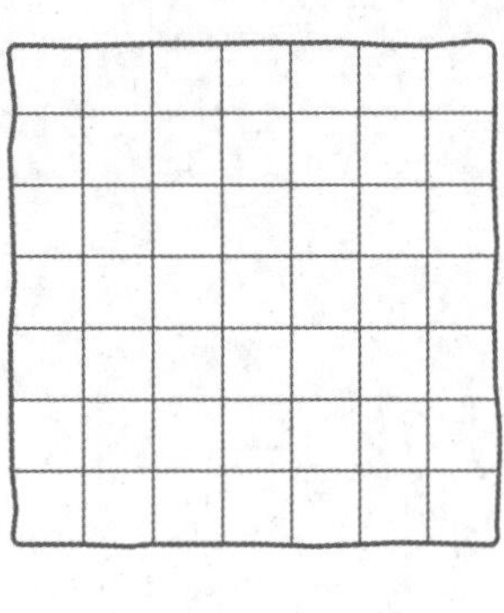

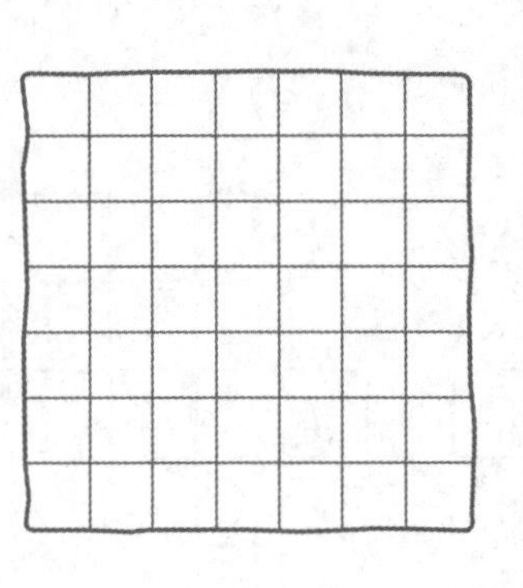

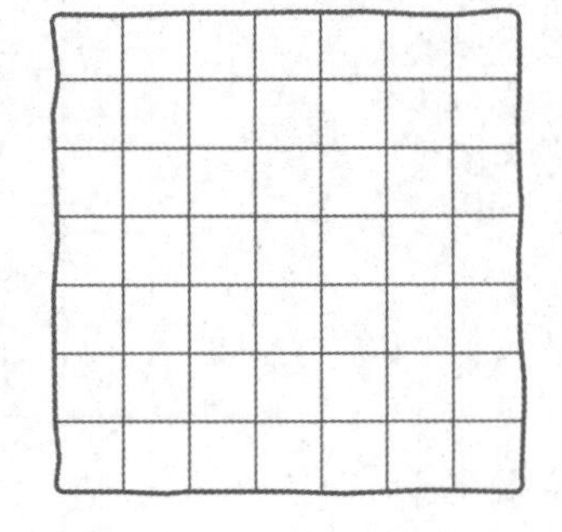

Got it?

Match each set of information with the correct linear equation.

1. line that passes through (2, 0) and (0, 1)
2. line with a slope of 0.5 and a y-intercept of 1
3. line that passes through (4, 2) and the origin
4. line that has a slope of 0 and passes through (5, 4)
5. line that has an undefined slope and passes through (5, 4)

a. $y = 0.5x$

b. $x = 5$

c. $y = 0.5x + 1$

d. $y = -0.5x + 1$

e. $y = 4$

Problem Solving

1. **Financial Literacy** Felix has $100 in his savings account, He plans to add $25 each week. The equation for the amount of money y Felix has in his savings account is $y = 100 + 25x$, where x is the number of weeks. Graph the equation. (Lesson 3)

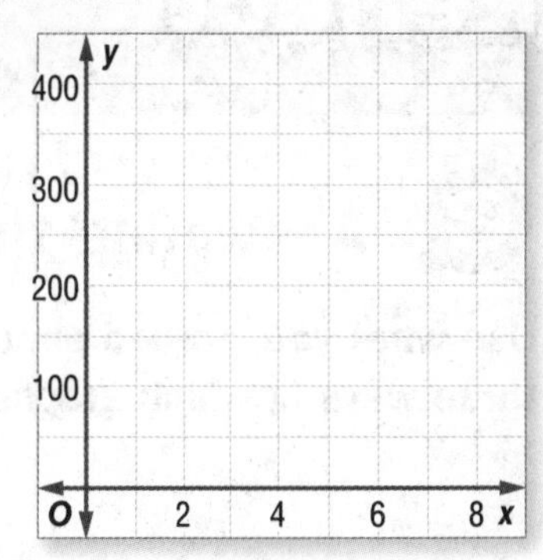

2. Hot dog buns are sold in packages of 8 and 12. Taro bought 192 total buns for a party. This is represented by the function $8x + 12y = 192$. (Lesson 5)

   a. Graph the function.

   b. Interpret the x-and y-intercept.

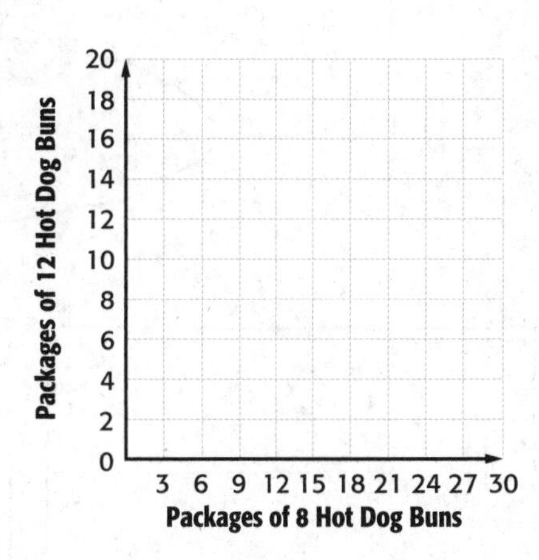

3. Twenty-five teenagers were surveyed about food. Five more preferred pizza than steak. Write and solve a system of equations to find out how many preferred pizza. (Lesson 8)

4. Lena bought a total of 20 postcards. She bought 6 more large postcards than small. Write a system of equations that represents the postcards Lena purchased. Solve the system by substitution. Interpret the solution. (Lesson 8)

Reflect

Answering the Essential Question

Use what you learned about graphs to complete the graphic organizer. List three ways in which graphs are helpful. Then give an example for each way.

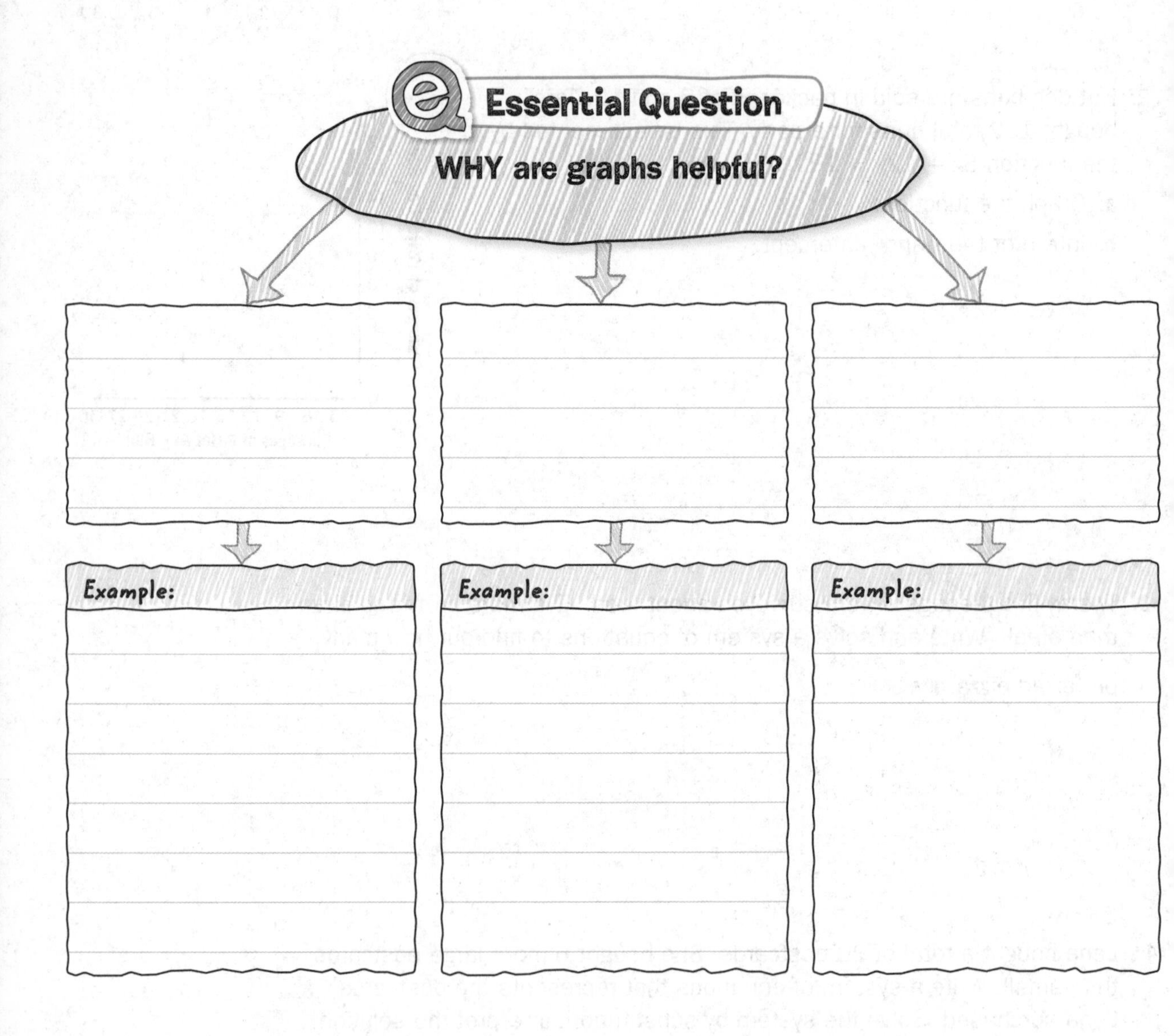

Answer the Essential Question. WHY are graphs helpful?

UNIT PROJECT

Watch

Web Design 101 When designing a good Web page, there are many details to consider in order to make your Web page stand out. In this project, you will:

- **Collaborate** with your classmates as you research an animal and design a Web page.
- **Share** the results of your research in a creative way
- **Reflect** on how you communicate mathematical ideas effectively.

By the end of this Project, you will be ready to design a live Web page about your favorite animal!

Collaborate

Go Online Work with your group to research and complete each activity. You will use your results in the Share section on the following page.

1. Choose your favorite animal. Research information about that animal, such as the population over the past 10 years, the kinds of food it eats, sleeping habits, its average lifespan, average size, and average speed. Present this information using tables and graphs.

2. Use the distance formula, distance = rate × time, to write an equation that represents the distance your animal can travel at its average speed. Find the average speed of two other animals and write equations using the distance formula. Graph all three equations on the same coordinate plane. Then describe the graphs.

3. Research the elements needed to make a good Web page. Then make a sketch of your own Web page about the favorite animal that you selected in Exercise 1. Be sure to include tables, equations, graphs, and photos.

4. Find another animal in the same animal kingdom as your favorite animal. On the sketch of your Web page, include a link to this other animal and an equation that describes one of its characteristics.

5. Research the cost of taking a Web design class. Write an equation that represents the time it will take you to save enough money for the class. Share this equation as you write a few paragraphs that explain your plan on how to save enough money

Share

With your group, decide on a way to share what you have learned about your animal and Web pages. Some suggestions are listed below, but you can also think of other creative ways to present your information. Remember to show how you used mathematics to complete each of the activities in this project!

- If possible, use Web page creation software to turn your design into a live Web page.
- Imagine you are going to be interviewed by a reporter about your work on this project. Write down what will be discussed in the interview. You may wish to actually record an interview.

Check out the note on the right to connect this project with other subjects.

Business Literacy Research Web design jobs in your area. Find out the following:

- What type of education is required?
- What skills should a Web designer possess?

Reflect

6. **Answer the Essential Question** HOW can you communicate mathematical ideas effectively?

a. How did you use what you learned in the Equations in One Variable chapter to communicate mathematical ideas effectively in this project?

____

____

____

b. How did you use what you learned in the Equations in Two Variables chapter to communicate mathematical ideas effectively in this project?

____

____

____

UNIT 3 Functions

Essential Question

HOW can you find and use patterns to model real-world situations?

Chapter 4 Functions

Functions can be represented using equations, graphs, tables, and verbal descriptions. In this chapter, you will use functions to model linear relationships. You will also investigate nonlinear functions.

Unit Project Preview

Green Thumb Do you have a green thumb for gardening? A community garden is a great way to meet your neighbors, beautify your neighborhood, and strengthen the sense of community.

There are many types of community gardens. Food pantry gardens donate the produce they grow to local food pantries. School gardens help to educate students in science and math, while entrepreneurial gardens generate income by selling the produce.

At the end of Chapter 4, you'll complete a project to discover the costs involved in creating a community garden. But for now, it's time to do an activity in your book. Complete the table shown by estimating the cost of selling various vegetables and fruits.

My Garden	
Item	**Cost Per Item or Per Pound**
Carrots	
Cucumbers	
Peas	
Strawberries	
Tomatoes	

When Will You Use This?

Your Turn! **You will solve this problem in the chapter.**

Try the Quick Check below.
Or, take the Online Readiness Quiz.

Quick Review

Common Core Review 6.NS.6, 7.NS.3

Example 1

Name the ordered pair for point *Q*.

Start at the origin. Move right along the *x*-axis until you reach 1.5. Then move up until you reach the *y*-coordinate, 2. Point *Q* is located at (1.5, 2).

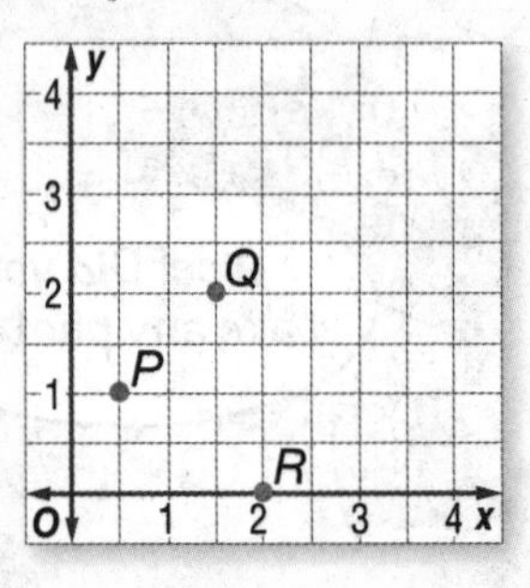

Example 2

Evaluate $6x + 1$ if $x = -4$.

$6x + 1 = 6(-4) + 1$ Replace *x* with −4.
$= -24 + 1$ Multiply 6 by −4.
$= -23$ Add.

Quick Check

Coordinate Graphing **Name the ordered pair for each point.**

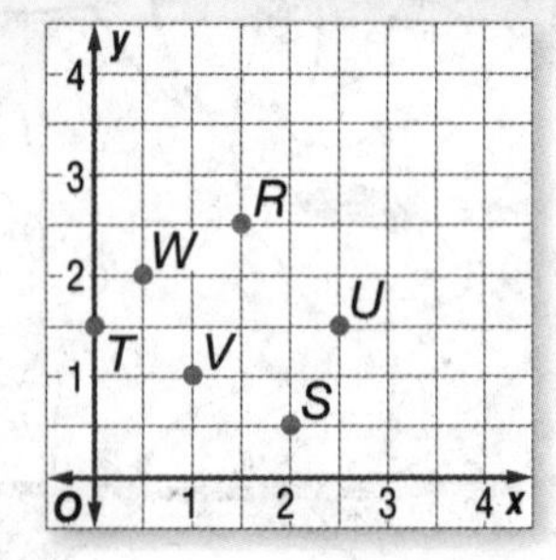

1. *R* ______
2. *S* ______
3. *T* ______
4. *U* ______
5. *V* ______
6. *W* ______

Evaluate Expressions **Evaluate each expression if $x = -6$.**

7. $3x$ ______
8. $4x + 9$ ______
9. $\frac{x}{2}$ ______
10. $\frac{3x}{9}$ ______

11. The weekly profit of a certain company is $48x - 875$, where *x* represents the number of units sold. Find the weekly profit if the company sells 37 units. ______

How Did You Do?

Which problems did you answer correctly in the Quick Check? Shade those exercise numbers below.

1 2 3 4 5 6 7 8 9  

Inquiry Lab

Relations and Functions

HOW can I determine if a relation is a function?

Content Standards
8.F.1
Mathematical Practices
1, 3, 4

Colors Mrs. Heinl asked three members of her class their favorite color. The mapping diagrams below show some possible results.

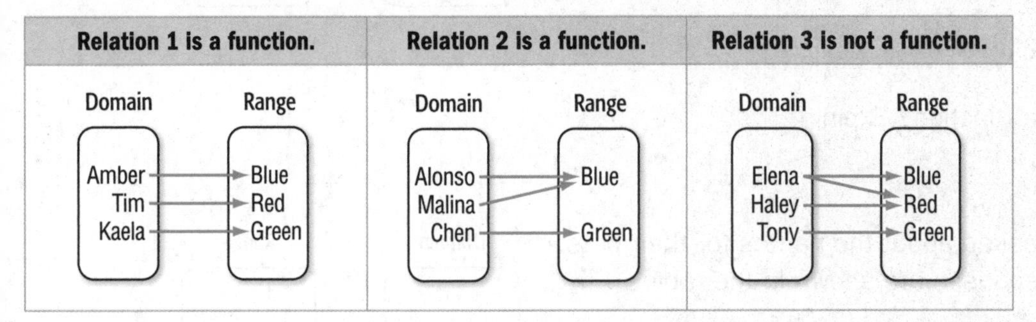

A *function* is a special relation in which each member of the domain is paired with *exactly* one member in the range. In the mapping above, Relation 3 is *not* a function because Elena chose two favorite colors, blue and red.

Investigation

Mr. Morgan asked his students how many pets they have. Some of the student responses are shown in the table.

Student Number	1	3	6
Number of Pets	2	5	7

Complete the mapping diagram shown.

Domain: 1, 3, 6
Range:

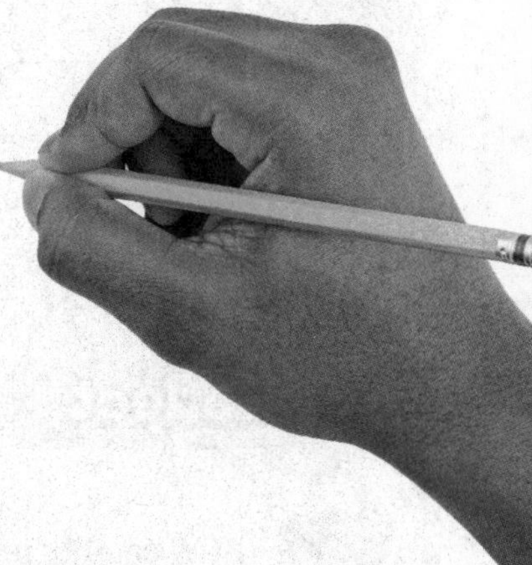

Is the relation a function? Explain.

Suppose Student 8 has 2 pets. Make a mapping diagram of this situation. Is this relation a function? Explain.

Domain: 1, 3, 6, 8
Range: 2, 5, 7

Collaborate

Model with Mathematics Work with a partner. Make a mapping diagram for each relation. Then determine whether each relation is a function.

1. Students were asked about the number of cell phone minutes they use. Some of the responses are shown in the table.

Domain | Range

Student	Sarah	Max	Jacob	Rebekah
Number of Minutes	275	220	350	275

Is this relation a function? Explain. ______________________________

____

2. Students were asked about the names for their pets. Some of the responses are shown in the table.

Domain | Range

Student	Klara	Adrienne	Simon
Pet Names	Tiny	Rover Betty	Mimi

Is this relation a function? Explain. ______________________________

____

Analyze

3. **Model with Mathematics** Make a table and a mapping diagram for the relation $\{(0, -2), (1, -2), (1, 3), (1, 8)\}$.

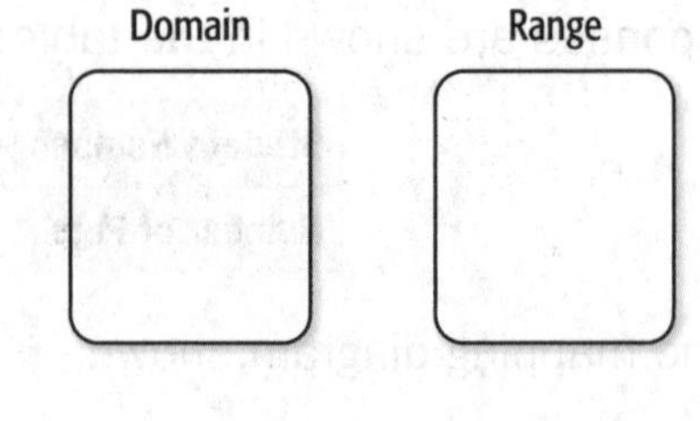

Domain				
Range				

Is this relation a function? Explain. ______________________________

____

Reflect

4. Inquiry HOW can I determine if a relation is a function?

____

____

Problem-Solving Investigation
Make a Table

Content Standards
8.F.4
Mathematical Practices
1, 2, 4

Case #1 Play Catch Up

Emilio's family going on vacation. His mom and sister leave at 7:00 in the morning, driving an average of 45 miles per hour. Emilio and his dad leave at 8:00. His dad drives an average of 60 miles per hour.

Will Emilio and his dad catch up to his mom and sister?

1 Understand *What are the facts?*

You know the times they left and their rates. You need to know if Emilio and his dad will catch up to his mom and sister.

2 Plan *What is your strategy to solve this problem?*

Make a table that shows how many miles each driver has driven.

3 Solve *How can you apply the strategy?*

Hours Since 7:00 A.M.	Distance Traveled (mi)	
	Emilio's Mom	Emilio's Dad
0	0	0
1	45	0
2		60
3		
4		

At ______ A.M., Emilio and Emilio's dad will catch up to his mom and sister.

4 Check *Does the answer make sense?*

45 mph × ____ h = ____ mi   60 mph × ____ h = ____ mi

The distances are equal. ✓

Reason Abstractly Suppose Emilio's mom drives at an average speed of 50 miles per hour. At what time will Emilio and his dad catch up to her?

Company	Deposit	Cost Per Day
Mike's Music	\$5	\$1.25
Karaoke Korner	\$4	\$1.50

Case #2 Karaoke Kid

Rina wants to rent a karaoke machine for a family reunion. The prices to rent the machine from two different companies are shown.

For how many days must she rent the machine for the cost from each place to be the same?

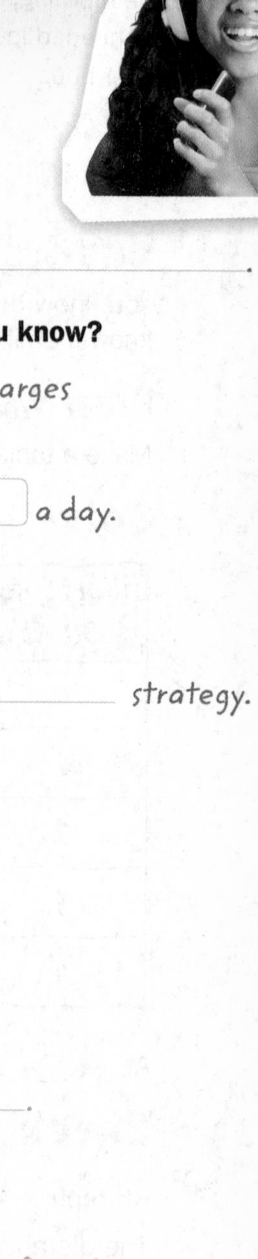

1 Understand

Read the problem. What are you being asked to find?

I need to find ______________________.

Underline key words and values. What information do you know?

I know that Mike's Music has a deposit of ☐ and charges ☐ a day.

Karaoke Korner's deposit is ☐ and they charge ☐ a day.

2 Plan

Choose a problem-solving strategy.

I will use the ______________________ strategy.

3 Solve

Use your problem-solving strategy to solve the problem.

	Day 1	Day 2	Day 3	Day 4
Mike's Music				
Karaoke Korner				

So, the cost at both companies is the same at ________.

4 Check

Use information from the problem to check your answer.

Mike's Music charges ☐ or ☐ for the first day.

Each day adds another ☐.

Karaoke Korner charges ☐ or ☐ for the first day.

Each day adds another ☐.

At ☐ days, both companies charge ☐.

Collaborate Work with a small group to solve the following cases. Show your work on a separate piece of paper.

Case #3 Plants

The table shows the height of a giant bamboo plant.

Assuming the bamboo grew at a steady rate, what was the height of the bamboo on the fifth day? ______

Bamboo Growth

Number of Days	Total Growth (ft)
5	?
6	?
7	10
8	13.5
9	17

Case #4 A Penny Saved

Sophia and Scott each open a bank account. Sophia's first deposit is \$27, and Scott's first deposit is \$62. Scott plans to save \$15 a week and Sophia plans to save \$20 a week.

Provided they make no withdrawals, will Sophia's balance be greater than Scott's balance? If so, when?

______

Case #5 Fitness

Marcie increases the distance she jogs each week by 0.25 mile.

If she jogs 3.05 miles in week 7, how far did she jog the fourth week?

______

Circle a strategy below to solve the problem.

- Look for a pattern.
- Use logical reasoning.
- Guess, check, and revise.
- Work backward.

Case #6 Animals

The graph shows the maximum length of several animals. The maximum length of a walrus is twice the maximum length of a lion, which is 0.4 meter longer than the maximum length of a giant panda.

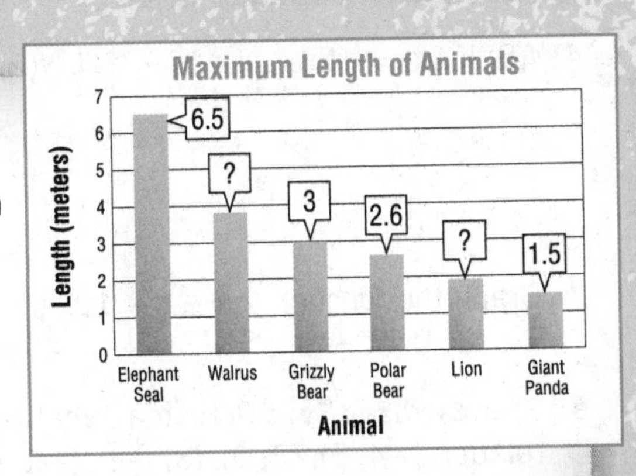

Find the maximum length of a walrus. ______

Mid-Chapter Check

Vocabulary Check

1. **CCSS Be Precise** Define *linear equation*. Give an example of a linear equation. (Lesson 1)

2. Describe the difference between the graph of a set of discrete data and the graph of a set of continuous data. (Lesson 4)

Skills Check and Problem Solving

3. There are 20 nickels in one dollar. (Lesson 2)
   a. Write an equation to find the number of nickels n in any number of dollars d.

   Show your work.

   b. Make a table to find the number of nickels in 5, 10, 15, or 20 dollars. Then graph the ordered pairs.

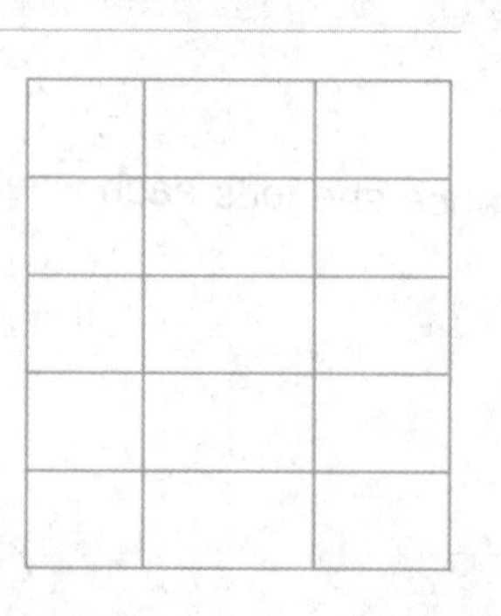

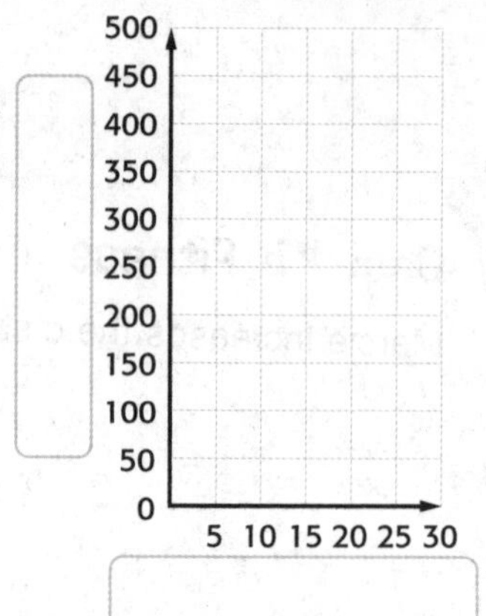

Find each function value. (Lesson 3)

4. $f(8)$ if $f(x) = 15x$

5. $f(2)$ if $f(x) = 2x - 5$

6. $f(4)$ if $f(x) = -3x + 15$

7. Graph the function $y = \frac{1}{3}x + 12$. (Lesson 4)

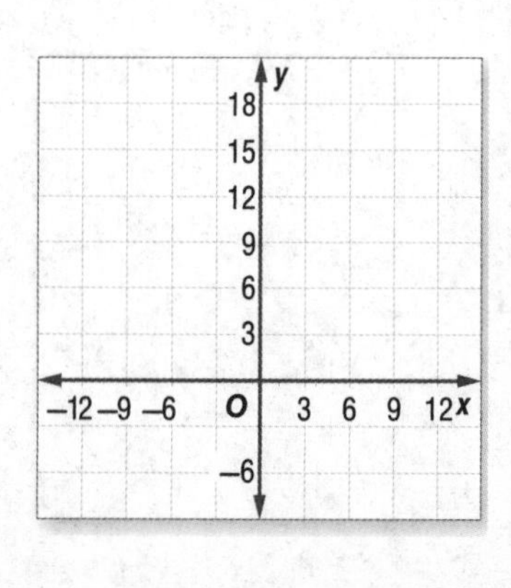

8. **Standardized Test Practice** What is the domain of the relation $\{-4, 5), (2, 0), (3, -2), (-1, 0)\}$? (Lesson 2)

   Ⓐ $\{-4, 2\}$
   Ⓑ $\{-4, -1, 2, 3\}$
   Ⓒ $\{-2, 0, 5\}$
   Ⓓ $\{(-4, 5), (2, 0), (3, -2), (-1, 0)\}$

Name ______________________ My Homework ______________________

Extra Practice

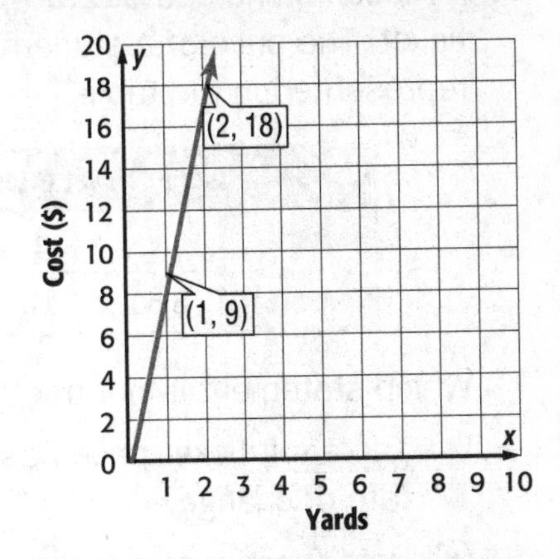

9. A fabric store sells cotton for $7.00 a yard. The price of special occasion fabric is shown in the graph. Compare the functions' rates of change.

Cotton fabric: $7.00 per yard

Special occasion fabric: $\frac{18-9}{2-1} = \frac{9}{1}$ or $9.00 per yard.

The special occasion fabric has the greater rate of change.

10. Two players played a game. The first player's score is represented by the function $p = 5c - 3$, where p is the number of points scored and c is the number of correct answers. The second player's score is shown in the table.

Questions Answered	Score
1	5
2	10
3	15
4	20

a. Compare the functions by comparing their y-intercepts and rates of change. ______________________

____

____

b. How many points will the first player have if he or she correctly answers 30 questions? ______________________

11. CCSS **Justify Conclusions** Jesse and Juan each open savings accounts. The amounts in Jesse's account are shown in the table. Juan saves $5 per week. Who will have more saved in 8 weeks? Explain.

Jesse's Savings	
Week	**Amount Saved ($)**
1	16
2	19
3	22
4	25
5	28

____

____

12. Canada Olympic Park features sports training and entertainment facilities. The Monster zip line produces average speeds of 120 kilometers per hour. A smaller line produces speeds represented by the function $d = 50h$ where d is the distance in kilometers after h hours. How much farther could you travel on the Monster zip line in 0.25 hours?

____

13. Raj gets a 1.5 mile head start and runs at a rate of 4.5 miles per hour. Jacinda's progress is represented by a graph that goes through the points (1, 10), (2, 20), and (3, 30). How long will Jacinda need to run to catch up with Raj? ______________________

Standardized Test Practice

14. A museum charges $12.50 per adult ticket. The price of a student ticket is represented in the table.

Student Ticket Price			
Tickets	1	2	3
Price ($)	8.50	17	25.50

Which statement is *not* true?

Ⓐ The adult ticket price has a greater rate of change.

Ⓑ Both functions have the same *y*-intercept.

Ⓒ The student ticket price has a greater rate of change.

Ⓓ Both functions show a direct variation.

15. Short Response Julie swam, biked, and ran in a 22.2 mile triathlon. She completed the race in 2.15 hours. The function $m = 13.8h$ represents the miles m Julie biked in h hours. Was her average speed biking less than or greater than her average speed for the entire race? Justify your answer. Round to the nearest tenth if necessary.

Common Core Review

Write an equation in slope-intercept form for each table of values. 8.F.3

16.

x	−1	0	1	2
y	−7	−3	1	5

17.

x	−3	−1	1	3
y	7	5	3	1

Evaluate each expression if $a = 12$ and $b = 8$. 7.EE.3

18. $4a - b$ ______

19. $3b - 2a$ ______

20. $3a + 2b$ ______

21. $a + 5b$ ______

Inquiry Lab

Graphing Technology: Families of Non-Linear Functions

HOW are families of nonlinear functions the same as the parent function? How are families of nonlinear functions different from the parent function?

Content Standards
8.F.3, 8.F.5

Mathematical Practices
1, 3, 7

Gardening Sasha is digging a vegetable garden and wants to know how much fertilizer to buy. The equation $y = x^2$ represents the area in square feet of the garden shown.

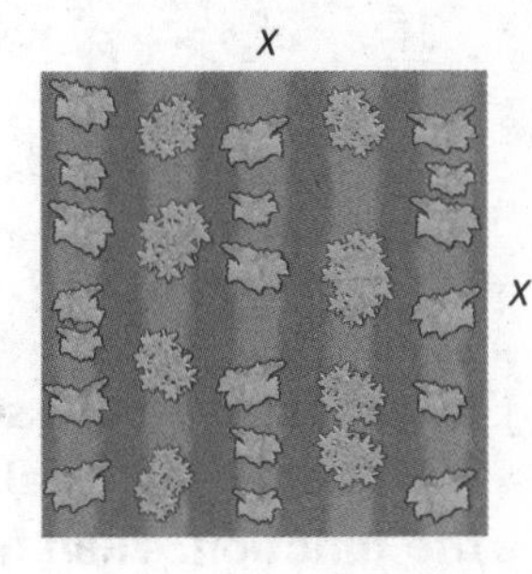

Investigation 1

Families of nonlinear functions share a common characteristic based on a parent function. The parent function, or simplest function, of a family of quadratic functions is $y = x^2$. You can use a graphing calculator to investigate families of quadratic functions.

Graph $y = x^2$, $y = x^2 + 5$, and $y = x^2 - 3$ on the same screen.

Step 1 Clear any existing equations from the Y= list by pressing [Y=] [CLEAR].

Step 2 Enter each equation. Press [X,T,θ,n] [x^2] [ENTER], [X,T,θ,n] [x^2] [+] 5 [ENTER], and [X,T,θ,n] [x^2] [−] 3 [ENTER].

Step 3 Press [ZOOM] 6.

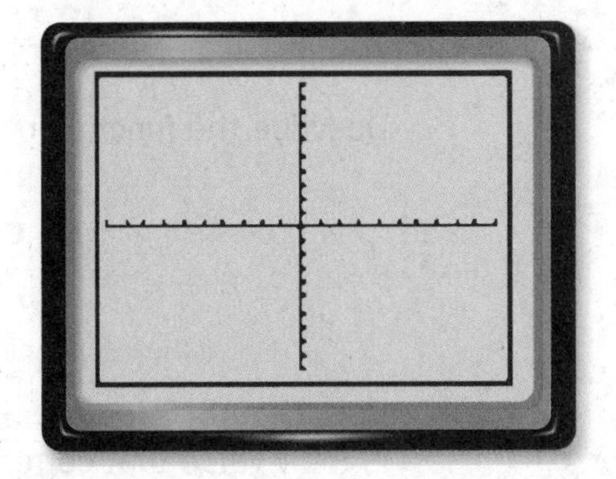

Copy your calculator screen on the blank screen shown.

How are the three equations related?

____

____

Describe how the graphs of the three equations are related.

____

____

Investigation 2

An **exponential function** is a nonlinear function in which the base is a constant and the exponent is an independent variable, x. The parent function for an exponential function is shown.

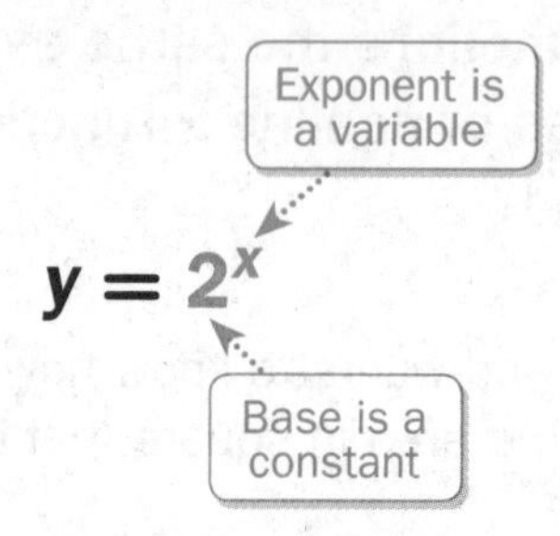

A certain type of bacteria doubles every hour. The function $y = 2^x$ represents the total number of bacteria y at the end of every hour x. Graph the function. Then find the number of bacteria at the end of 5 hours.

Step 1 Clear any existing equations from the Y= list by pressing [Y=] [CLEAR].

Step 2 Enter the equation. Press [Y=] 2 [^] [X,T,θ,n]

Step 3 Graph the equation in the standard viewing window. Press [ZOOM] 6.

Copy your calculator screen on the blank screen shown.

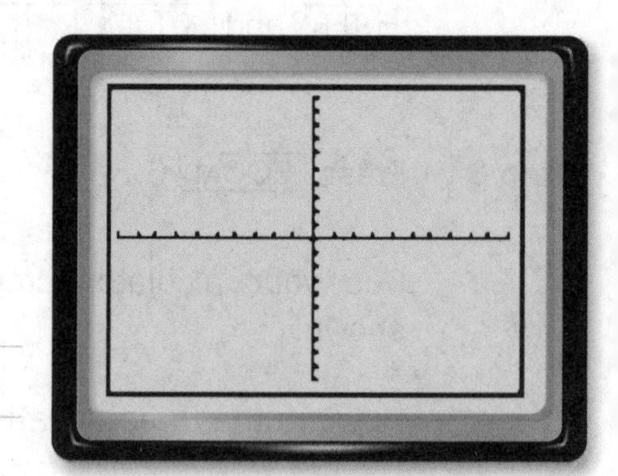

Describe the function by analyzing the graph.

____

____

Step 4 Use the TABLE feature. Press [2nd] [GRAPH]. What is the y value that corresponds to the x value of 5? ☐

So, there are ☐ bacteria at the end of 5 hours.

Collaborate

Work with a partner.

1. Use a graphing calculator to graph $y = x^2$, $y = x^2 - 6$, and $y = x^2 + 4$. Copy your calculator screen on the blank screen shown.

How does changing the value of c in the equation $y = x^2 + c$ affect the graph?

2. Use a graphing calculator to graph $y = 0.5x^2$, $y = x^2$, and $y = 2x^2$. Copy your calculator screen on the blank screen shown.

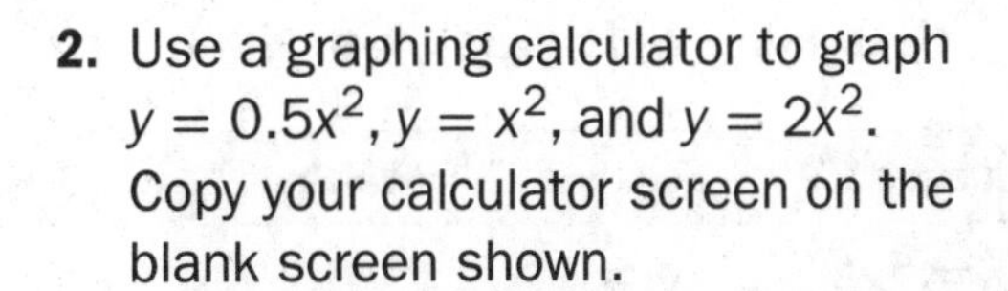

How does changing the value of a in the equation $y = ax^2$ affect the graph?

3. Use a graphing calculator to graph $y = 2^x$, $y = 2^x + 1$, and $y = 2^x - 3$. Copy your calculator screen on the blank screen shown.

How does changing the value of c in the equation $y = 2^x + c$ affect the graph?

4. Use a graphing calculator to graph $y = 0.5^x$, $y = 0.25^x$ and $y = 2^x$. Copy your calculator screen on the blank screen shown.

How does changing the value of a to a fraction in the equation $y = a^x$ affect the graph?

Analyze

CCSS Identify Structure Work with a partner to complete the table. Without graphing, determine which graph is wider.

	Equation 1	Equation 2	Which graph is wider?
5.	$y = 5x^2$	$y = x^2$	
6.	$y = \frac{1}{3}x^2$	$y = 3x^2$	
7.	$y = 2^x$	$y = 4^x$	
8.	$y = 0.25^x$	$y = 0.75^x$	

9. Miley's parents started a savings account when she was born by depositing $100 into the account. The account has an annual interest rate of 3%. The balance y in the account can be represented by the function $y = 100(1.03)^x$, where x is the number of years.

a. Graph the function on your graphing calculator. Copy your calculator screen on the blank screen shown. (*Hint:* Use the x-scale -50 to 50 and the y-scale 0 to 500.)

b. Use the TABLE feature. How much money is in the account after 13 years? ______________

c. Describe the function by analyzing the graph.

____

____

Reflect

10. Inquiry HOW are families of nonlinear functions the same as the parent function? How are families of nonlinear functions different from the parent function?

____

____

____

21ST CENTURY CAREER in Physical Therapy

Physical Therapist

Are you a compassionate person? Do you have a strong desire to help others? If so, a career as a physical therapist might be a good choice for you. Physical therapists help restore function, improve mobility, and relieve pain of patients suffering from injuries or disease. One of their jobs is to teach exercises or recommend activities to help patients regain balance, flexibility, endurance, and strength.

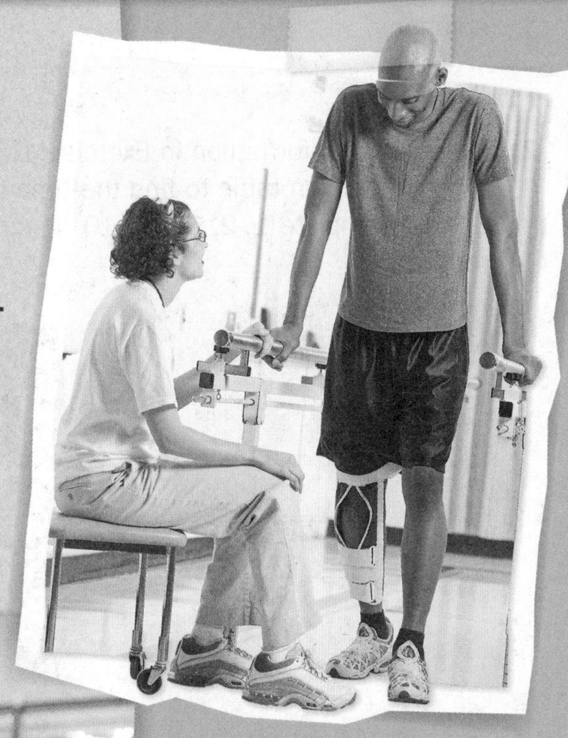

Explore college and careers at ccr.mcgraw-hill.com

Is This the Career for You?

Are you interested in a career as a physical therapist? Take some of the following courses in high school.

- Algebra
- Biology
- Chemistry
- Introduction To Physical Therapy

Turn the page to find out how math relates to a career in Physical Therapy.

Focusing on Recovery

Use the information in the table below to solve each problem.

1. The function $t(r) = 12r$, where r is the number of repetitions, represents the total time $t(r)$ in seconds to complete a flexibility exercise. Find $t(8)$. Then interpret the solution. ______________________

____

2. Refer to the information in Exercise 1. Make a function table to find the time it will take to complete 1, 2, 5, and 10 repetitions.

r	$12r$	$t(r)$

3. Write a function to represent the distance d in miles a runner will travel in t minutes.

____

4. Refer to the function that you wrote in Exercise 3. How far will a runner travel after 80 minutes? ______________________

5. Graph the function from Exercise 3. Then use the graph to estimate the distance a runner will travel after 90 minutes.

____

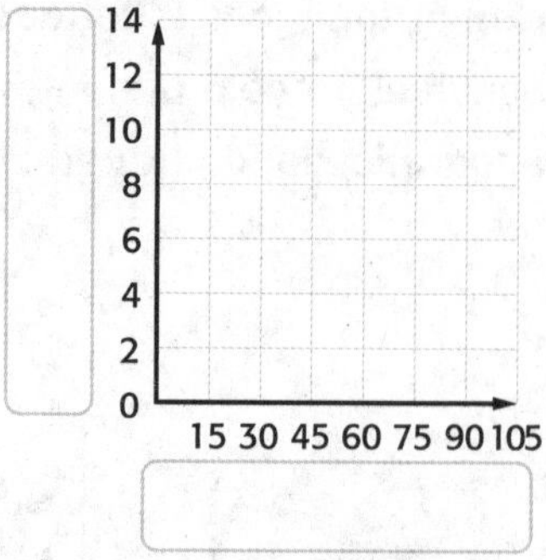

Endurance Exercise: Cross-Country Running	
Time (min)	**Distance (mi)**
15	2.25
30	4.5
45	6.75
60	9.0

Career Project

It's time to update your career portfolio! Make a list of questions that you would like to know about a career in physical therapy. Then interview a physical therapist in your area. Include all the interview questions and answers in your portfolio.

____

____

____

____

____

List other careers that someone with an interest in physical therapy could pursue.

-
-
-
-
-

UNIT PROJECT

Watch

Green Thumb If you have a knack for gardening, volunteering in a community garden is a great way to get involved with your community and also earn a little money. In this project you will:

- **Collaborate** with your classmates as you research the costs involved with growing vegetables and predict possible profits.
- **Share** the results of your research in a creative way.
- **Reflect** on how you find and use patterns to model real-world situations.

By the end of this Project, you just might be a young entrepreneur!

Collaborate

Go Online Work with your group to research and complete each activity. You will use your results in the Share section on the following page.

1. Choose a vegetable that is sold individually, and find its cost at a grocery store. Write an equation to represent the total cost as a function of the number of vegetables. Make a function table to find the cost of 1, 2, 3, 4, 5, and 6 vegetables. Then graph the ordered pairs.

2. Research a vegetable you would like to grow in a community garden. Find the costs involved such as buying seeds and gardening tools. Then determine how much you will charge per vegetable (or per pound) based on grocery store or farm market prices.

3. Based on the information you found in Exercise 2, write a linear function to represent your profit. Describe what the variables represent. Then graph and describe the function.

4. Research the following terms: *gross profit, total revenue, and gross profit margin.* Make a diagram explaining these terms. Then find your gross profit margin based on estimated gross profit and total revenue. What does your gross profit tell you?

5. Research the average temperatures in your area for the growing season of the vegetable you chose. Then sketch a qualitative graph that shows the change in temperature over the growing season. Include a brief explanation of your graph.

With your group, decide on a way to share what you have learned about growing and selling vegetables. Some suggestions are listed below, but you can also think of other creative ways to present your information. Remember to show how you used mathematics to complete each of the activities in this project!

- Imagine you sell your vegetables at a farmer's market. Describe your experience in a blog.
- Use a budget spreadsheet to show how your vegetable can generate a profit. Include tables, equations, and graphs.

Environmental Literacy Research information about Earth's soil and the qualities needed to grow plants. Some questions to consider are:

- What type of soil allows fruits and vegetables to grow well?
- What type of soil is typically found in your area?
- What could you add to the soil to make it better for growing your plants?

Check out the note on the right to connect this project with other subjects.

6. **Answer the Essential Question** How can you find and use patterns to model real-world situations?

   a. How did you use what you learned about constructing functions in this chapter to find and use patterns to model real-world situations in this project?

   ____

   ____

   ____

   b. How did you use what you learned about different representations of functions in this chapter to find and use patterns to model real-world situations in this project?

   ____

   ____

   ____

Glossary/Glosario

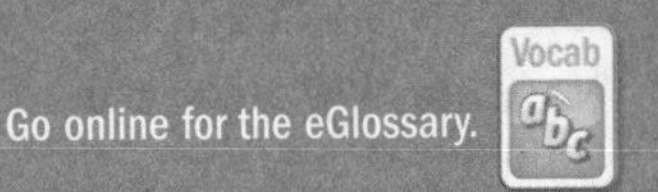

The eGlossary contains words and definitions in the following 13 languages:

Arabic	**Cantonese**	**Hmong**	**Spanish**	**Urdu**
Bengali	**English**	**Korean**	**Tagalog**	**Vietnamese**
Brazilian Portuguese	**Haitian Creole**	**Russian**		

English

Español

Aa

accuracy The degree of closeness of a measurement to the true value.

exactitud Cercanía de una medida a su valor verdadero.

acute angle An angle whose measure is less than 90°.

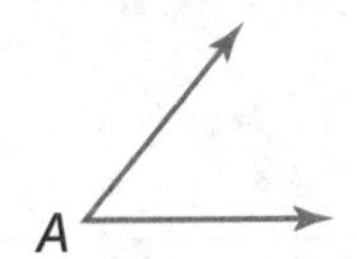

ángulo agudo Ángulo que mide menos de 90°.

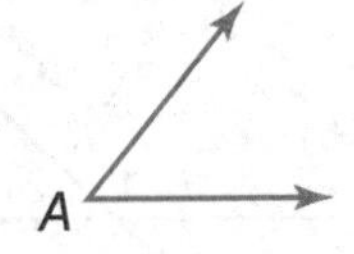

acute triangle A triangle with all acute angles.

triángulo acutángulo Triángulo con todos los ángulos agudos.

Addition Property of Equality If you add the same number to each side of an equation, the two sides remain equal.

propiedad de adición de la igualdad Si sumas el mismo número a ambos lados de una ecuación, los dos lados permanecen iguales.

adjacent angles Angles that share a common vertex, a common side, and do not overlap. In the figure, the adjacent angles are $\angle 5$ and $\angle 6$.

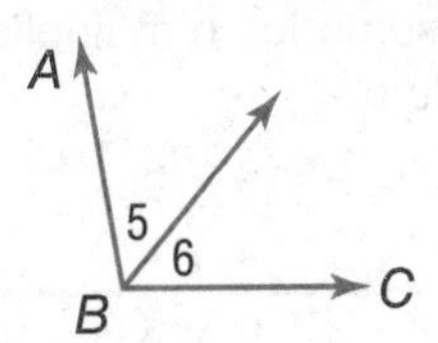

ángulos adyacentes Ángulos que comparten un vértice, un lado común y no se traslapan. En la figura, los ángulos adyacentes son $\angle 5$ y $\angle 6$.

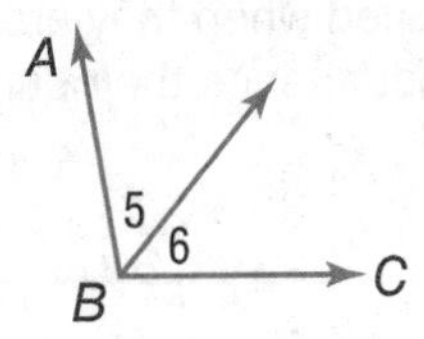

algebra A branch of mathematics that involves expressions with variables.

álgebra Rama de las matemáticas que trabaja con expresiones con variables.

algebraic expression A combination of variables, numbers, and at least one operation.

expresión algebraica Una combinación de variables, números y por lo menos una operación.

alternate exterior angles Exterior angles that lie on opposite sides of the transversal. In the figure, transversal t intersects lines ℓ and m. $\angle 1$ and $\angle 7$, and $\angle 2$ and $\angle 8$ are alternate exterior angles. If line ℓ and m are parallel, then these pairs of angles are congruent.

ángulos alternos externos Ángulos externos que se encuentran en lados opuestos de la transversal. En la figura, la transversal t interseca las rectas ℓ y m. $\angle 1$ y $\angle 7$, y $\angle 2$ y $\angle 8$ son ángulos alternos externos. Si las rectas ℓ y m son paralelas, entonces estos ángulos son pares de ángulos congruentes.

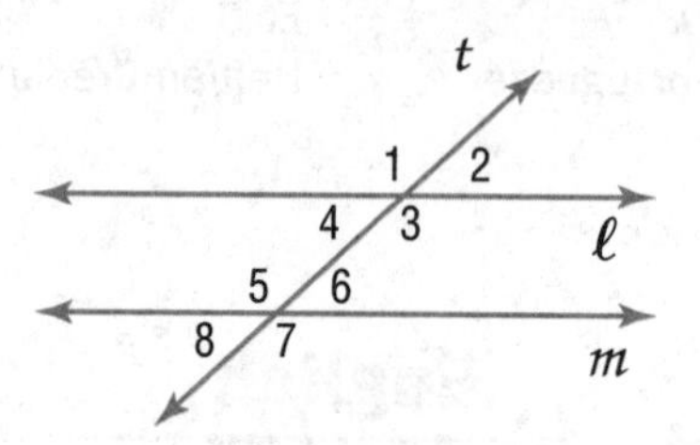

alternate interior angles Interior angles that lie on opposite sides of the transversal. In the figure below, transversal t intersects lines ℓ and m. $\angle 3$ and $\angle 5$, and $\angle 4$ and $\angle 6$ are alternate interior angles. If lines ℓ and m are parallel, then these pairs of angles are congruent.

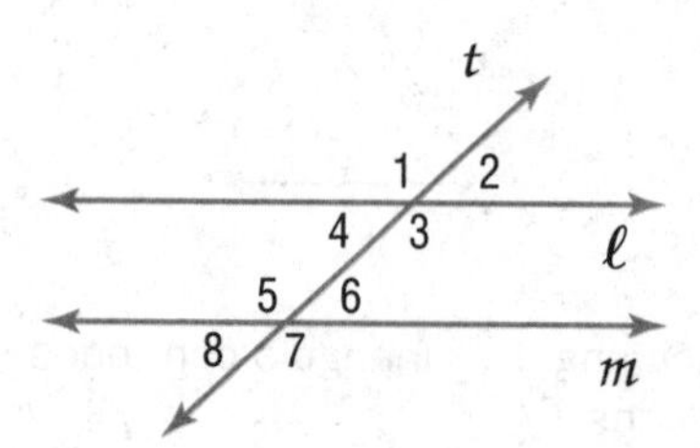

ángulos alternos internos Ángulos internos que se encuentran en lados opuestos de la transversal. En la figura, la transversal t interseca las rectas ℓ y m. $\angle 3$ y $\angle 5$, y $\angle 4$ y $\angle 6$ son ángulos alternos internos. Si las rectas ℓ y m son paralelas, entonces estos ángulos son pares de ángulos congruentes.

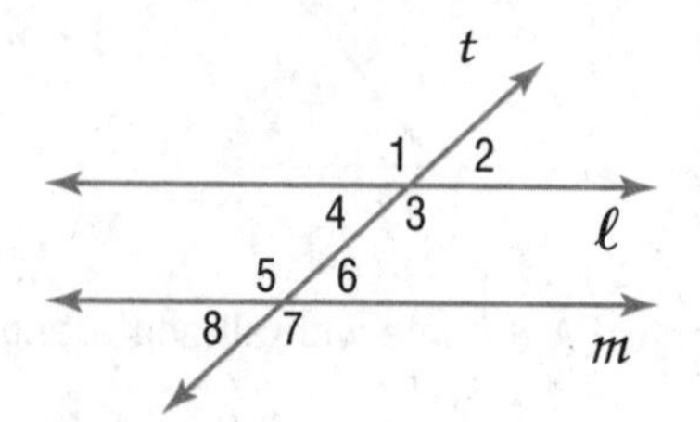

angle of rotation The degree measure of the angle through which a figure is rotated.

ángulo de rotación Medida en grados del ángulo sobre el cual se rota una figura.

arc One of two parts of a circle separated by a central angle.

arco Una de dos partes de un círculo separadas por un ángulo central.

Associative Property The way in which three numbers are grouped when they are added or multiplied does not change their sum or product.

propiedad asociativa La forma en que se agrupan tres números al sumarlos o multiplicarlos no altera su suma o producto.

base In a power, the number that is the common factor. In 10^3, the base is 10. That is, $10^3 = 10 \times 10 \times 10$.

base En una potencia, número que es el factor común. En 10^3, la base es 10. Es decir, $10^3 = 10 \times 10 \times 10$.

base One of the two parallel congruent faces of a prism.

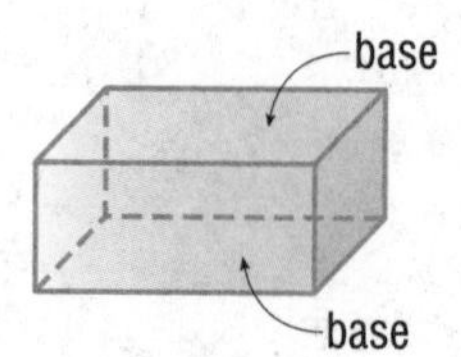

base Una de las dos caras paralelas congruentes de un prisma.

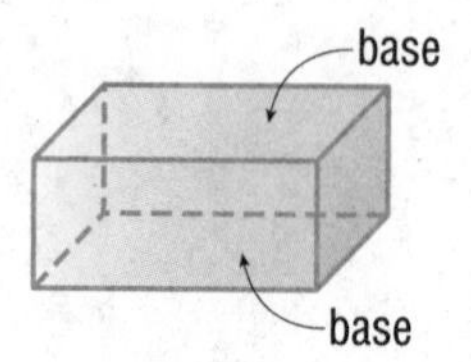

biased sample A sample drawn in such a way that one or more parts of the population are favored over others.

muestra sesgada Muestra en que se favorece una o más partes de una población.

bivariate data Data with two variables, or pairs of numerical observations.

datos bivariantes Datos con dos variables, o pares de observaciones numéricas.

box plot A method of visually displaying a distribution of data values by using the median, quartiles, and extremes of the data set. A box shows the middle 50% of the data.

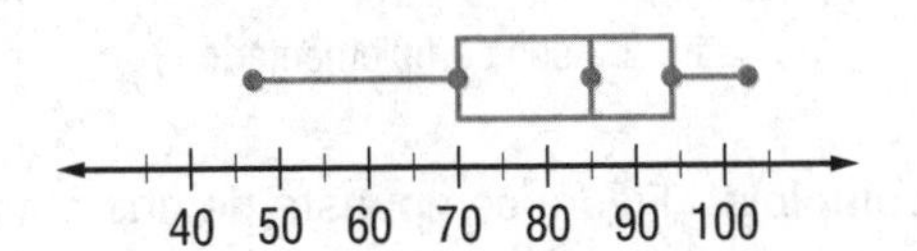

diagrama de caja Un método de mostrar visualmente una distribución de valores usando la mediana, cuartiles y extremos del conjunto de datos. Una caja muestra el 50% del medio de los datos.

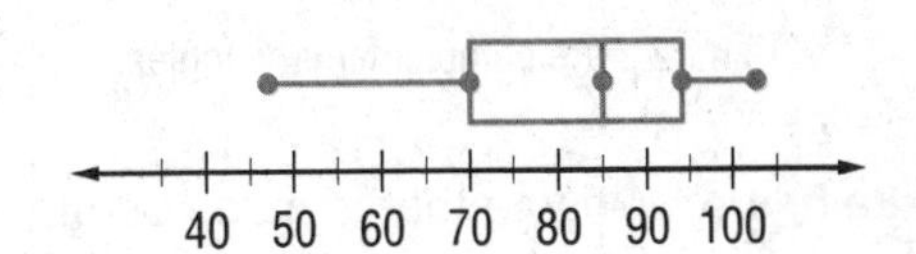

Cc

center The given point from which all points on a circle are the same distance.

centro Un punto dado del cual equidistan todos los puntos de un círculo.

center of dilation The center point from which dilations are performed.

centro de la homotecia Punto fijo en torno al cual se realizan las homotecias.

center of rotation A fixed point around which shapes move in a circular motion to a new position.

centro de rotación Punto fijo alrededor del cual se giran las figuras en movimiento circular alrededor de un punto fijo.

central angle An angle that intersects a circle in two points and has its vertex at the center of the circle.

ángulo central Ángulo que interseca un círculo en dos puntos y cuyo vértice es el centro del círculo.

circle The set of all points in a plane that are the same distance from a given point called the center.

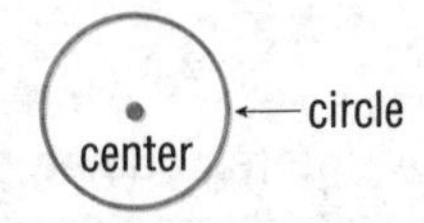

círculo Conjunto de todos los puntos en un plano que equidistan de un punto dado llamado centro.

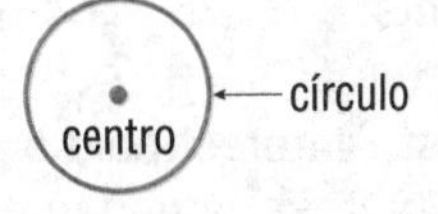

circumference The distance around a circle.

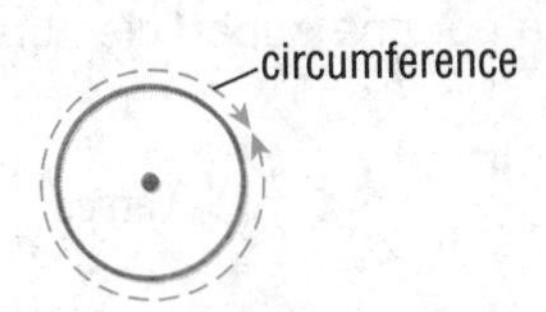

circunferencia La distancia alrededor de un círculo.

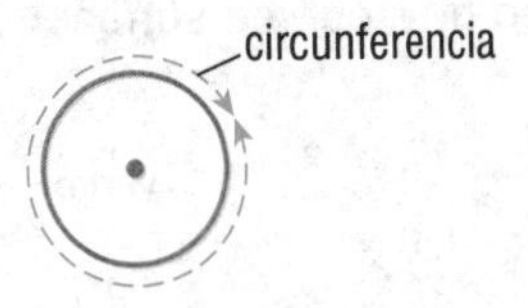

chord A segment with endpoints that are on a circle.

cuerda Segmento cuyos extremos están sobre un círculo.

coefficient The numerical factor of a term that contains a variable.

coeficiente Factor numérico de un término que contiene una variable.

common difference The difference between any two consecutive terms in an arithmetic sequence.

diferencia común La diferencia entre cualquier par de términos consecutivos en una sucesión aritmética.

Commutative Property The order in which two numbers are added or multiplied does not change their sum or product.

complementary angles Two angles are complementary if the sum of their measures is 90°.

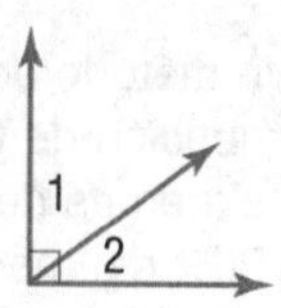

∠1 and ∠2 are complementary angles.

composite figure A figure that is made up of two or more shapes.

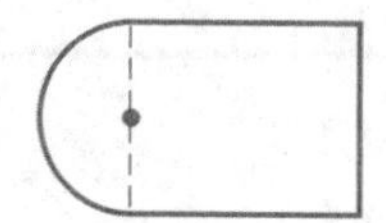

composite solid An object made up of more than one type of solid.

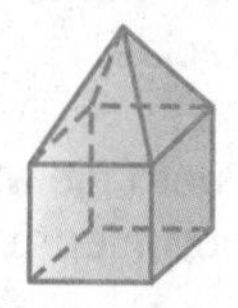

composition of transformations The resulting transformation when a transformation is applied to a figure and then another transformation is applied to its image.

compound event An event that consists of two or more simple events.

compound interest Interest paid on the initial principal and on interest earned in the past.

cone A three-dimensional figure with one circular base connected by a curved surface to a single vertex.

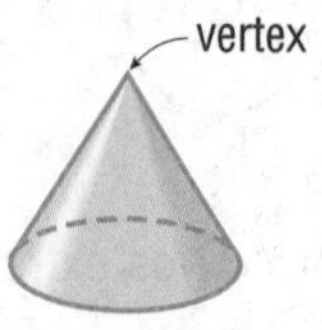

congruent Having the same measure; if one image can be obtained by another by a sequence of rotations, reflections, or translations.

constant A term without a variable.

propiedad conmutativa La forma en que se suman o multiplican dos números no altera su suma o producto.

ángulos complementarios Dos ángulos son complementarios si la suma de sus medidas es 90°

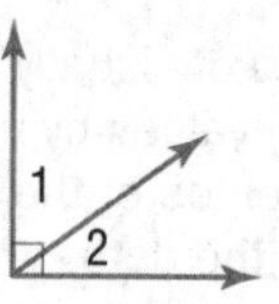

∠1 y ∠2 son complementarios.

figura compleja Figura compuesta de dos o más formas.

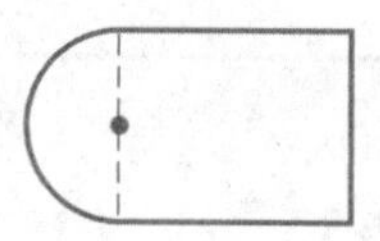

sólido complejo Cuerpo compuesto de más de un tipo de sólido.

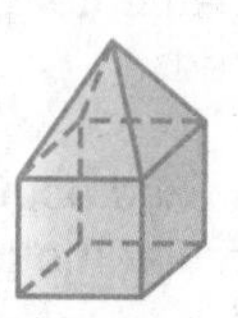

composición de transformaciones Transformación que resulta cuando se aplica una transformación a una figura y luego se le aplica otra transformación a su imagen.

evento compuesto Evento que consta de dos o más eventos simples.

interés compuesto Interés que se paga por el capital inicial y sobre el interés ganado en el pasado.

cono Una figura tridimensional con una circular base conectada por una superficie curva para un solo vértice.

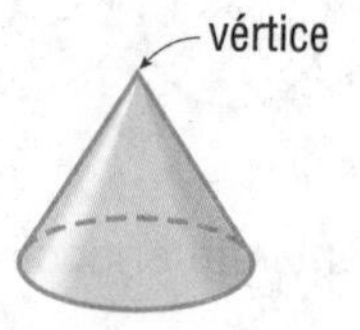

congruente Que tienen la misma medida; si una imagen puede obtenerse de otra por una secuencia de rotaciones, reflexiones o traslaciones.

constante Término sin variables.

constant of proportionality The constant ratio in a proportional linear relationship.

constante de proporcionalidad La razón constante en una relación lineal proporcional.

constant of variation A constant ratio in a direct variation.

constante de variación Razón constante en una relación de variación directa.

constant rate of change The rate of change between any two points in a linear relationship is the same or *constant.*

tasa constante de cambio La tasa de cambio entre dos puntos cualesquiera en una relación lineal permanece igual o *constante*.

continuous data Data that can take on any value. There is no space between data values for a given domain. Graphs are represented by solid lines.

datos continuos Datos que pueden tomar cualquier valor. No hay espacio entre los valores de los datos para un dominio dado. Las gráficas se representan con rectas sólidas.

convenience sample A sample which includes members of the population that are easily accessed.

muestra de conveniencia Muestra que incluye miembros de una población fácilmente accesibles.

converse The converse of a theorem is formed when the parts of the theorem are reversed. The converse of the Pythagorean Theorem can be used to test whether a triangle is a right triangle. If the sides of the triangle have lengths *a, b,* and *c,* such that $c^2 = a^2 + b^2$, then the triangle is a right triangle.

recíproco El recíproco de un teorema se forma cuando se invierten las partes del teorema. El recíproco del teorema de Pitágoras puede usarse para averiguar si un triángulo es un triángulo rectángulo. Si las longitudes de los lados de un triángulo son *a, b* y *c,* tales que $c^2 = a^2 + b^2$, entonces el triángulo es un triángulo rectángulo.

coordinate plane A coordinate system in which a horizontal number line and a vertical number line intersect at their zero points.

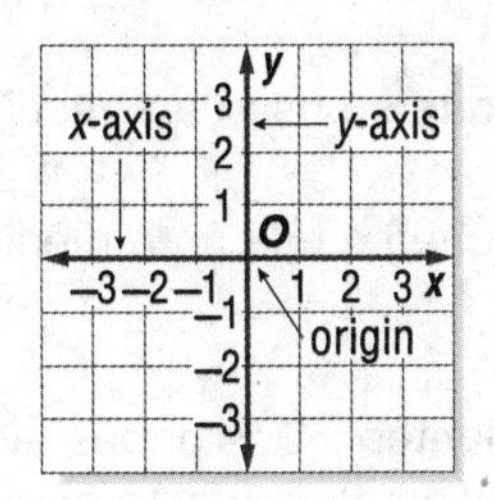

plano de coordenadas Sistema de coordenadas en que una recta numérica horizontal y una recta numérica vertical se intersecan en sus puntos cero.

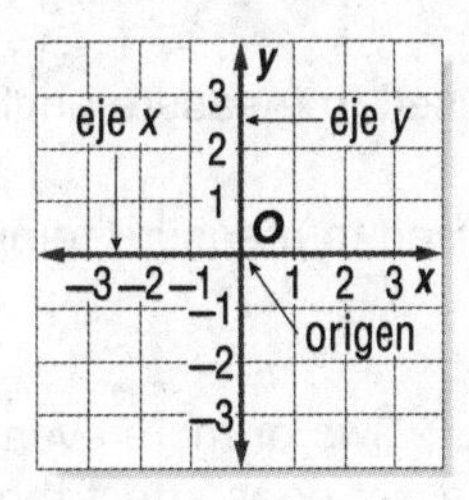

coplanar Lines that lie in the same plane.

coplanario Rectas que yacen en el mismo plano.

corresponding angles Angles that are in the same position on two parallel lines in relation to a transversal.

ángulos correspondientes Ángulos que están en la misma posición sobre dos rectas paralelas en relación con la transversal.

corresponding parts Parts of congruent or similar figures that match.

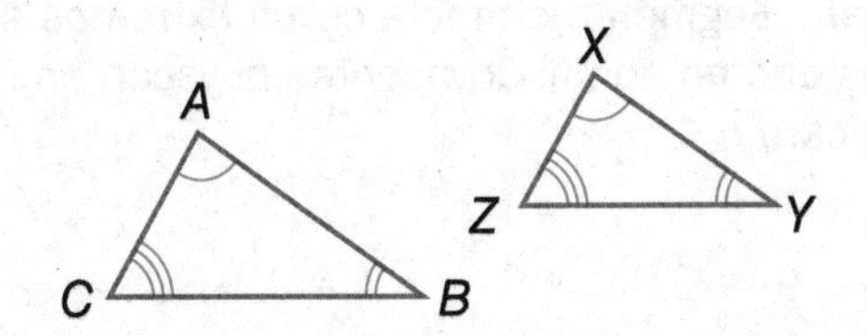

partes correspondientes Partes de figuras congruentes o semejantes que coinciden.

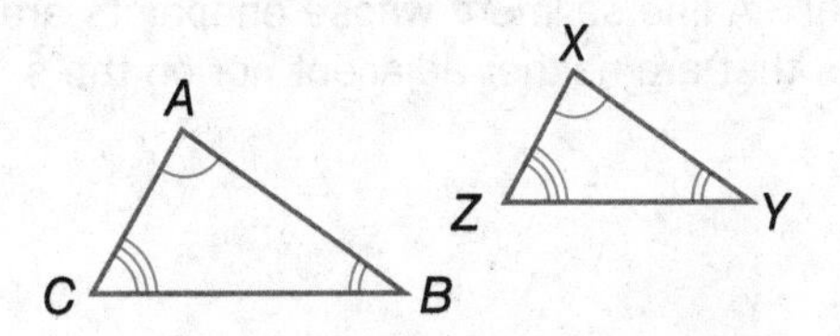

counterexample A statement or example that shows a conjecture is false.

contraejemplo Ejemplo o enunciado que demuestra que una conjetura es falsa.

cross section The intersection of a solid and a plane.

sección transversal Intersección de un sólido y un plano.

cube root One of three equal factors of a number. If $a^3 = b$, then a is the cube root of b. The cube root of 64 is 4 since $4^3 = 64$.

raíz cúbica Uno de tres factores iguales de un número. Si $a^3 = b$, entonces a es la raíz cúbica de b. La raíz cúbica de 64 es 4, dado que $4^3 = 64$.

cylinder A three-dimensional figure with two parallel congruent circular bases connected by a curved surface.

cilindro Una figura tridimensional con dos paralelas congruentes circulares bases conectados por una superficie curva.

Dd

deductive reasoning A system of reasoning that uses facts, rules, definitions, or properties to reach logical conclusions.

razonamiento deductivo Sistema de razonamiento que emplea hechos, reglas, definiciones o propiedades para obtener conclusions lógicas.

defining a variable Choosing a variable and a quantity for the variable to represent in an expression or equation.

definir una variable El elegir una variable y una cantidad que esté representada por la variable en una expresión o en una ecuación.

degree A unit used to measure angles.

grado Unidad que se usa para medir ángulos.

degree A unit used to measure temperature.

grado Unidad que se usa para medir la temperatura.

dependent events Two or more events in which the outcome of one event does affect the outcome of the other event or events.

eventos dependientes Dos o más eventos en que el resultado de uno de ellos afecta el resultado de los otros eventos.

dependent variable The variable in a relation with a value that depends on the value of the independent variable.

variable dependiente La variable en una relación cuyo valor depende del valor de la variable independiente.

derived unit A unit that is derived from a measurement system base unit, such as length, mass, or time.

unidad derivada Unidad derivada de una unidad básica de un sistema de medidas como por ejemplo, la longitud, la masa o el tiempo.

diagonal A line segment whose endpoints are vertices that are neither adjacent nor on the same face.

diagonal Segmento de recta cuyos extremos son vértices que no son ni adyacentes ni yacen en la misma cara.

diameter The distance across a circle through its center.

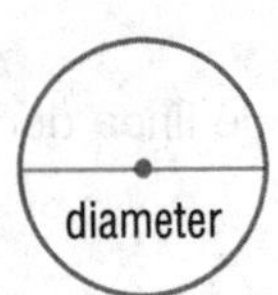

diámetro La distancia a través de un círculo pasando por el centro.

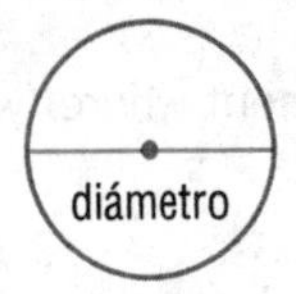

dilation A transformation that enlarges or reduces a figure by a scale factor.

homotecia Transformación que produce la ampliación o reducción de una imagen por un factor de escala.

dimensional analysis The process of including units of measurement when you compute.

análisis dimensional Proceso que incorpora las unidades de medida al hacer cálculos.

direct variation A relationship between two variable quantities with a constant ratio.

variación directa Relación entre dos cantidades variables con una razón constante.

discount The amount by which a regular price is reduced.

descuento La cantidad de reducción del precio normal.

discrete data Data with space between possible data values. Graphs are represented by dots.

datos discretos Datos con espacios entre posibles valores de datos. Las gráficas están representadas por puntos.

disjoint events Events that cannot happen at the same time.

eventos disjuntos Eventos que no pueden ocurrir al mismo tiempo.

Distance Formula The distance d between two points with coordinates (x_1, y_1) and (x_2, y_2) is given by the formula

$$d = \sqrt{(x_1 - x_2)^2 + (y_1 - y_2)^2}.$$

fórmula de la distancia La distancia d entre dos puntos con coordenadas (x_1, y_1) and (x_2, y_2) viene dada por la fórmula

$$d = \sqrt{(x_1 - x_2)^2 + (y_1 - y_2)^2}.$$

distribution A way to show the arrangement of data values.

distribución Una manera de mostrar la agrupación de valores.

Distributive Property To multiply a sum by a number, multiply each addend by the number outside the parentheses.

$$5(x + 3) = 5x + 15$$

propiedad distributiva Para multiplicar una suma por un número, multiplica cada sumando por el número fuera de los paréntesis.

$$5(x + 3) = 5x + 15$$

Division Property of Equality If you divide each side of an equation by the same nonzero number, the two sides remain equal.

propiedad de división de la igualdad Si cada lado de una ecuación se divide entre el mismo número no nulo, los dos lados permanecen iguales.

domain The set of x-coordinates in a relation.

dominio Conjunto de coordenadas x en una relación.

double box plot Two box plots graphed on the same number line.

doble diagrama de puntos Dos diagramas de caja sobre la misma recta numérica.

Ee

edge The line segment where two faces of a polyhedron intersect.

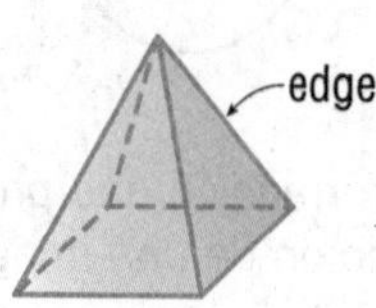

arista El segmento de línea donde se cruzan dos caras de un poliedro.

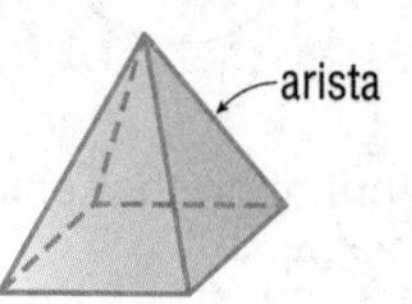

equation A mathematical sentence stating that two quantities are equal.

ecuación Enunciado matemático que establece que dos cantidades son iguales.

equiangular A polygon in which all angles are congruent.

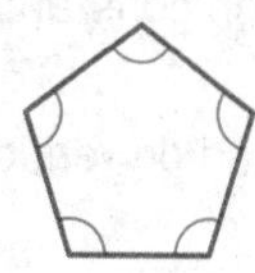

equiangular Polígono en el cual todos los ángulos son congruentes.

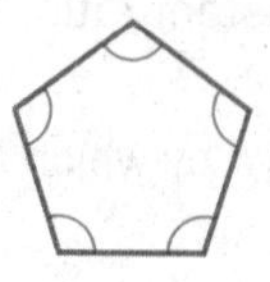

equilateral triangle A triangle with three congruent sides.

triángulo equilátero Triángulo con tres lados congruentes.

equivalent expressions Expressions that have the same value regardless of the value(s) of the variable(s).

expresiones equivalentes Expresiones que poseen el mismo valor, sin importar los valores de la(s) variable(s).

event An outcome is a possible result.

evento Un resultado posible.

experimental probability An estimated probability based on the relative frequency of positive outcomes occurring during an experiment.

probabilidad experimental Probabilidad estimada que se basa en la frecuencia relativa de los resultados positivos que ocurren durante un experimento.

exponent In a power, the number of times the base is used as a factor. In 10^3, the exponent is 3.

exponente En una potencia, el número de veces que la base se usa como factor. En 10^3, el exponente es 3.

exponential function A nonlinear function in which the base is a constant and the exponent is an independent variable.

función exponencial Función no lineal en la cual la base es una constante y el exponente es una variable independiente.

exterior angles The four outer angles formed by two lines cut by a transversal.

ángulo externo Los cuatro ángulos exteriores que se forman cuando una transversal corta dos rectas.

Ff

face A flat surface of a polyhedron.

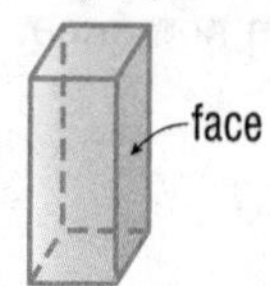

cara Una superficie plana de un poliedro.

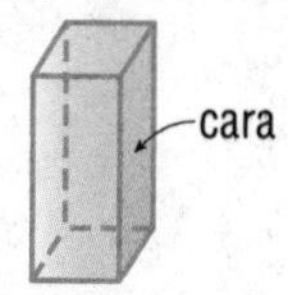

Glossary/Glosario

fair game A game where each player has an equally likely chance of winning.

juego justo Juego donde cada jugador tiene igual posibilidad de ganar.

five-number summary A way of characterizing a set of data that includes the minimum, first quartile, median, third quartile, and the maximum.

resumen de los cinco números Una manera de caracterizar un conjunto de datos que incluye el mínimo, el primer cuartil, la mediana, el tercer cuartil y el máximo.

formal proof A two-column proof containing statements and reasons.

demostración formal Demostración endos columnas contiene enunciados y razonamientos.

function A relation in which each member of the domain (input value) is paired with exactly one member of the range (output value).

función Relación en la cual a cada elemento del dominio (valor de entrada) le corresponde exactamente un único elemento del rango (valor de salida).

function table A table organizing the domain, rule, and range of a function.

tabla de funciones Tabla que organiza la regla de entrada y de salida de una función.

Fundamental Counting Principle Uses multiplication of the number of ways each event in an experiment can occur to find the number of possible outcomes in a sample space.

principio fundamental de contar Método que usa la multiplicación del número de maneras en que cada evento puede ocurrir en un experimento, para calcular el número de resultados posibles en un espacio muestral.

geometric sequence A sequence in which each term after the first is found by multiplying the previous term by a constant.

sucesión geométrica Sucesión en la cual cada término después del primero se determina multiplicando el término anterior por una constante.

half-plane The part of the coordinate plane on one side of the boundary.

semiplano Parte del plano de coordenadas en un lado de la frontera.

hemisphere One of two congruent halves of a sphere.

hemisferio Una de dos mitades congruentes de una esfera.

hypotenuse The side opposite the right angle in a right triangle.

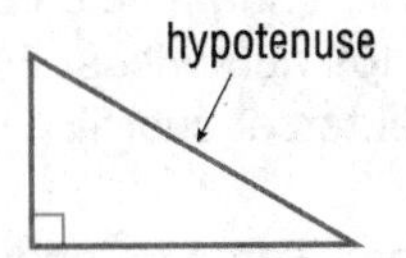

hipotenusa El lado opuesto al ángulo recto de un triángulo rectángulo.

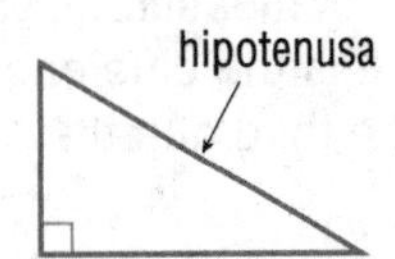

identity An equation that is true for every value for the variable.

identidad Ecuación que es verdad para cada valor de la variable.

image The resulting figure after a transformation.

independent events Two or more events in which the outcome of one event does not affect the outcome of the other event(s).

independent variable The variable in a function with a value that is subject to choice.

indirect measurement A technique using properties of similar polygons to find distances or lengths that are difficult to measure directly.

inductive reasoning Reasoning that uses a number of specific examples to arrive at a plausible generalization or prediction. Conclusions arrived at by inductive reasoning lack the logical certainty of those arrived at by deductive reasoning.

inequality A mathematical sentence that contains $<, >, \neq, \leq$, or $\geq$.

inscribed angle An angle that has its vertex on the circle. Its sides contain chords of the circle.

informal proof A paragraph proof.

interest The amount of money paid or earned for the use of money.

interior angle An angle inside a polygon.

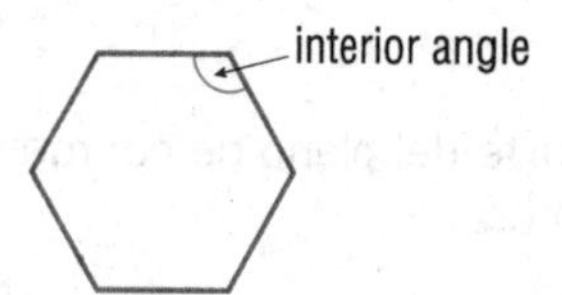

interior angles The four inside angles formed by two lines cut by a transversal.

interquartile range A measure of variation in a set of numerical data. It is the difference between the first quartile and the third quartile.

inverse operations Pairs of operations that undo each other. Addition and subtraction are inverse operations. Multiplication and division are inverse operations.

irrational number A number that cannot be expressed as the quotient $\frac{a}{b}$, where a and b are integers and $b \neq 0$.

imagen Figura que resulta después de una transformación.

eventos independientes Dos o más eventos en los cuales el resultado de un evento no afecta el resultado de los otros eventos.

variable independiente Variable en una función cuyo valor está sujeto a elección.

medición indirecta Técnica que usa las propiedades de polígonos semejantes para calcular distancias o longitudes difíciles de medir directamente.

razonamiento inductivo Razonamiento que usa varios ejemplos especificos para lograr una generalización o una predicción plausible. Las conclusions obtenidas por razonamiento inductivo carecen de la certeza lógica de aquellas obtenidas por razonamiento deductivo.

desigualdad Enunciado matemático que contiene $<, >, \neq, \leq$, o $\geq$.

ángulo inscrito Ángulo cuyo vértice está en el círculo y cuyos lados contienen cuerdas del círculo.

demostración informal Demonstración en forma de párrafo.

interés Cantidad que se cobra o se paga por el uso del dinero.

ángulo interno Ángulo dentro de un polígono.

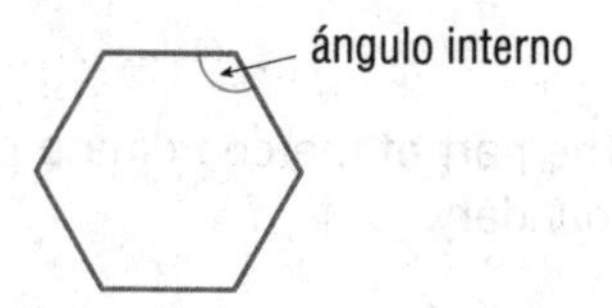

ángulo interno Los cuatro ángulos internos formados por dos rectas intersecadas por una transversal.

rango intercuartílico Una medida de la variación en un conjunto de datos numéricos. Es la diferencia entre el primer y el tercer cuartil.

peraciones inversas Pares de operaciones que se anulan mutuamente. La adición y la sustracción son operaciones inversas. La multiplicación y la división son operaciones inversas.

números irracionales Número que no se puede expresar como el cociente $\frac{a}{b}$, donde a y b son enteros y $b \neq 0$.

sosceles triangle A triangle with at least two congruent sides.

triángulo isósceles Triángulo con por lo menos dos lados congruentes.

lateral area The sum of the areas of the lateral faces of a solid.

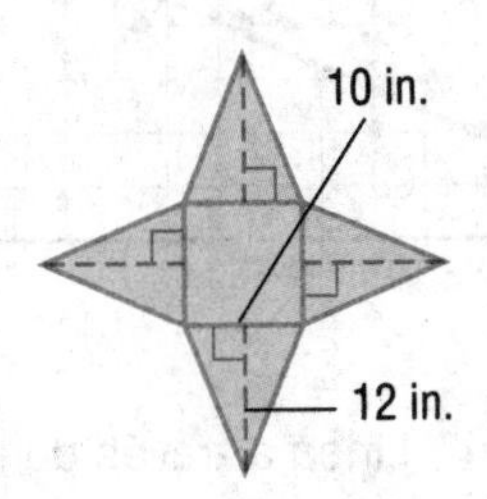

lateral area $= 4\left(\frac{1}{2} \times 10 \times 12\right) = 240$ square inches

área lateral La suma de las áreas de las caras laterales de un sólido.

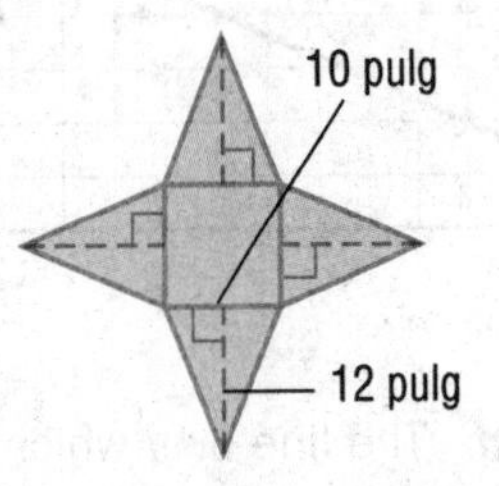

área lateral $= 4\left(\frac{1}{2} \times 10 \times 12\right) = 240$ pulgadas cuadradas

lateral face Any flat surface that is not a base.

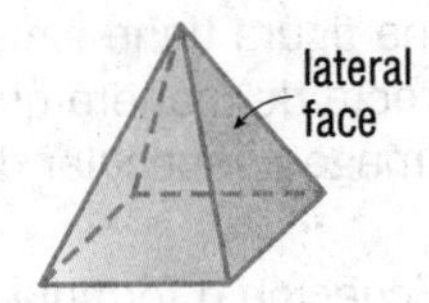

cara lateral Cualquier superficie plana que no es la base.

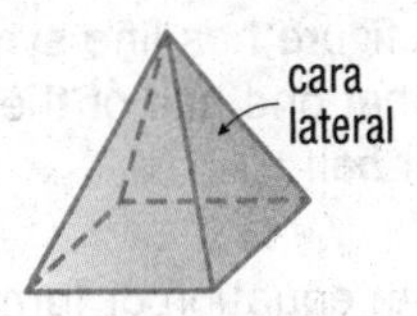

legs The two sides of a right triangle that form the right angle.

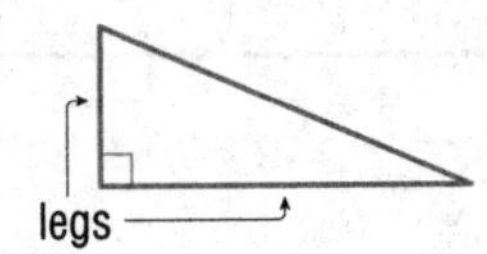

catetos Los dos lados de un triángulo rectángulo que forman el ángulo recto.

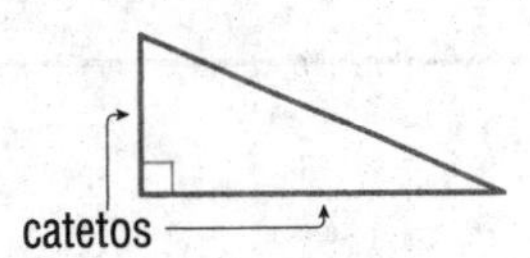

like fractions Fractions that have the same denominators.

fracciones semejantes Fracciones que tienen el mismo denominador.

like terms Terms that contain the same variable(s) to the same powers.

términos semejantes Términos que contienen la misma variable o variables elevadas a la misma potencia.

linear To fall in a straight line.

lineal Que cae en una línea recta.

linear equation An equation with a graph that is a straight line.

ecuación lineal Ecuación cuya gráfica es una recta.

linear function A function in which the graph of the solutions forms a line.

función lineal Función en la cual la gráfica de las soluciones forma un recta.

linear relationship A relationship that has a straight-line graph.

relación lineal Relación cuya gráfica es una recta.

line of best fit A line that is very close to most of the data points in a scatter plot.

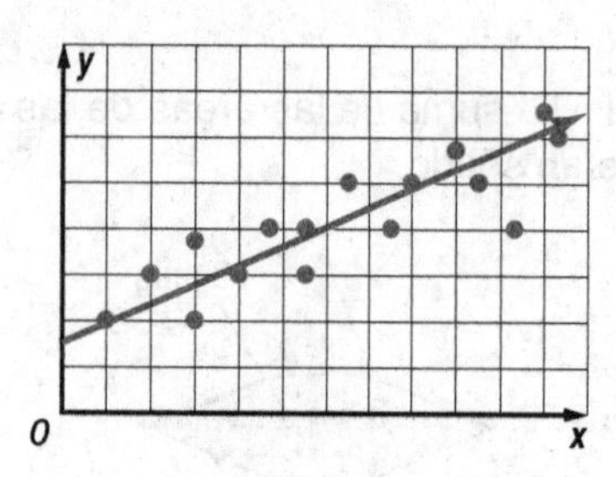

recta de mejor ajuste Recta que más se acerca a la mayoría de puntos de los datos en un diagrama de dispersión.

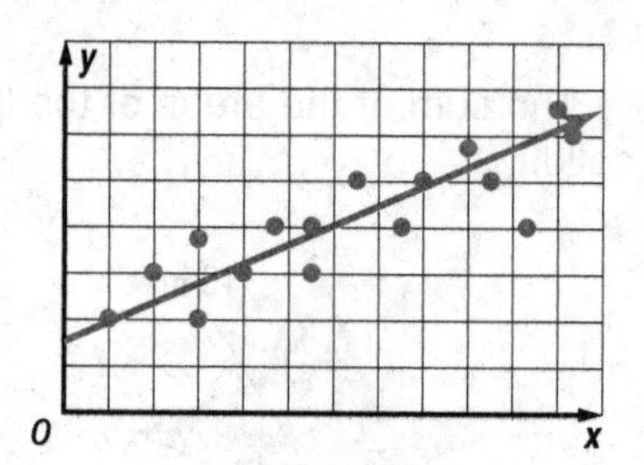

line of reflection The line over which a figure is reflected.

línea de reflexión Línea a través de la cual se refleja una figura.

line of symmetry Each half of a figure is a mirror image of the other half when a line of symmetry is drawn.

eje de simetría Recta que divide una figura en dos mitades especulares.

line symmetry A figure has line symmetry if a line can be drawn so that one half of the figure is a mirror image of the other half.

simetría lineal Una figura tiene simetría lineal si se puede trazar una recta de manera que una mitad de la figura sea una imagen especular de la otra mitad.

literal equation An equation or formula that has more than one variable.

ecuación literal Ecuación o fórmula con más de una variable.

Mm

markup The amount the price of an item is increased above the price the store paid for the item.

margen de utilidad Cantidad de aumento en el precio de un artículo por encima del precio que paga la tienda por dicho artículo.

mean The sum of the data divided by the number of items in the set.

media La suma de datos dividida entre el número total de artículos.

mean absolute deviation The average of the absolute values of differences between the mean and each value in a data set.

desviación media absoluta El promedio de los valores absolutos de diferencias entre el medio y cada valor de un conjunto de datos.

measures of center Numbers that are used to describe the center of a set of data.These measures include the mean, median, and mode.

medidas del centro Números que describen el centro de un conjunto de datos. Estas medidas incluyen la media, la mediana y la moda.

measures of variation Numbers used to describe the distribution or spread of a set of data.

medidas de variación Números que se usan para describir la distribución o separación de un conjunto de datos.

median A measure of center in a set of numerical data. The median of a list of values is the value appearing at the center of a sorted version of the list—or the mean of the two central values, if the list contains an even number of values.

mediana Una medida del centro en un conjunto de datos numéricos. La mediana de una lista de valores es el valor que aparece en el centro de una versión ordenada de la lista, o la media de los dos valores centrales si la lista contiene un número par de valores.

mode The number(s) or item(s) that appear most often in a set of data.

moda El número(s) o artículo(s) que aparece con más frecuencia en un conjunto de datos.

monomial A number, a variable, or a product of a number and one or more variables.

monomio Un número, una variable o el producto de un número por una o más variables.

Multiplication Property of Equality If you multiply each side of an equation by the same number, the two sides remain equal.

propiedad de multiplicación de la igualdad Si cada lado de una ecuación se multiplica por el mismo número, los lados permanecen iguales.

multiplicative inverses Two numbers with a product of 1. The multiplicative inverse of $\frac{2}{3}$ is $\frac{3}{2}$.

inversos multiplicativo Dos números cuyo producto es 1. El inverso multiplicativo de $\frac{2}{3}$ es $\frac{3}{2}$.

Nn

net A two-dimensional pattern of a three-dimensional figure.

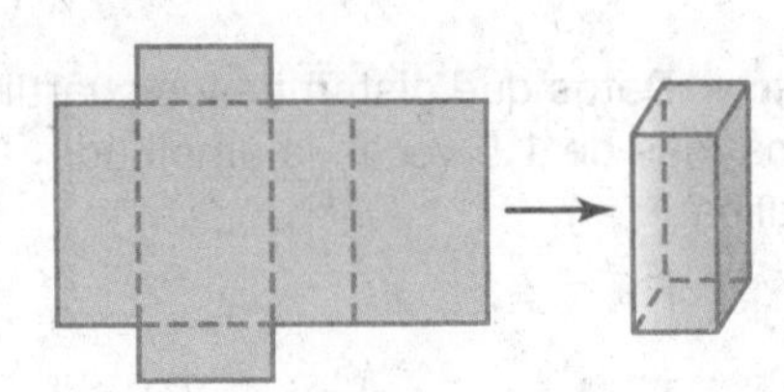

red Patrón bidimensional de una figura tridimensional.

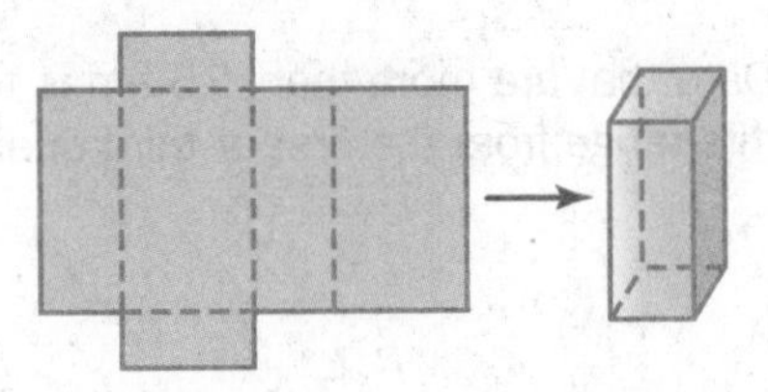

nonlinear function A function whose rate of change is not constant. The graph of a nonlinear function is not a straight line.

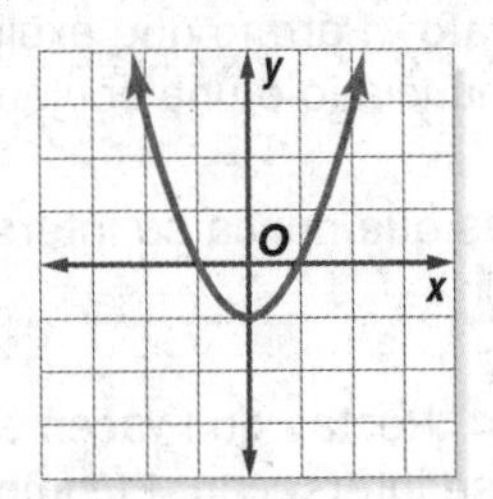

función no lineal Función cuya tasa de cambio no es constante. La gráfica de una función no lineal no es una recta.

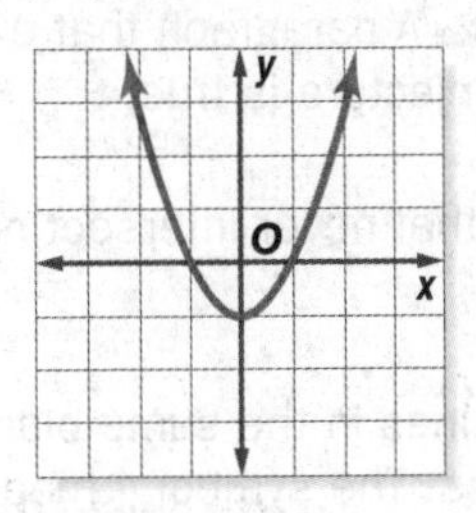

null set The empty set.

conjunto nulo El conjunto vacío.

obtuse angle An angle whose measure is between 90° and 180°.

ángulo obtuso Ángulo cuya medida está entre 90° y 180°.

obtuse triangle A triangle with one obtuse angle.

triángulo obtusángulo Triángulo con un ángulo obtuso.

ordered pair A pair of numbers used to locate a point in the coordinate plane. The ordered pair is written in this form: (x-coordinate, y-coordinate).

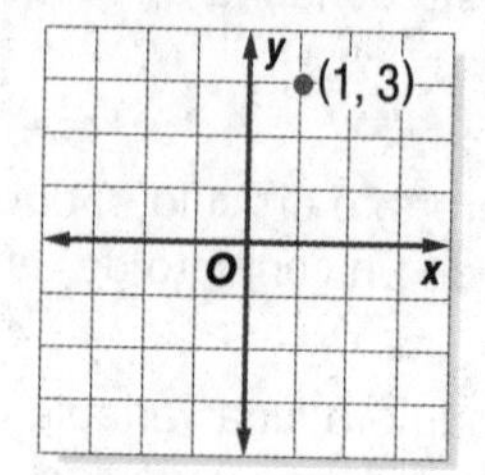

origin The point of intersection of the x-axis and y-axis in a coordinate plane.

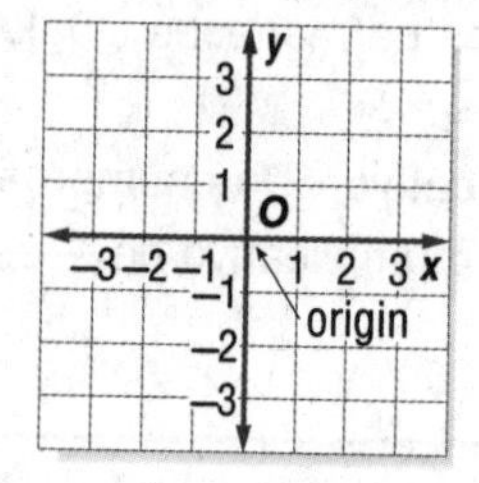

outcome One possible result of a probability event. For example, 4 is an outcome when a number cube is rolled.

outlier Data that are more than 1.5 times the interquartile range from the first or third quartiles.

par ordenado Par de números que se utiliza para ubicar un punto en un plano de coordenadas. Se escribe de la siguiente forma: (coordenada x, coordenada y).

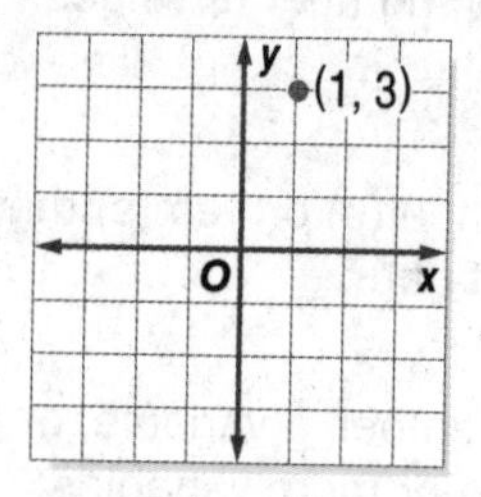

origen Punto en que el eje x y el eje y se intersecan en un plano de coordenadas.

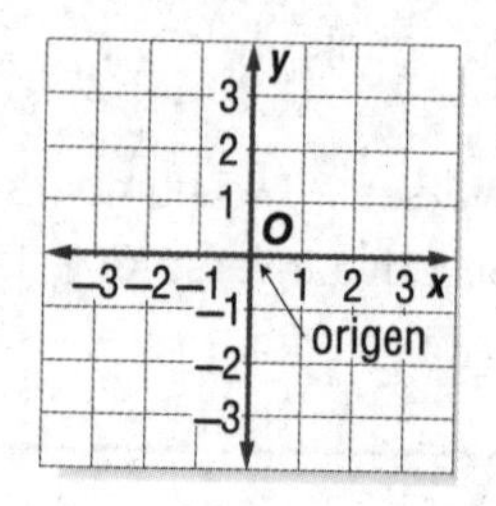

resultado Una consecuencia posible de un evento de probabilidad. Por ejemplo, 4 es un resultado posible al lanzar un cubo numérico.

valor atípico Datos que distan de los cuartiles respectivos más de 1.5 veces la amplitud intercuartílica.

paragraph proof A paragraph that explains why a statement or conjecture is true.

parallel Lines that never intersect no matter how far they extend.

parallel lines Lines in the same plane that never intersect or cross. The symbol || means parallel.

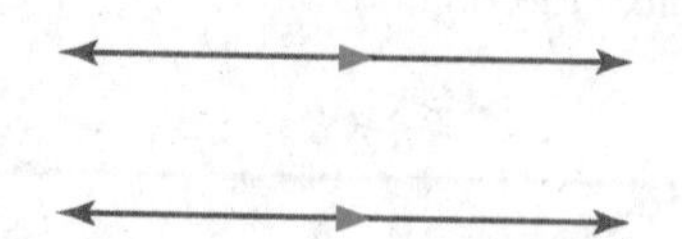

parallelogram A quadrilateral with both pairs of opposite sides parallel and congruent.

prueba por párrafo Párrafo que explica por qué es verdadero un enunciado o una conjetura.

paralelo Rectas que nunca se intersecan sea cual sea su extensión.

rectas paralelas Rectas que yacen en un mismo plano y que no se intersecan. El símbolo || significa paralela a.

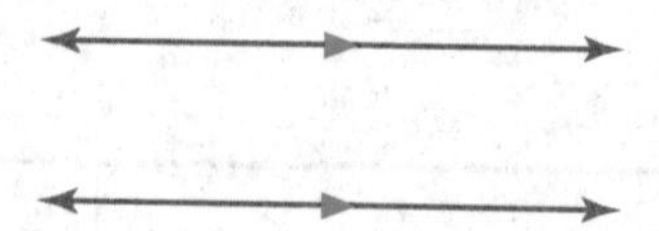

paralelogramo Cuadrilátero con ambos pares de lados opuestos, paralelos y congruentes.

percent equation An equivalent form of a percent proportion in which the percent is written as a decimal.

part = percent • whole

ecuación porcentual Forma equivalente de proporción porcentual en la cual el por ciento se escribe como un decimal.

parte = por ciento • entero

percent of change A ratio that compares the change in quantity to the original amount.

$$\text{percent of change} = \frac{\text{amount of change}}{\text{original amount}}$$

porcentaje de cambio Razón que compara el cambio en una cantidad a la cantidad original.

$$\text{procentaje de cambio} = \frac{\text{cantidad de cambio}}{\text{cantidad original}}$$

percent of decrease When the percent of change is negative.

porcentaje de disminución Cuando el porcentaje de cambio es negativo.

percent of increase When the percent of change is positive.

porcentaje de aumento Cuando el porcentaje de cambio es positivo.

percent proportion Compares part of a quantity to the whole quantity using a percent.

$$\frac{\text{part}}{\text{whole}} = \frac{\text{percent}}{100}$$

proporción porcentual Compara parte de una cantidad con la cantidad total mediante un por ciento.

$$\frac{\text{parte}}{\text{entero}} = \frac{\text{por ciento}}{100}$$

perfect cube A rational number whose cube root is a whole number. 27 is a perfect cube because its cube root is 3.

cubo perfecto Número racional cuya raíz cúbica es un número entero. 27 es un cubo perfecto porque su raíz cúbica es 3.

perfect square A rational number whose square root is a whole number. 25 is a perfect square because its square root is 5.

cuadrados perfectos Número racional cuya raíz cuadrada es un número entero. 25 es un cuadrado perfecto porque su raíz cuadrada es 5.

permutation An arrangement or listing in which order is important.

permutación Arreglo o lista donde el orden es importante.

perpendicular lines Two lines that intersect to form right angles.

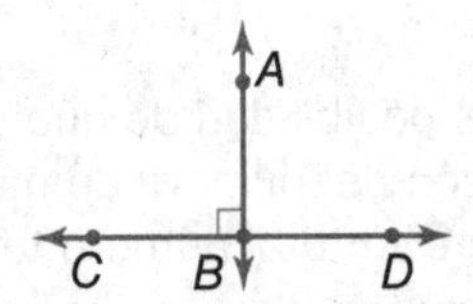

rectas perpendiculares Dos rectas que se intersecan formando ángulos rectos.

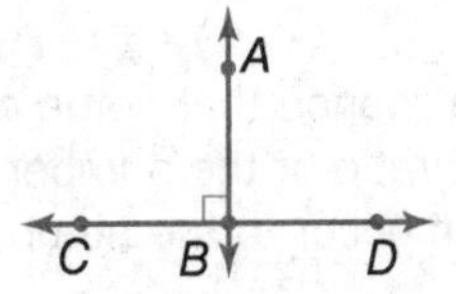

pi The ratio of the circumference of a circle to its diameter. The Greek letter π represents this number. The value of pi is always 3.1415926… .

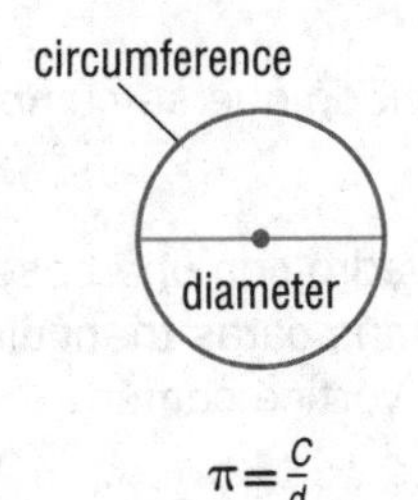

$$\pi = \frac{C}{d}$$

pi Razón de la circunferencia de un círculo al diámetro del mismo. La letra griega π representa este número. El valor de pi es siempre 3.1415926… .

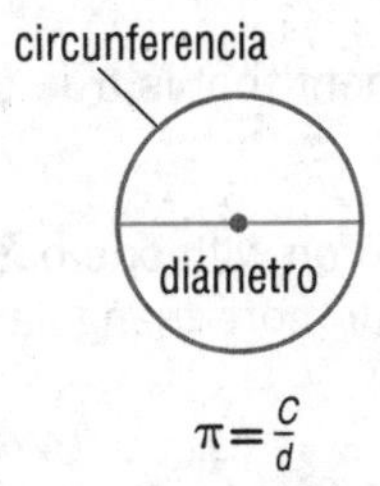

$$\pi = \frac{C}{d}$$

point-slope form An equation of the form $y - y_1 = m(x - x_1)$, where m is the slope and (x_1, y_1) is a given point on a nonvertical line.

forma punto-pendiente Ecuación de la forma $y - y_1 = m(x - x_1)$ donde m es la pendiente y (x_1, y_1) es un punto dado de una recta no vertical.

polygon A simple, closed figure formed by three or more line segments.

polígono Figura simple y cerrada formada por tres o más segmentos de recta.

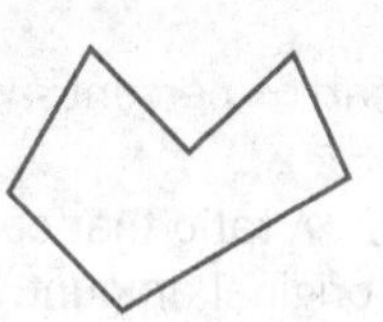

polyhedron A three-dimensional figure with faces that are polygons.

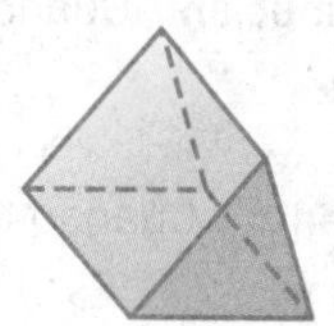

poliedro Una figura tridimensional con caras que son polígonos.

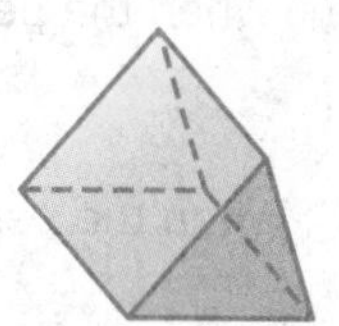

power A product of repeated factors using an exponent and a base. The power 7^3 is read *seven to the third power,* or *seven cubed.*

potencia Producto de factores repetidos con un exponente y una base. La potencia 7^3 se lee *siete a la tercera potencia* o *siete al cubo*.

precision The ability of a measurement to be consistently reproduced.

precisión Capacidad de una medida a ser reproducida consistentemente.

preimage The original figure before a transformation.

preimagen Figura original antes de una transformación.

principal The amount of money invested or borrowed.

capital Cantidad de dinero que se invierte o que se toma prestada.

prism A polyhedron with two parallel congruent faces called bases.

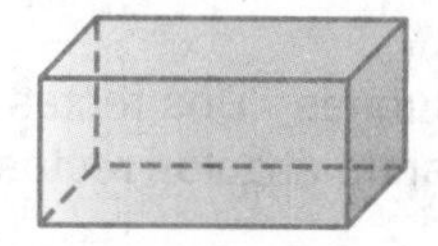

prisma Poliedro con dos caras congruentes y paralelas llamadas bases.

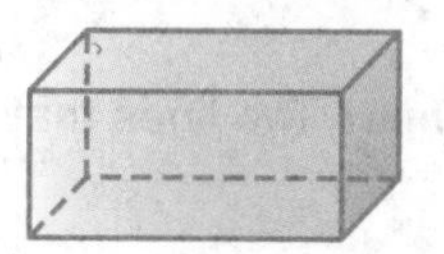

probability The chance that some event will happen. It is the ratio of the number of ways a certain event can occur to the number of possible outcomes.

probabilidad La posibilidad de que suceda un evento. Es la razón del número de maneras en que puede ocurrir un evento al número total de resultados posibles.

proof A logical argument in which each statement that is made is supported by a statement that is accepted as true.

prueba Argumento lógico en el cual cada enunciado hecho se respalda con un enunciado que se acepta como verdadero.

property A statement that is true for any numbers.

propiedad Enunciado que se cumple para cualquier número.

pyramid A polyhedron with one base that is a polygon and three or more triangular faces that meet at a common vertex.

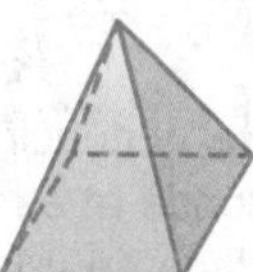

pirámide Un poliedro con una base que es un polígono y tres o más caras triangulares que se encuentran en un vértice común.

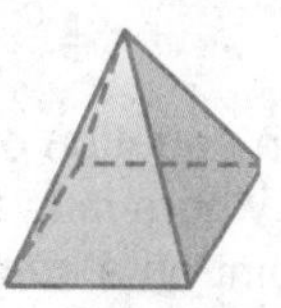

Pythagorean Theorem In a right triangle, the square of the length of the hypotenuse c is equal to the sum of the squares of the lengths of the legs a and b. $a^2 + b^2 = c^2$

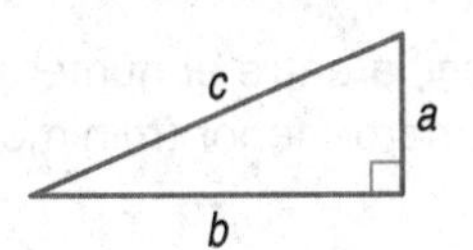

Qq

quadrants The four sections of the coordinate plane.

quadratic function A function in which the greatest power of the variable is 2.

quadrilateral A closed figure with four sides and four angles.

qualitative graph A graph used to represent situations that do not necessarily have numerical values.

quantitative data Data that can be given a numerical value.

quartiles Values that divide a set of data into four equal parts.

Rr

radical sign The symbol used to indicate a positive square root, $\sqrt{}$.

radius The distance from the center of a circle to any point on the circle.

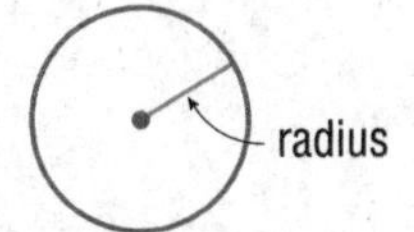

Teorema de Pitágoras En un triángulo rectángulo, el cuadrado de la longitud de la hipotenusa es igual a la suma de los cuadrados de las longitudes de los catetos. $a^2 + b^2 = c^2$

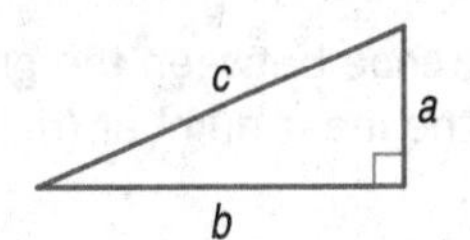

cuadrantes Las cuatro secciones del plano de coordenadas.

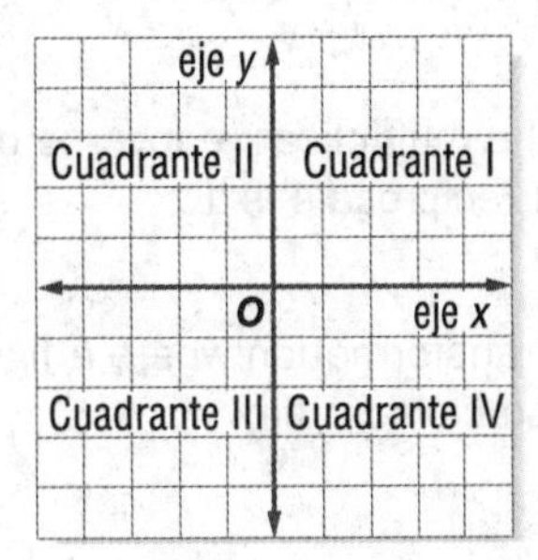

función cuadrática Función en la cual la potencia mayor de la variable es 2.

cuadrilátero Figura cerrada con cuatro lados y cuatro ángulos.

gráfica cualitativa Gráfica que se usa para representar situaciones que no tienen valores numéricos necesariamente.

datos cualitativos Datos que se pueden dar un valor numérico.

cuartiles Valores que dividen un conjunto de datos en cuatro partes iguales.

signo radical Símbolo que se usa para indicar una raíz cuadrada no positiva, $\sqrt{}$.

radio Distancia desde el centro de un círculo hasta cualquier punto del mismo.

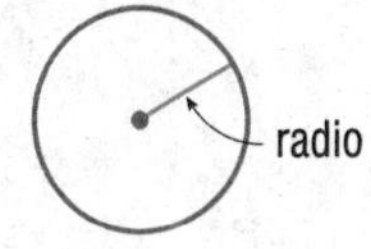

random Outcomes occur at random if each outcome is equally likely to occur.

azar Los resultados ocurren al azar si todos los resultados son equiprobables.

range The set of *y*-coordinates in a relation.

rango Conjunto de coordenadas y en una relación.

range The difference between the greatest number (maximum) and the least number (minimum) in a set of data.

rango La diferencia entre el número mayor (máximo) y el número menor (mínimo) en un conjunto de datos.

rational number Numbers that can be written as the ratio of two integers in which the denominator is not zero. All integers, fractions, mixed numbers, and percents are rational numbers.

número racional Números que pueden escribirse como la razón de dos enteros en los que el denominador no es cero. Todos los enteros, fracciones, números mixtos y porcentajes son números racionales.

real numbers The set of rational numbers together with the set of irrational numbers.

número real El conjunto de números racionales junto con el conjunto de números irracionales.

reciprocals The multiplicative inverse of a number. The product of reciprocals is 1.

recíproco El inverso multiplicativo de un número. El producto de recíprocos es 1.

reflection A transformation where a figure is flipped over a line. Also called a flip.

reflexión Transformación en la cual una figura se voltea sobre una recta. También se conoce como simetría de espejo.

regular polygon A polygon that is equilateral and equiangular.

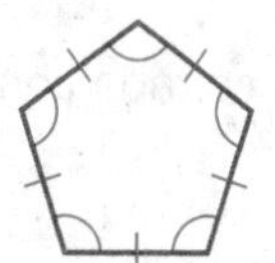

polígono regular Polígono equilátero y equiangular.

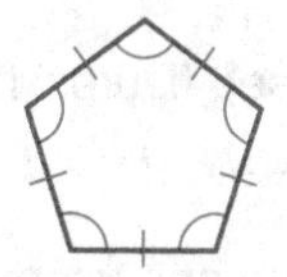

regular pyramid A pyramid whose base is a regular polygon.

pirámide regular Pirámide cuya base es un polígono regular.

relation Any set of ordered pairs.

relación Cualquier conjunto de pares ordenados.

relative frequency The ratio of the number of experimental successes to the total number of experimental attempts.

frecuencia relativa Razón del número de éxitos experimentales al número total de intentos experimentales.

remote interior angles The angles of a triangle that are not adjacent to a given exterior angle.

ángulos internos no adyacentes Ángulos de un triángulo que no son adya centes a un ángulo exterior dado.

repeating decimal Decimal form of a rational number.

decimal periódico Forma decimal de un número racional.

rhombus A parallelogram with four congruent sides.

rombo Paralelogramo con cuatro lados congruentes.

right angle An angle whose measure is exactly 90°.

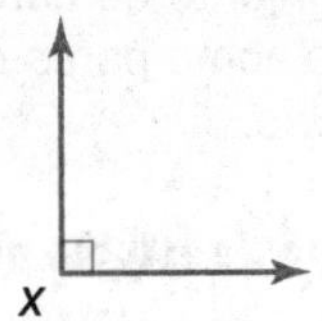

right triangle A triangle with one right angle.

rise The vertical change between any two points on a line.

rotation A transformation in which a figure is turned about a fixed point.

rotational symmetry A type of symmetry a figure has if it can be rotated less than 360° about its center and still look like the original.

run The horizontal change between any two points on a line.

ángulo recto Ángulo que mide exactamente 90°.

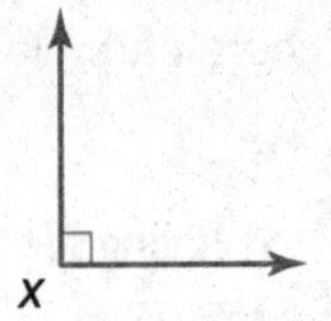

triángulo rectángulo Triángulo con un ángulo recto.

elevación El cambio vertical entre cualquier par de puntos en una recta.

rotación Transformación en la cual una figura se gira alrededor de un punto fijo.

simetría rotacional Tipo de simetría que tiene una figura si se puede girar menos que 360° en torno al centro y aún sigue viéndose como la figura original.

carrera El cambio horizontal entre cualquier par de puntos en una recta.

Ss

sales tax An additional amount of money charged on certain goods and services.

sample A randomly-selected group chosen for the purpose of collecting data.

sample space The set of all possible outcomes of a probability experiment.

scale factor The ratio of the lengths of two corresponding sides of two similar polygons.

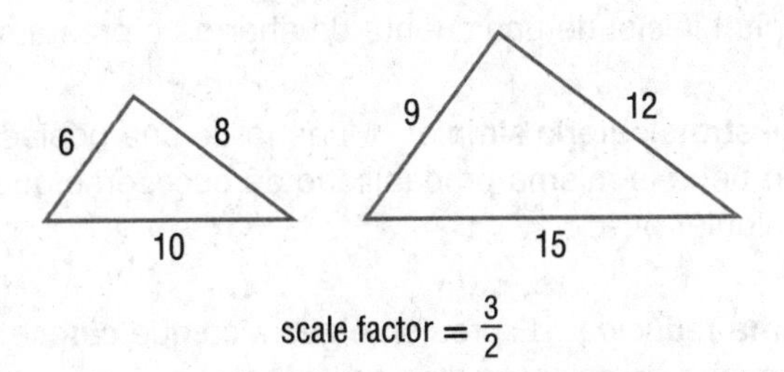

scalene triangle A triangle with no congruent sides.

impuesto sobre las ventas Cantidad de dinero adicional que se cobra por ciertos artículos y servicios.

muestra Subconjunto de una población que se usa con el propósito de recoger datos.

espacio muestral Conjunto de todos los resultados posibles de un experimento de probabilidad.

factor de escala La razón de las longitudes de dos lados correspondientes de dos polígonos semejantes.

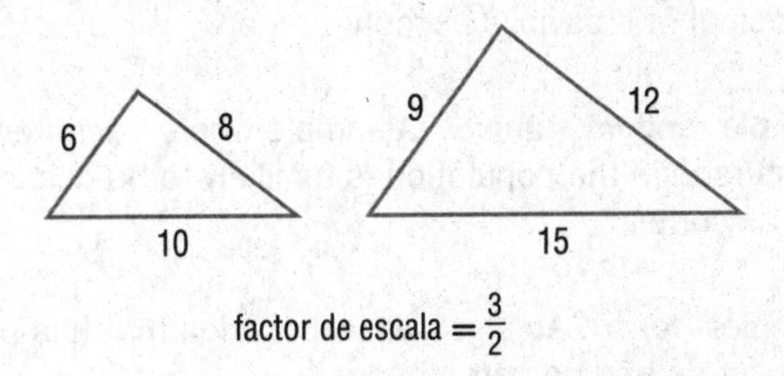

triángulo escaleno Triángulo sin lados congruentes.

scatter plot A graph that shows the relationship between a data set with two variables graphed as ordered pairs on a coordinate plane.

diagrama de dispersión Gráfica que muestra la relación entre un conjunto de datos con dos variables graficadas como pares ordenados en un plano de coordenadas.

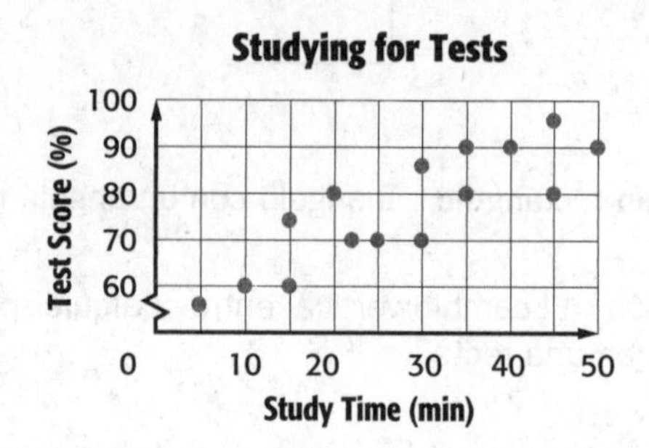

scientific notation A compact way of writing numbers with absolute values that are very large or very small. In scientific notation, 5,500 is 5.5×10^3.

notación científica Manera abreviada de escribir números con valores absolutos que son muy grandes o muy pequeños. En notación científica, 5,500 es 5.5×10^3.

selling price The amount the customer pays for an item.

precio de venta Cantidad de dinero que paga un consumidor por un artículo.

semicircle An arc measuring 180°.

semicírculo Arco que mide 180°.

sequence An ordered list of numbers, such as 0, 1, 2, 3 or 2, 4, 6, 8.

sucesión Lista ordenada de números, tales como 0, 1, 2, 3 o 2, 4, 6, 8.

similar If one image can be obtained from another by a sequence of transformations and dilations.

similar Si una imagen puede obtenerse de otra mediante una secuencia de transformaciones y dilataciones.

similar polygons Polygons that have the same shape.

polígonos semejantes Polígonos con la misma forma.

similar solids Solids that have exactly the same shape, but not necessarily the same size.

sólidos semejantes Sólidos que tienen exactamente la misma forma, pero no necesariamente el mismo tamaño.

simple interest Interest paid only on the initial principal of a savings account or loan.

interés simple Interés que se paga sólo sobre el capital inicial de una cuenta de ahorros o préstamo.

simple random sample A sample where each item or person in the population is as likely to be chosen as any other.

muestra aleatoria simple Muestra de una población que tiene la misma probabilidad de escogerse que cualquier otra.

simplest form An algebraic expression that has no like terms and no parentheses.

forma reducida Expresión algebraica que carece de términos semejantes y de paréntesis.

simplify To perform all possible operations in an expression.

simplificar Realizar todas las operaciones posibles en una expresión.

simulation An experiment that is designed to model the action in a given situation.

simulacro Un experimento diseñado para modelar la acción en una situación dada.

slant height The altitude or height of each lateral face of a pyramid.

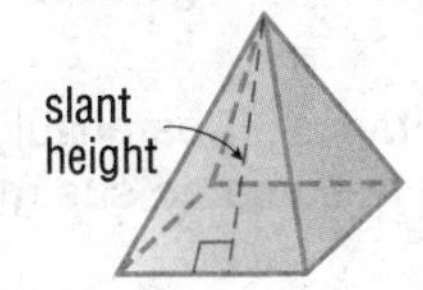

altura oblicua La longitud de la altura de cada cara lateral de una pirámide.

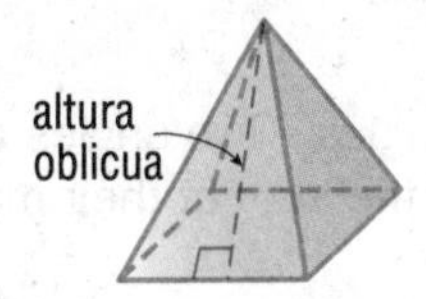

slope The rate of change between any two points on a line. The ratio of the rise, or vertical change, to the run, or horizontal change.

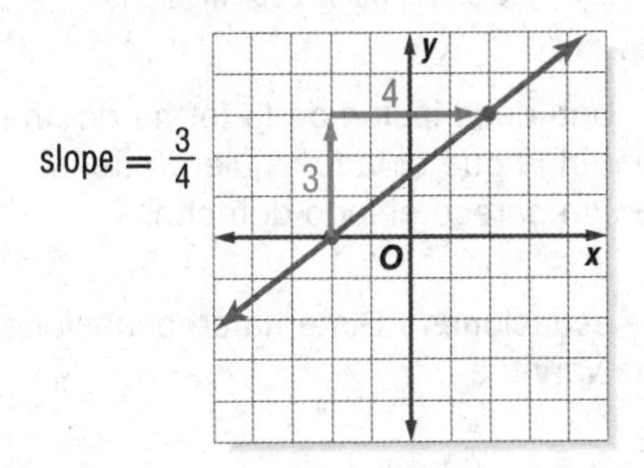

pendiente Razón de cambio entre cualquier par de puntos en una recta. La razón de la altura, o cambio vertical, a la carrera, o cambio horizontal.

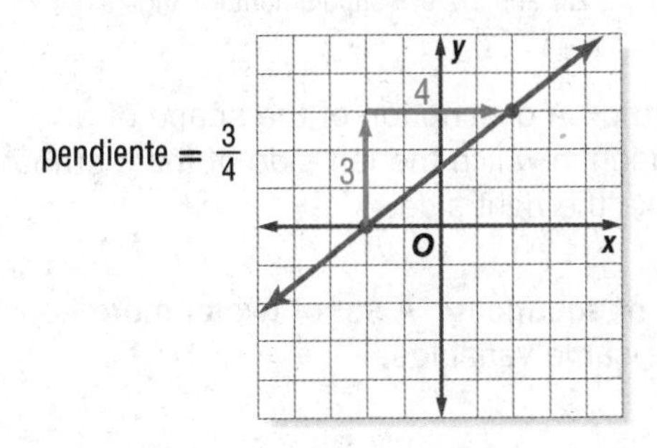

slope-intercept form An equation written in the form $y = mx + b$, where m is the slope and b is the y-intercept.

forma pendiente intersección Ecuación de la forma $y = mx + b$, donde m es la pendiente y b es la intersección y.

solid A three-dimensional figure formed by intersecting planes.

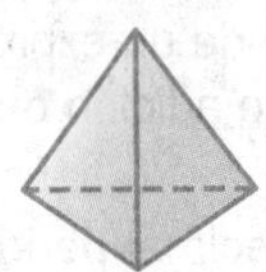

sólido Figura tridimensional formada por planos que se intersecan.

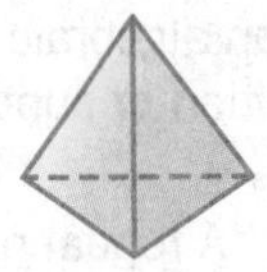

sphere The set of all points in space that are a given distance from a given point called the center.

esfera Conjunto de todos los puntos en el espacio que están a una distancia dada de un punto dado llamado centro.

square root One of the two equal factors of a number. If $a^2 = b$, then a is the square root of b. A square root of 144 is 12 since $12^2 = 144$.

raíz cuadrada Uno de dos factores iguales de un número. Si $a^2 = b$, la a es la raíz cuadrada de b. Una raíz cuadrada de 144 es 12 porque $12^2 = 144$.

standard deviation A measure of variation that describes how the data deviates from the mean of the data.

desviación estándar Una medida de variación que describe cómo los datos se desvía de la media de los datos.

standard form An equation written in the form $Ax + By = C$.

forma estándar Ecuación escrita en la forma $Ax + By = C$.

straight angle An angle whose measure is exactly 180°.

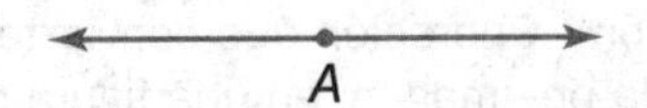

ángulo llano Ángulo que mide exactamente 180°.

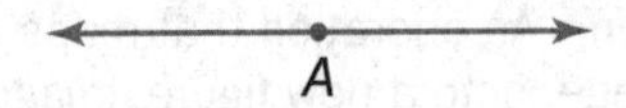

substitution An algebraic model that can be used to find the exact solution of a system of equations.

sustitución Modelo algebraico que se puede usar para calcular la solución exacta de un sistema de ecuaciones.

Subtraction Property of Equality If you subtract the same number from each side of an equation, the two sides remain equal.

propiedad de sustracción de la igualdad Si sustraes el mismo número de ambos lados de una ecuación, los dos lados permanecen iguales.

supplementary angles Two angles are supplementary if the sum of their measures is 180°.

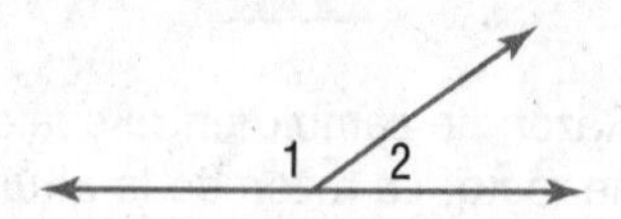

∠1 and ∠2 are supplementary angles.

ángulos suplementarios Dos ángulos son suplementarios si la suma de sus medidas es 180°.

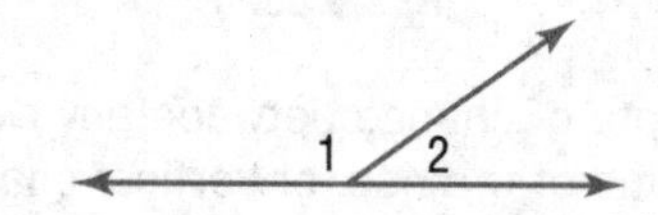

∠1 y ∠2 son ángulos suplementarios.

symmetric A description of the shape of a distribution in which the left side of the distribution looks like the right side.

simétrico Una descripción de la forma de una distribución en la que el lado izquierdo de la distribución se parece el lado derecho.

system of equations A set of two or more equations with the same variables.

sistema de ecuaciones Sistema de ecuaciones con las mismas variables.

Tt

term A number, a variable, or a product of numbers and variables.

término Un número, una variable o un producto de números y variables.

term Each part of an algebraic expression separated by an addition or subtraction sign.

término Cada parte de un expresión algebraica separada por un signo adición o un signo sustracción.

terminating decimal A repeating decimal where the repeating digit is zero.

decimal finito Un decimal periódico donde el dígito que se repite es cero.

theorem A statement or conjecture that can be proven.

teorema Un enunciado o conjetura que puede probarse.

theoretical probability Probability based on known characteristics or facts.

probabilidad teórica Probabilidad que se basa en características o hechos conocidos.

third quartile For a data set with median M, the third quartile is the median of the data values greater than M.

tercer cuartil Para un conjunto de datos con la mediana M, el tercer cuartil es la mediana de los valores mayores que M.

three-dimensional figure A figure with length, width, and height.

figura tridimensional Figura que tiene largo, ancho y alto.

total surface area The sum of the areas of the surfaces of a solid.

área de superficie total La suma del área de las superficies de un sólido.

transformation An operation that maps a geometric figure, preimage, onto a new figure, image.

transformación Operación que convierte una figura geométrica, la pre-imagen, en una figura nueva, la imagen.

translation A transformation that slides a figure from one position to another without turning.

traslación Transformación en la cual una figura se desliza de una posición a otra sin hacerla girar.

transversal A line that intersects two or more other lines.

transversal Recta que interseca dos o más rectas.

trapezoid A quadrilateral with exactly one pair of parallel sides.

trapecio Cuadrilátero con exactamente un par de lados paralelos.

tree diagram A diagram used to show the total number of possible outcomes in a probability experiment.

diagrama de árbol Diagrama que se usa para mostrar el número total de resultados posibles en un experimento de probabilidad.

triangle A figure formed by three line segments that intersect only at their endpoints.

triángulo Figura formada por tres segmentos de recta que se intersecan sólo en sus extremos.

two-column proof A formal proof that contains statements and reasons organized in two columns. Each step is called a statement, and the properties that justify each step are called reasons.

demostración de dos columnas Demonstración formal que contiene enunciados y razones organizadas en dos columnas. Cada paso se llama enunciado y las propiedades que lo justifican son las razones.

two-step equation An equation that contains two operations.

ecuación de dos pasos Ecuación que contiene dos operaciones.

two-step inequality An inequality that contains two operations.

desigualdad de dos pasos Desigualdad que contiene dos operaciones.

two-way table A table that shows data that pertain to two different categories.

tabla de doble entrada Una tabla que muestra datos que pertenecen a dos categorías diferentes.

Uu

unbiased sample A sample that is selected so that it is representative of the entire population.

muestra no sesgada Muestra que se selecciona de modo que sea representativa de la población entera.

unit rate/ratio A rate or ratio with a denominator of 1.

tasa/razón unitaria Una tasa o razón con un denominador de 1.

univariate data Data with one variable.

datos univariante Datos con una variable.

unlike fractions Fractions whose denominators are different.

fracciones con distinto denominador Fracciones cuyos denominadores son diferentes.

variable A symbol, usually a letter, used to represent a number in mathematical expressions or sentences.

variable Un símbolo, por lo general, una letra, que se usa para representar números en expresiones o enunciados matemáticos.

vertex The point where the sides of an angle meet.

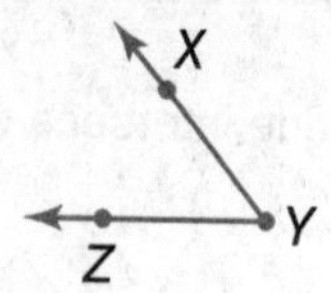

vértice Punto donde se encuentran los lados.

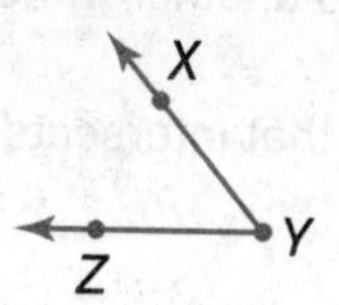

vertex The point where three or more faces of a polyhedron intersect.

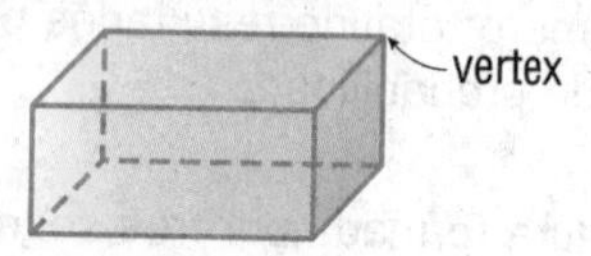

vértice El punto donde tres o más caras de un poliedro se cruzan.

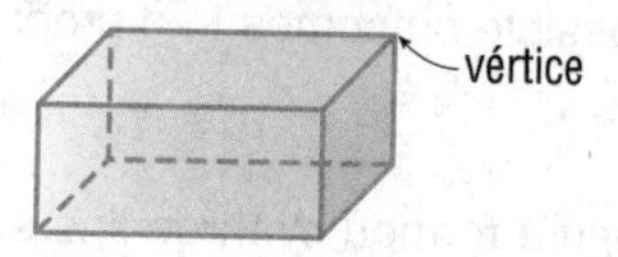

vertex The point at the tip of a cone.

vértice El punto en la punta de un cono.

vertical angles Opposite angles formed by the intersection of two lines. Vertical angles are congruent. In the figure, the vertical angles are $\angle 1$ and $\angle 3$, and $\angle 2$ and $\angle 4$.

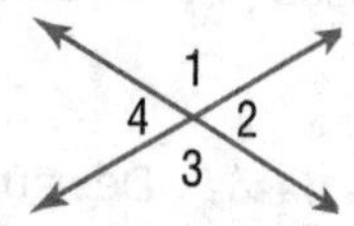

ángulos opuestos por el vértice Ángulos congruentes que se forman de la intersección de dos rectas. En la figura, los ángulos opuestos por el vértice son $\angle 1$ y $\angle 3$, y $\angle 2$ y $\angle 4$.

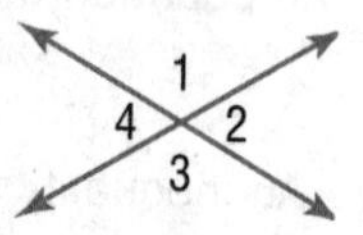

volume The measure of the space occupied by a solid. Standard measures are cubic units such as in^3 or ft^3.

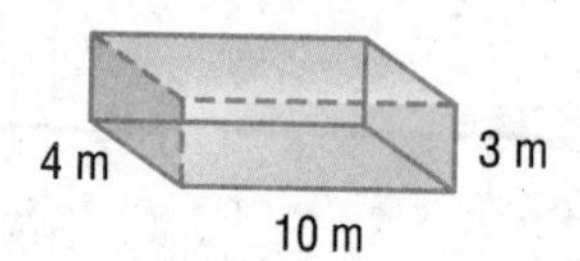

$V = 10 \times 4 \times 3 = 120$ cubic meters

volumen Medida del espacio que ocupa un sólido. Las medidas estándares son las unidades cúbicas, como $pulg^3$ o $pies^3$.

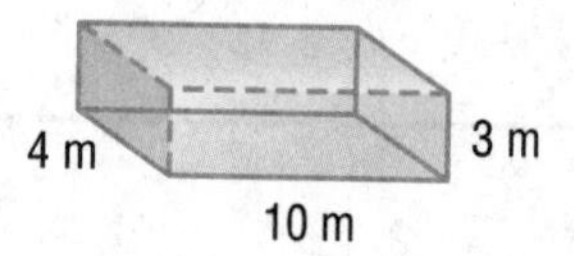

$V = 10 \times 4 \times 3 = 120$ metros cúbicos

voluntary response sample A sample which involves only those who want to participate in the sampling.

muestra de respuesta voluntaria Muestra que involucra sólo aquellos que quieren participar en el muestreo.

***x*-axis** The horizontal number line that helps to form the coordinate plane.

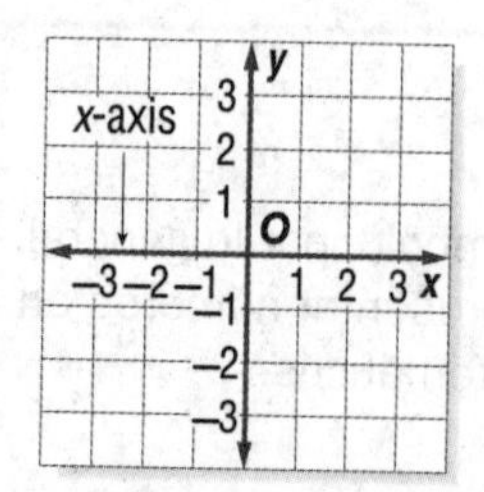

eje *x* La recta numérica horizontal que ayuda a formar el plano de coordenadas.

x-coordinate The first number of an ordered pair.

coordenada x El primer número de un par ordenado.

x-intercept The x-coordinate of the point where the line crosses the x-axis.

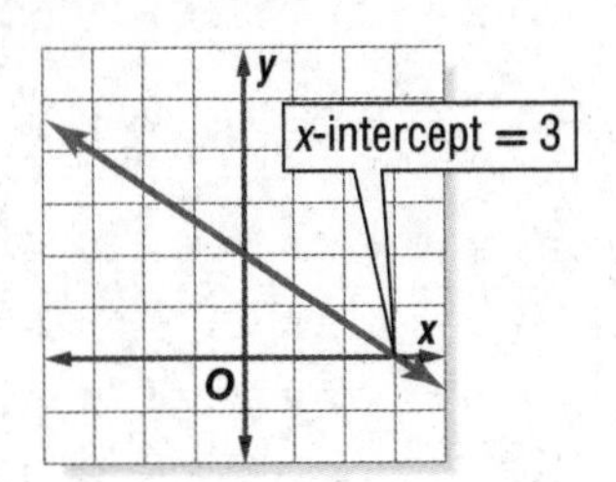

intersección x La coordenada x del punto donde cruza la gráfica el eje x.

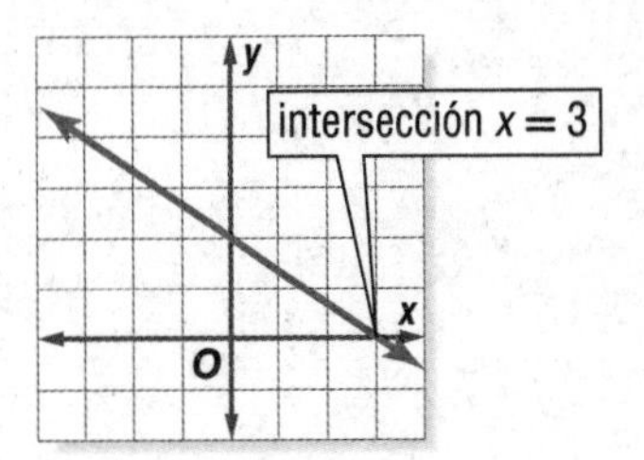

Yy

y-axis The vertical number line that helps to form the coordinate plane.

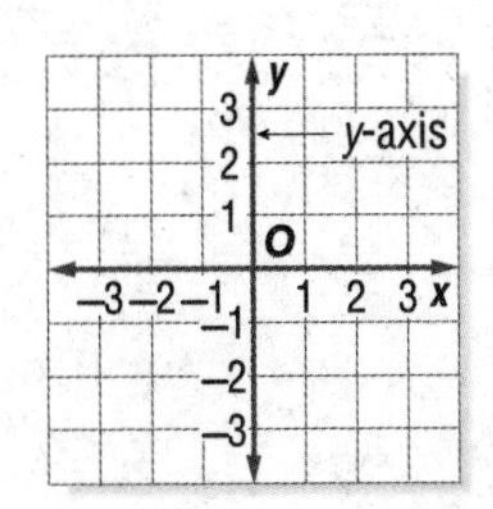

eje y La recta numérica vertical que ayuda a formar el plano de coordenadas.

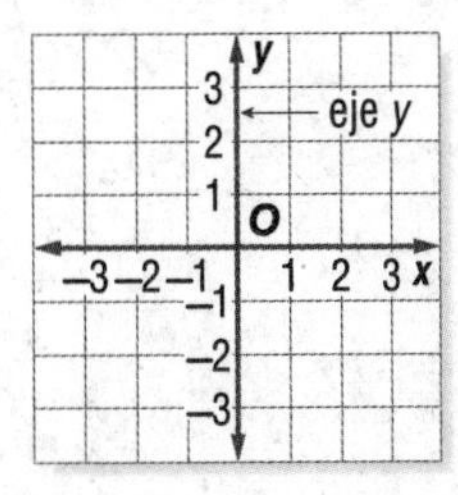

y-coordinate The second number of an ordered pair.

coordenada y El segundo número de un par ordenado.

y-intercept The y-coordinate of the point where the line crosses the y-axis.

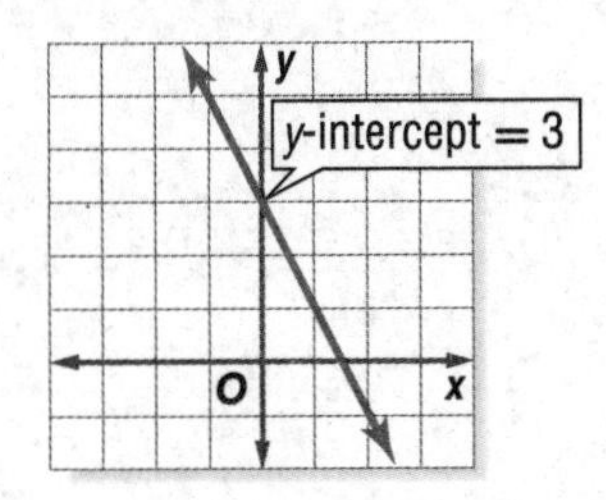

intersección y La coordenada y del punto donde cruza la gráfica el eje y.

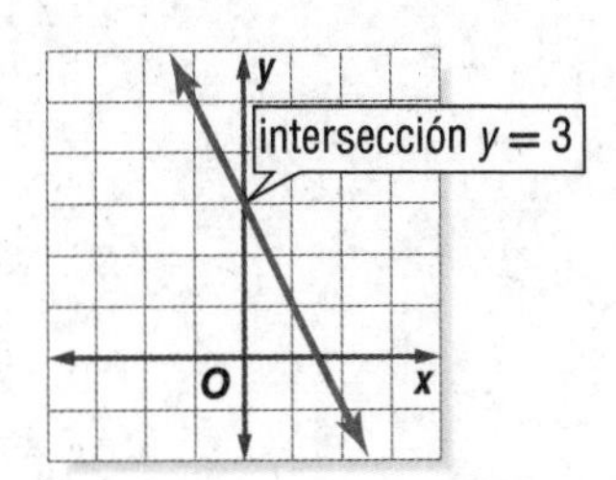

Chapter 1 Real Numbers

Page 6 Chapter 1 Are You Ready?

1. 256 **3.** $2,048 **5.** 2 × 2 × 2 × 3 **7.** 2 × 2 × 5 × 5 **9.** −1 × 2 × 3 × 7

Pages 11–12 Lesson 1-1 Independent Practice

1. 0.4 **3.** 0.825 **5** $-0.\overline{54}$ **7a.** $0.0\overline{6}$ **7b.** $0.1\overline{6}$ **7c.** 0.333 **7d.** 0.417 **9** $-7\frac{8}{25}$ **11.** $-\frac{5}{11}$ **13.** $5\frac{11}{20}$

15. $1\frac{1}{16}$ in.; 1.0625 in. **17.** Sample answer: When dividing, there are two possibilities for the remainder. If the remainder is 0, the decimal terminates. If the remainder is not 0, then at the point where the remainder repeats or equals the original dividend the decimal begins to repeat. **19.** D

Pages 13–14 Lesson 1-1 Extra Practice

21. $7\frac{5}{33}$ **23.** 5.3125 **25.** $-1\frac{11}{20}$ **27.** $-\frac{1}{11}$ **29.** $2\frac{2}{5}$ in. **31.** 0.45 **33.** Felisa: 0.9; Morgan: 0.542; Yasmine: 0.682; Gail: 0.714 **35.** I **37.** > **39.** =

Pages 19–20 Lesson 1-2 Independent Practice

1. $(-5)^4$ **3.** m^5 **5.** $\frac{1}{81}$ **7** 8,000,000,000 or 8 billion **9** −311 **11.** 16 **13a.** 10^6 **13b.** 10^9 **13c.** 10^{15} **15.** Sample answer: As the exponent decreases by 1, the simplified answer is divided by 3; $\frac{1}{2}$

Pages 21–22 Lesson 1-2 Extra Practice

17. $3^3 \cdot p^3$ **19.** $\left(-\frac{5}{6}\right)^3$ **21.** $4^2 \cdot b^4$ **23.** 224 **25.** =

27a.

Side Length (in.)	Perimeter (in.)	Area (in²)
1	4	1
2	8	4
3	12	9
4	16	16
5	20	25
6	24	36
7	28	49
8	32	64
9	36	81
10	40	100

27b.

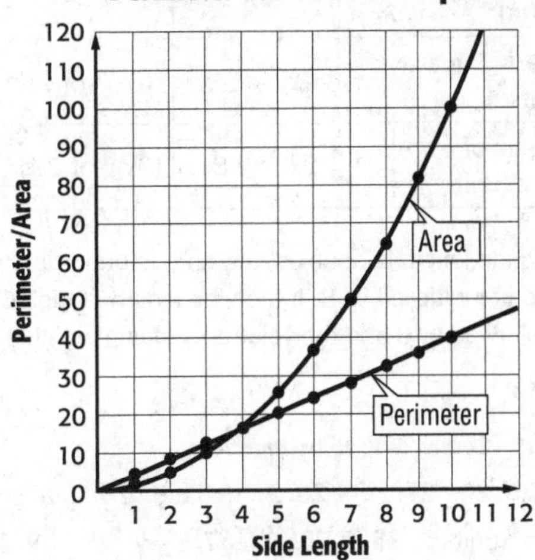

27c. Sample answer: The graph representing perimeter of a square is linear because each side length is multiplied by 4. The graph representing area of a square is nonlinear because each side length is squared and does not increase at a constant rate. **29.** 1,331 **31a.** 10 **31b.** 5,120 **33.** −31 **35.** 1

Pages 27–28 Lesson 1-3 Independent Practice

1. $(-6)^7$ or −279,936 **3.** $-35a^5b^5c^5$ **5** $2t^3$ **7.** 3^3x^2 or $27x^2$ **9.** 6^5 or 7,776 **11** 10^{14} instructions **13a.** 10^9 times greater **13b.** 10^6 or one million **15.** 9 **17.** 6 **19.** 7

21. Equal; sample answer: Using the quotient of powers, $\frac{3^{100}}{3^{99}} = 3^{100-99}$ or 3^1, which is 3. **23.** A

Pages 29–30 Lesson 1-3 Extra Practice

25. h^1 or h **27.** $-8w^{11}$ **29.** 2^8 or 256 **31.** $5^2 \cdot 7^0 \cdot 10$ or 250 **33a.** $2r$ **33b.** $\frac{\pi}{4}$

33c.

Radius (units)	2	3	4	$2r$
Area of Circle (units²)	$\pi(2)^2$ or 4π	9π	16π	$4\pi r^2$
Length of 1 Side of the Square	4	6	8	$4r$
Area of Square (units²)	4^2 or 16	36	64	$16r^2$
Ratio (Area of circle / Area of square)	$\frac{\pi}{4}$	$\frac{\pi}{4}$	$\frac{\pi}{4}$	$\frac{\pi}{4}$

33d. The ratio is $\frac{\pi}{4}$. **35.** G **37.** −28 **39.** −35 **41.** −9 **43.** $\frac{1}{8}$

Pages 35–36 Lesson 1-4 Independent Practice

1. 4^6 **3.** d^{42} **5.** 3^8 **7.** $625j^{24}$ **9.** $216a^6b^{18}$ **11.** $-243w^{15}z^{40}$ **13.** $27c^{18}d^6$ cubic units **15.** $729x^{12}y^{18}$ **17.** $-2{,}048v^{29}$

19a.

Side Length (units)	x	$2x$	$3x$
Area of Square (units²)	x^2	$(2x)^2$ or $4x^2$	$(3x)^2$ or $9x^2$
Volume of Cube (units³)	x^3	$(2x)^3$ or $8x^3$	$(3x)^3$ or $27x^3$

19b. If the side length is doubled, the area is quadrupled and the volume is multiplied by 8. If the side length is tripled, the area is multiplied by 9 and the volume is multiplied by 27. **21.** 3

Pages 37–38 Lesson 1-4 Extra Practice

23. 2^{14} **25.** 3^8 **27.** z^{55} **29.** 2^{18} **31.** $64g^6h^2$ units^2 **33.** $125r^6s^9$ units^3 **35.** $0.25k^{10}$ **37.** $\frac{1}{16}w^{10}z^6$ **39.** D **41.** D **43.** 6^{11} **45.** $18x^{14}$ **47.** Bridalveil: 620 ft; Fall Creek: 256 ft; Shoshone: 212 ft

Page 41 Problem-Solving Investigation The Four-Step Plan

Case 3. 18 tour guides **Case 5.** 21 toothpicks

Pages 47–48 Lesson 1-5 Independent Practice

1. $\frac{1}{7^{10}}$ **3.** $\frac{1}{g^7}$ **5.** 12^{-4} **7.** 5^{-3} **9.** $10^{-1}, 10^{-2}, 10^{-3}, 10^{-6}$ **11.** $\frac{1}{128}$ **13.** y^3 **15.** 81 **17.** y^4 **19.** 10^5 or 100,000 times **21.** $11^{-3}, 11^0, 11^2$; Sample answer: The exponents in order from least to greatest are −3, 0, 2. **23.** Sample answer: $\left(\frac{1}{2}\right)^{-1} = 2$, $\left(\frac{34}{43}\right)^{-1} = \frac{43}{34}$, $\left(\frac{56}{65}\right)^{-1} = \left(\frac{65}{56}\right)$; When you raise a fraction to the −1 power, it is the same as finding the reciprocal of the fraction.

Pages 49–50 Lesson 1-5 Extra Practice

25. $\frac{1}{3^5}$ **27.** $\frac{1}{6^8}$ **29.** $\frac{1}{s^9}$ **31.** z^{-1} or $\frac{1}{z}$ **33.** b^{-12} or $\frac{1}{b^{12}}$ **35.** $\frac{1}{16}$ **37.** $\frac{1}{10{,}000}$ **39.** 12 **41.** −11 **43.** D **45.** 100 **47.** 1,000,000 **49.** 1,000 **51.** 10,000 **53.** 1,000

Pages 55–56 Lesson 1-6 Independent Practice

1. 3,160 **3.** 0.0000252 **5.** 7.2×10^{-3} **7.** Arctic, Southern, Indian, Atlantic, Pacific **9.** 17.32 millimeters; the number is small so choosing a smaller unit of measure is more meaningful. **11.** $<$ **13.** 1.2×10^6; 1.2×10^5 is only 120,000, but 1.2×10^6 is just over one million. **15.** D

Pages 57–58 Lesson 1-6 Extra Practice

17. 7.07×10^{-6} **19.** 0.0078 **21.** 6.7×10^3 **23.** 3.7×10^{-2} **25.** 2.2×10^3, 310,000, 3.1×10^7, 216,000,000 **27.** 10 **29.** I **31.** 4.355 **33.** 4.44 **35.** 1.6 **37.** $50x^7$

Pages 63–64 Lesson 1-7 Independent Practice

1. 8.97×10^8 **3.** 8.19×10^{-2} **5.** 2.375×10^{11} **7.** 8,000 times **9.** 9.83×10^8 **11.** 8.70366×10^4

13. $\frac{6.63 \times 10^{-6}}{5.1 \times 10^{-2}} = \left(\frac{6.63}{5.1}\right)\left(\frac{10^{-6}}{10^{-2}}\right)$
$= 1.3 \times 10^{-6-(-2)}$
$= 1.3 \times 10^{-4}$

15. 10^{109} times

Pages 65–66 Lesson 1-7 Extra Practice

17. 4.44×10^1 **19.** 4×10^2 **21.** 1.334864×10^{10} **23.** $13\frac{5}{9}$ h **25.** B **27.** C

29.

x	x^2	x^3	x	x^2	x^3
1	1	1	7	49	343
2	4	8	8	64	512
3	9	27	9	81	729
4	16	64	10	100	1,000
5	25	125	11	121	1,331
6	36	216	12	144	1,728

Pages 75–76 Lesson 1-8 Independent Practice

1. 4 **3.** no real solution **5.** −1.6 **7.** ±9 **9.** ±0.13 **11.** −0.5 **13.** 13 students **15.** 44 in. **17.** 24 m **19.** $\frac{25}{81}$ **21.** x **23.** D

Pages 77–78 Lesson 1-8 Extra Practice

25. $-\frac{8}{15}$ **27.** ±1.2 **29.** −8 **31.** −7 **33.** $\pm\frac{3}{8}$ **35.** $\frac{1}{2}$ **37.** 20 **39.** 400 **41.** 60 chairs **43.** I **45.** 2,197 **47.** 3,375 **49.** 55 **51.** 20 **53.** $64r^9s^3$ units^3

Pages 85–86 Lesson 1-9 Independent Practice

1. 5 **3.** 4 **5.** 3 **7.** 10 **9.** Sample answer: 54 ft and 57 ft; 55.5 ft and 55.8 feet; 55.71 ft and 55.74 feet; 56 feet **11.** about 2.75 seconds **13.** $\sqrt[3]{105}, 5, \sqrt{38}, 7$ **15.** 10; Since 94 is less than 100, $\sqrt{94}$ is less than 10. **17.** She incorrectly estimated. She found half of 200, not the square root. Since $196 < 200 < 225$, the square root of 200 is between 14 and 15. Since 200 is closer to 196, the square root of 200 is about 14. **19.** B

Pages 87–88 Lesson 1-9 Extra Practice

21. 6 **23.** 5 **25.** 8 **27.** 10 or −10 **29.** 6 in. **31.** 70 feet on each side **33.** I **35.** $\frac{-36}{1}$ **37.** $\frac{-6}{125}$ **39.** $5^2, 3^3, 4(8)$ **41.** $12^2 + 4, 10^3, 25^2 \cdot 3$

Pages 93–94 Lesson 1-10 Independent Practice

1. rational **3.** rational **5.** $<$ **7** $<$
9 $\sqrt{5}, \frac{7}{3}, \sqrt{6}, 2.5, 2.55$

$\sqrt{5}$ $\frac{7}{3}$ $\sqrt{6}$ 2.5 2.55

2.1 2.2 2.3 2.4 2.5 2.6 2.7

11. about 1.9 m^2 **13.** $>$ **15.** $<$ **17.** always **19.** always

Pages 95–96 Lesson 1-10 Extra Practice

21. irrational **23.** natural, whole, integer, rational
25. irrational **27.** $>$ **29.** 18.52 m **31.** $<$ **33.** $>$ **35.** F
37. $\sqrt{32}, 6, 7, \sqrt{53}$ **39.** $\frac{1}{7}$ or $-\frac{1}{7}$ **41.** 7.92×10^{-2}
43. China, India, United States, Indonesia

Page 99 Chapter Review Vocabulary Check

Across
5. perfect cube **9.** rational number
Down
1. radical sign **3.** exponent **7.** cube root

Page 100 Chapter Review Key Concept Check

1. real **3.** Power of a Product rule **5.** 9^{28} **7.** less than

Page 101 Chapter Review Problem Solving

1. 216 calls **3.** 0.004375 lb **5.** $2\frac{1}{5}, 2.\overline{2}, \sqrt{5}, 2.25$

Chapter 2 Equations in One Variable

Page 110 Chapter 2 Are You Ready?

1. −17 **3.** 19 **5.** 7 **7.** $18 + h = 92$; 74 marbles

Pages 115–116 Lesson 2-1 Independent Practice

1. 72 **3** 24 **5.** −12 **7.** 2 **9.** $\frac{4}{5}$
11 q = total questions; $0.8q = 16$; 20 questions
13. Multiplicative Inverse: $1\frac{1}{3}, -\frac{1}{2}$; Division: 0.2, −5
15. true; Sample answer: The product of $\frac{3}{4}$ and $\frac{4}{3}$ is $\frac{12}{12}$, which simplifies to 1. **17.** 53; Since $10 = \frac{1}{5}x$, then $x = 50$ and $x + 3 = 53$.

Pages 117–118 Lesson 2-1 Extra Practice

19. $1\frac{1}{4}$ **21.** $3\frac{1}{2}$ **23.** −10.5 **25.** $6\frac{3}{10}$ **27.** $\frac{8}{9}$
29.

$$-\frac{7}{8}x = 24$$
$$\left(-\frac{8}{7}\right)\left(-\frac{7}{8}x\right) = 24\left(-\frac{8}{7}\right)$$
$$x = -27\frac{3}{7}$$

31. H **33.** −25 **35.** −5.55 **37.** 2
39. Simone: $s + 37.50 = 127.75$; \$90.25; Dan: $s - 65.35 = d$; \$24.90

Pages 125–126 Lesson 2-2 Independent Practice

1. 3 **3.** −4 **5** −52 **7** 5 bracelets **9.** 64
11. −26 **13a.** 146 messages **13b.** 135 messages
15. Sample answer: Andrea saved x dollars each week for 3 weeks. She spent \$25 and had \$125 left. How much did she save each week?; \$50

Pages 127–128 Lesson 2-2 Extra Practice

17. 6 **19.** −3 **21.** −8 **23.** 7 **25.** −10 **27.** \$1.50; Sample answer: Subtraction Property of Equality, Division Property of Equality **29.** 40 **31.** −22 **33.** −4
35. $j + 45 = 79.50$; \$34.50

Pages 133–134 Lesson 2-3 Independent Practice

1. $5n - 4 = 11$ **3** $7n - 6 = -20$
5 s = the number of songs; $0.25s + 9.99 = 113.74$; 415 songs **7.** s = height of the Statue of Liberty; $s + (s + 0.89) = 92.99$; 46.05 m **9a.** $175 = 3c - 20$; 65 mph **9b.** $s = \frac{1}{5} \cdot 175 - 1$; 34 mph
9c. $175 = 6h + 13$; 27 mph **11.** $n + 2n + (n + 3) = 27$; 6, 9, 12 **13.** D

Pages 135–136 Lesson 2-3 Extra Practice

15. $4n + 16 = -2$ **17.** $6 + 9n = 456$
19. x = the number of groups of pitches; $4 + 0.75x = 7$; 4 groups **21.** D **23.** $3m + 6 = 120$; 38 cards
25. −648 **27.** 6

Page 139 Problem-Solving Investigation Work Backward

Case 3. \$92 **Case 5.** \$1,238.50

Pages 149–150 Lesson 2-4 Independent Practice

1. −2 **3** 10 **5.** 48 **7** Let n = the number; $0.5n - 9 = 4n + 5$; −4 **9a.** Sample answer: Set the side lengths equal to each other and solve for x.
9b. $4x - 2 = 2x + 8$ **9c.** 18 units **11.** Sample answer: You have 20 crafts made and continue to make crafts at the rate of 3 per hour. How many hours will it take you and your friend to make the same amount of crafts, if she makes crafts at a rate of 5 per hour. **13.** A

Pages 151–152 Lesson 2-4 Extra Practice

15. 3 **17.** 5 **19** Let n = the number; $3n - 18 = 2n$; 18
21. $60x = 8x + 26$; 0.5 **23.** G **25.** $4x + 6$ **27.** −9
29. $6x + 30$ **31.** $15z - 36$

Pages 157–158 Lesson 2-5 Independent Practice

1. −9 **3** 6 **5.** null set or no solution **7.** −6
9 \$4.72 **11a.** $20 + 0.15m = 30 + 0.1m$ **11b.** m = 200; 200 messages **13.** 13 in. and 15 in.

Pages 159–160 Lesson 2-5 Extra Practice

15. 13 **17.** $4\frac{4}{7}$ **19** −9 **21.** identity or all numbers
23a. Sample answer: $3x + 5 = 3x - 2 + 7$ **23b.** Sample answer: $2(x - 1) = 2x + 2$ **23.** D **25.** G

27. $x \geq -3$

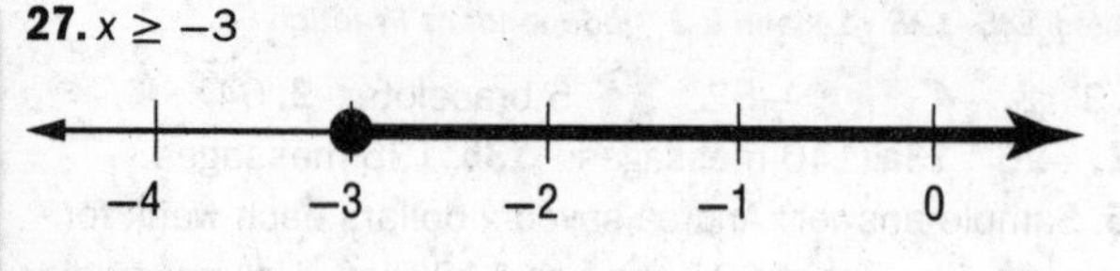

29. $n < 8$

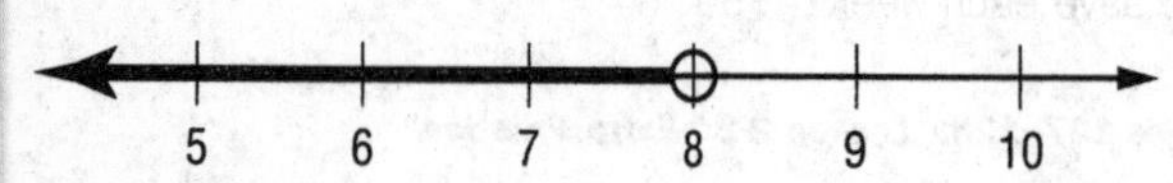

31. $m < -4$

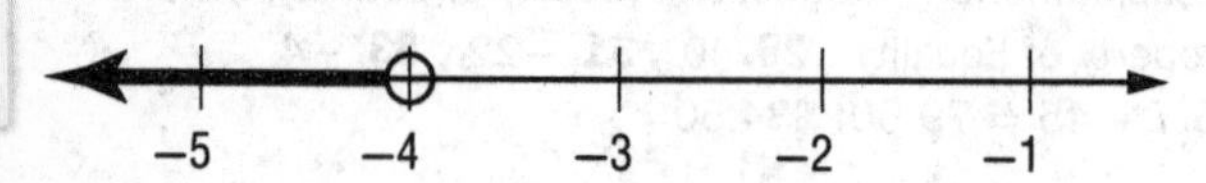

Page 163 Chapter Review Vocabulary Check

1. coefficient **3.** properties

Page 164 Chapter Review Key Concept Check

1. 31 **3.** Sample steps: 2; 3; 1; 4; $7\frac{2}{7}$

Page 165 Chapter Review Problem Solving

1. 14,000 mi^2 **3.** $8 + 4d = 28$; 5 more days **5.** $9.17

Chapter 3 Equations in Two Variables

Page 170 Chapter 3 Are You Ready?

1. 9 **3.** −7 **5.** 6 **7.** $\frac{2}{5}$ **9.** $\frac{1}{5}$ **11.** $-\frac{2}{5}$

Pages 175–176 Lesson 3-1 Independent Practice

1 Yes; the rate of change between cost and time for each hour is a constant 3¢ per hour. **3.** Yes; the rate of change between vinegar and oil for each cup of oil is a constant $\frac{3}{8}$ cup vinegar per cup of oil. **5** Yes; the rate of change between the actual distance and the map distance for each inch on the map is a constant 7.5 mi/in. **7.** Yes; the ratio of the cost to time is a constant 3¢ per hour, so the relationship is proportional. **9.** Yes; the ratio of actual distance to map distance is a constant $\frac{15}{2}$ miles per inch, so the relationship is proportional. **11.** Sample answer:

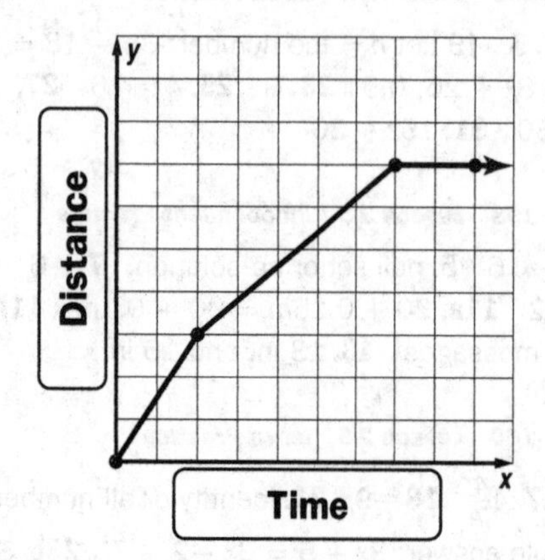

13. D

Pages 177–178 Lesson 3-1 Extra Practice

15. No; the rate of change from 1 to 2 hours, $\frac{24 - 12}{2 - 1}$ or 12 per hour, is not the same as the rate of change from 3 to 4 hours, $\frac{60 - 36}{4 - 3}$ or 24 per hour, so the rate of change is not constant. **17.** −50 mph; the distance decreased by 50 miles every hour. **19.** 0.5; $\frac{1}{2}$ of retail price. **21.** $15 per week **23.** 24 mi/gal **25.** $0.55/red pepper

Pages 185–186 Lesson 3-2 Independent Practice

1 $-\frac{5}{8}$ **3.** $-\frac{3}{4}$ **5.** 2 **7** −4 **9.** yes; $\frac{1}{15} < \frac{1}{12}$ **11.** Jacob did not use the *x*-coordinates in the same order as the *y*-coordinates.

$$m = \frac{3 - 2}{4 - 0}$$
$$m = \frac{1}{4}$$

13. D

Pages 187–188 Lesson 3-2 Extra Practice

15. 3 **17.** −3 **19.** 2 **21.** $\frac{1}{5}$ **23.** B **25.** $\frac{30}{180} = \frac{x}{240}$; 40 minutes **27.** 15 **29.** 7.5 **31.** −60

Pages 195–196 Lesson 3-3 Independent Practice

1. $0.50 per paper **3** Computers R Us; Sample answer: The unit cost for Computer Access is $25 per hour. The unit cost for Computers R Us is $23.50. 23.5 < 25 **5.** yes; 4 **7** 127 cm **9.** $y = \frac{2}{5}x$; 4 **11.** Sample answer: (4, 3), (8, 6), (0, 0) **13.** B

Pages 197–198 Lesson 3-3 Extra Practice

15. $y = 4.2x$; $4.20 per pound

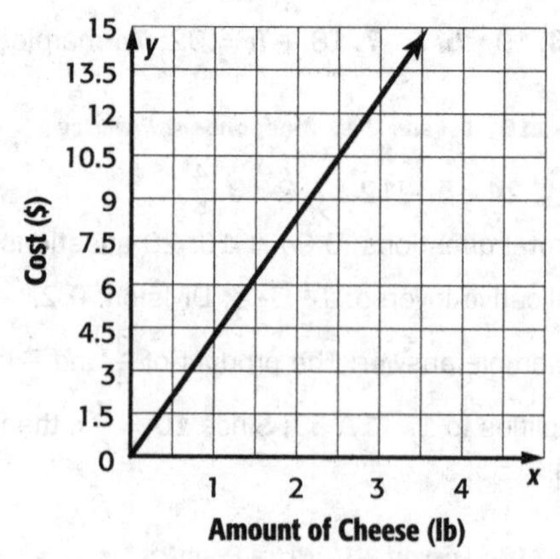

17. tickets to the play; Sample answer: The unit rate per raffle ticket is $5 and the unit rate per play ticket is $6.25. 6.25 > 5 **19.** 36 pages **21.** $\frac{2}{1}$ or 2 **23.** $-\frac{7}{3}$ **25.** undefined

Pages 203–204 ***Lesson 3-4*** ***Independent Practice***

1. 3; 4 **3.** −3; −4 **5** $y = \frac{5}{6}x + 8$ **7.** $y = \frac{5}{4}x - 12$

9.

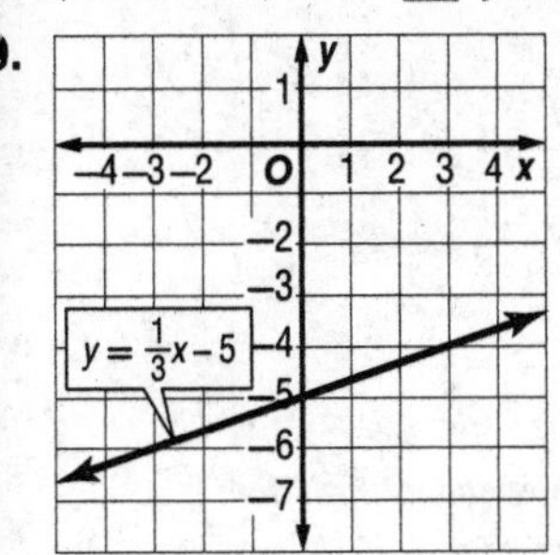

11

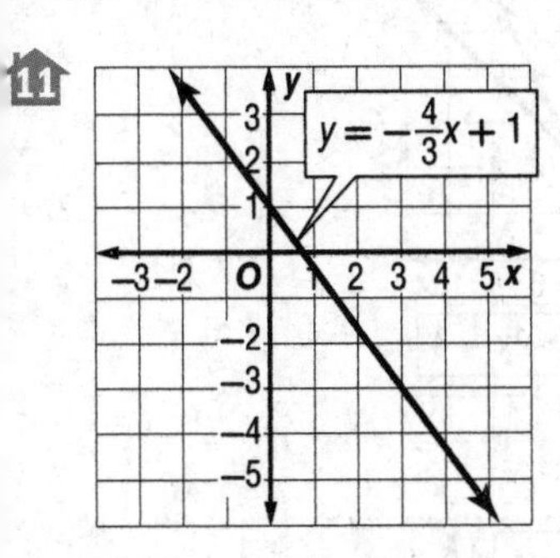

13. 0; Sample answer: A line that has a *y*-intercept but no *x*-intercept is a horizontal line. **15.** C

Pages 205–206 ***Lesson 3-4*** ***Extra Practice***

17. $\frac{1}{2}$; −6 **19.** $y = \frac{1}{2}x + 6$ **21.** $y = -\frac{3}{5}x - \frac{1}{5}$

23.

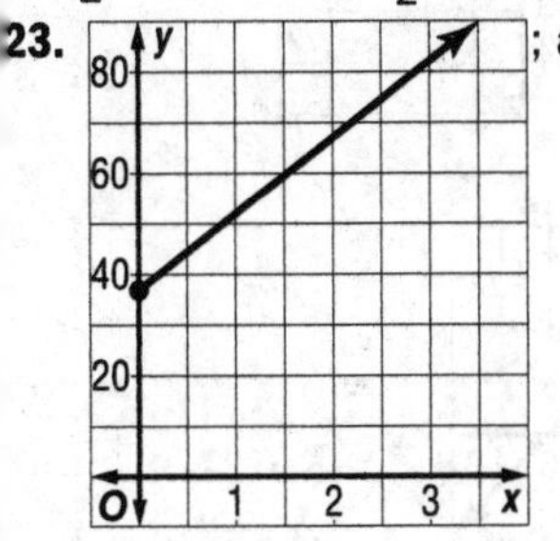

; about 3 chirps

25. A **27.** $y = -\frac{2}{3}x - 4$ **29.** −2 **31.** 4 **33.** −6

35. yes; 0.07

Pages 213–214 ***Lesson 3-5*** ***Independent Practice***

1. *x*-intercept: 3.5; *y*-intercept: 7

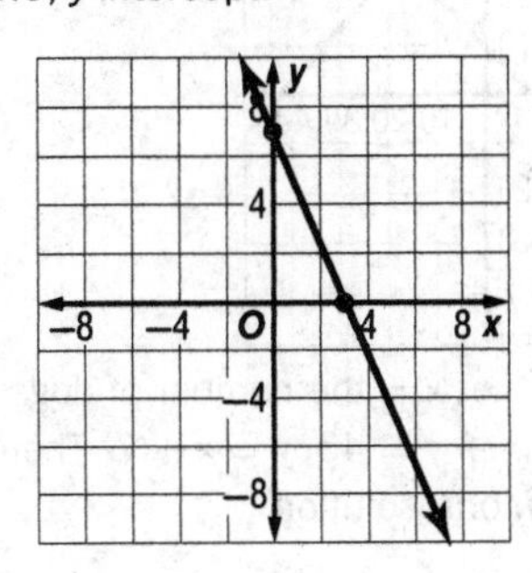

3 *x*-intercept: $1\frac{1}{4}$; *y*-intercept: $1\frac{2}{3}$

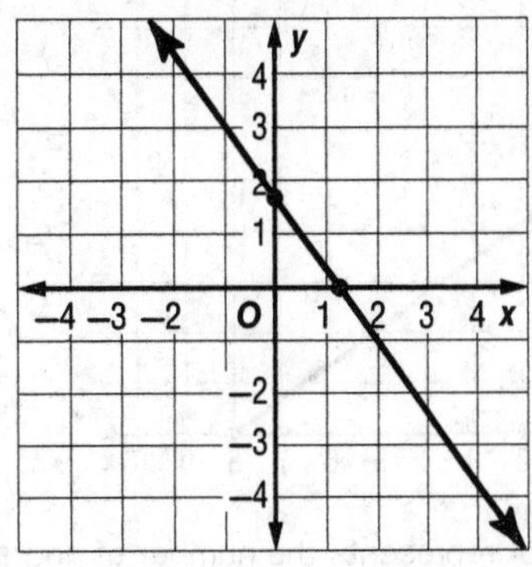

5

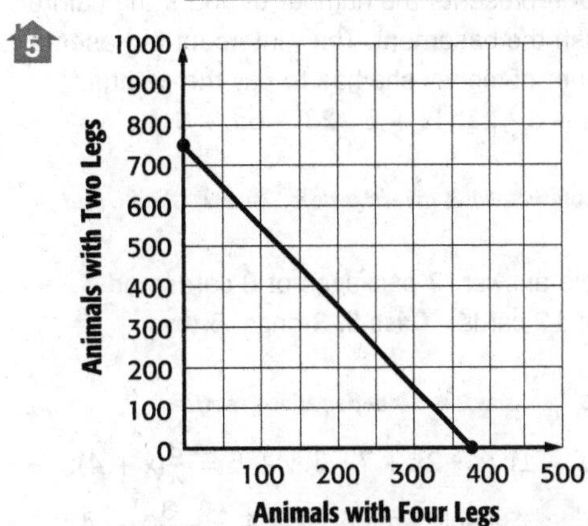

The *x*-intercept of 375 means that if the zoo had only four-legged animals, there would be 375 of them. The *y*-intercept of 750 means that if the zoo had only two-legged animals, there would be 750 of them. **7.** After 3*x* = 12, Carmen didn't divide both sides by 3 to get the *x*-intercept of 4. **9.** D

Pages 215–216 ***Lesson 3-5*** ***Extra Practice***

11. (12, 0), (0, 8)

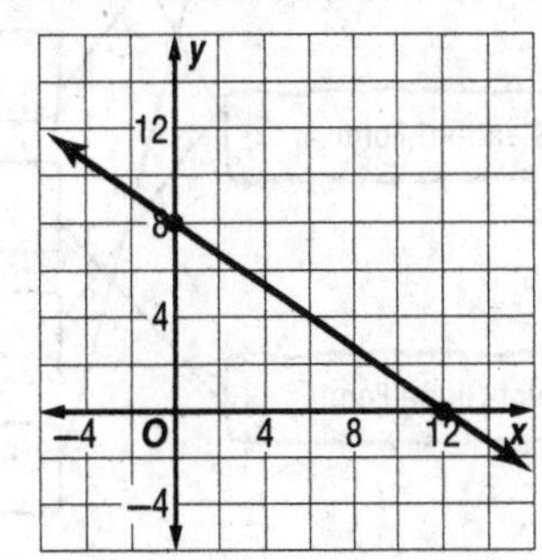

13. (6, 0), (0, 10)

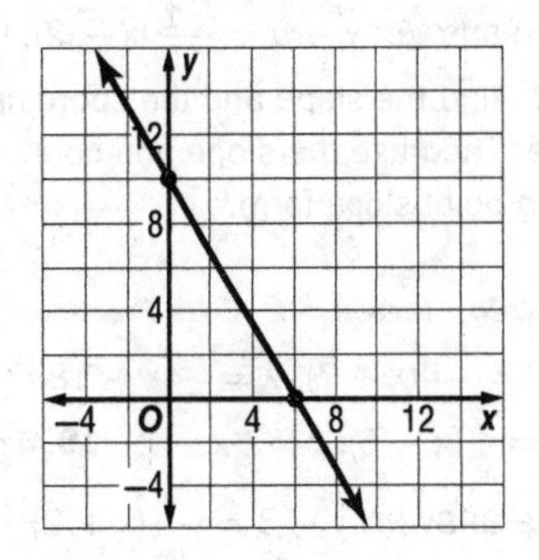

15.

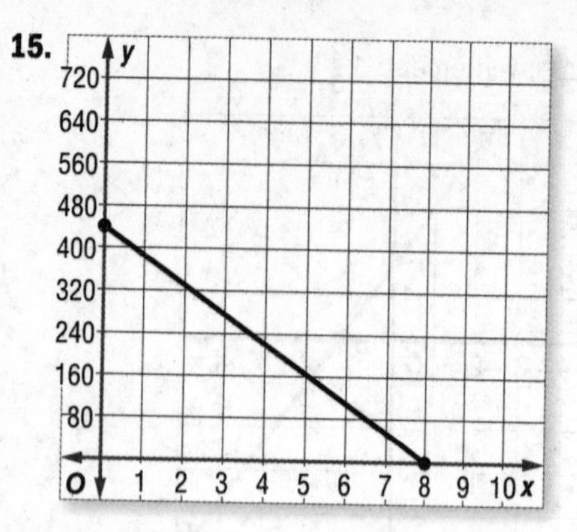

The x-intercept represents the number of hours the painter worked to finish the basement. The y-intercept represents the total amount of money she has to pay the painter.

17. G **19.** $2x + 1$ **21.** $4x + 6$ **23.** $-6a - 6$

Page 219 Problem-Solving Investigation Guess, Check, and Revise

Case 3. Sample answer: 3 packages of 8 cards and 4 packages of 12 cards **Case 5.** 3 rings, 5 toys

Pages 225–226 Lesson 3-6 Independent Practice

1. $y - 9 = 2(x - 1)$; $y = 2x + 7$ **3.** $y + 5 = \frac{3}{4}(x + 4)$; $y = \frac{3}{4}x - 2$ **5** Sample answer: $y + 4 = -\frac{3}{2}(x - 4)$; $y = -\frac{3}{2}x + 2$ **7.** Sample answer: $y - 14 = \frac{1}{5}(x - 10)$ **9** $3x + y = 13$

11.

Slope-Intercept Form — Standard Form — Point-Slope Form

$5x + 3y = 12$

$y = 2x - 8$

$7x = y$

$y - 8 = \frac{1}{2}(x - 9)$

$4x - 6y = 24$

$y = 10 - 3x$

13. Sample answer: $y - 5 = -\frac{1}{2}(x - 2)$; First, use the equation to find the slope and the coordinates of any point on the line. Then use the slope and coordinates to write an equation in point-slope form.

Pages 227–228 Lesson 3-6 Extra Practice

15. $y - 10 = -4(x + 7)$; $y = -4x - 18$

17. $y - 2 = \frac{2}{3}(x - 6)$; $y = \frac{2}{3}x - 2$ **19.** $4x - 5y = 17$

21. Sample answer: $y - 3 = -\frac{5}{2}(x + 2)$ **23.** C **25.** I

27.

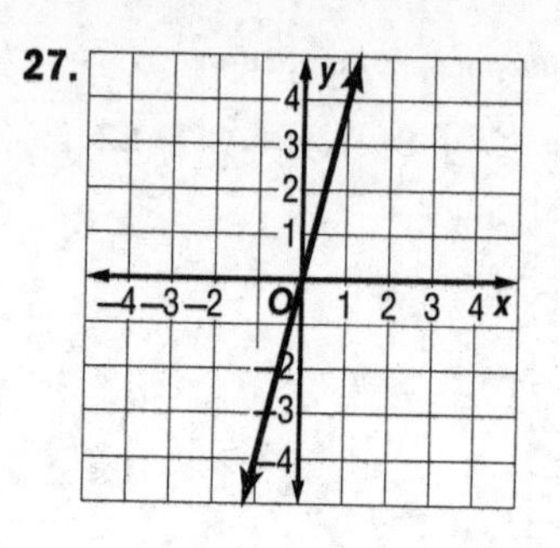

Pages 239–240 Lesson 3-7 Independent Practice

1. (4, 4)

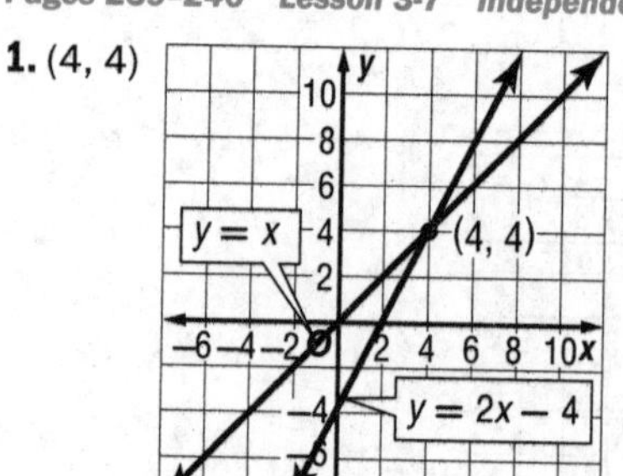

3 no solution

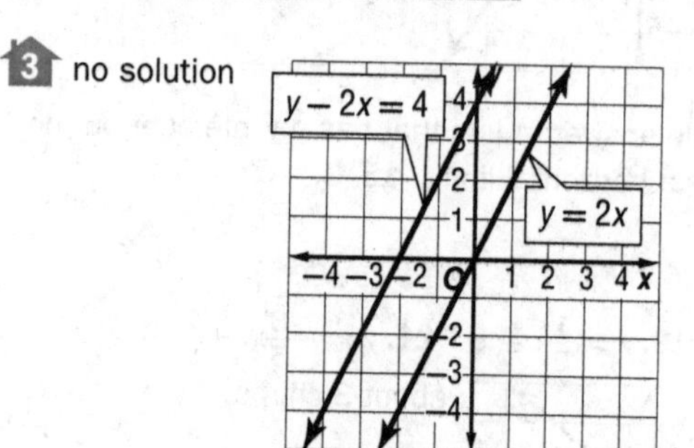

5. (0, 3)

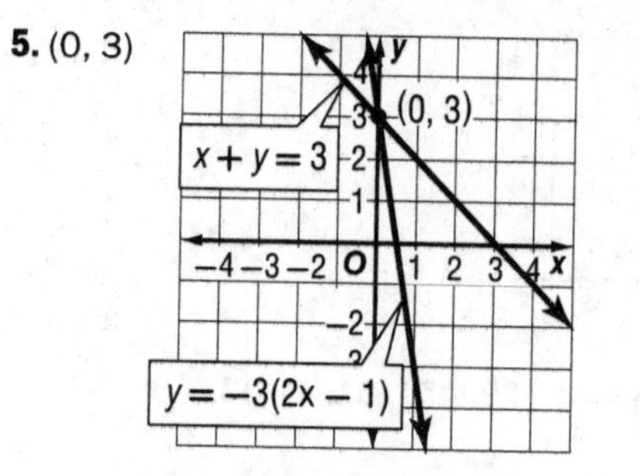

7.

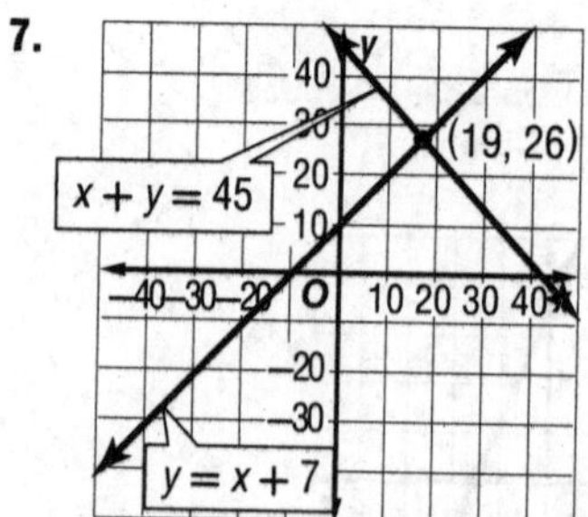

Sample answer: Let x = the number of dogs and y = the number of cats; $x + y = 45$, $y = x + 7$; There are 19 dogs and 26 cats. **9.** one solution

11 a.

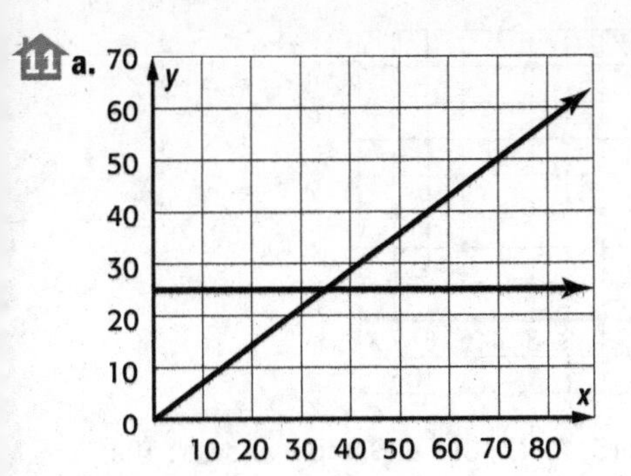

11b. 18 or more rides **13.** B

Pages 241–242 Lesson 3-7 Extra Practice

15. (1, 2)

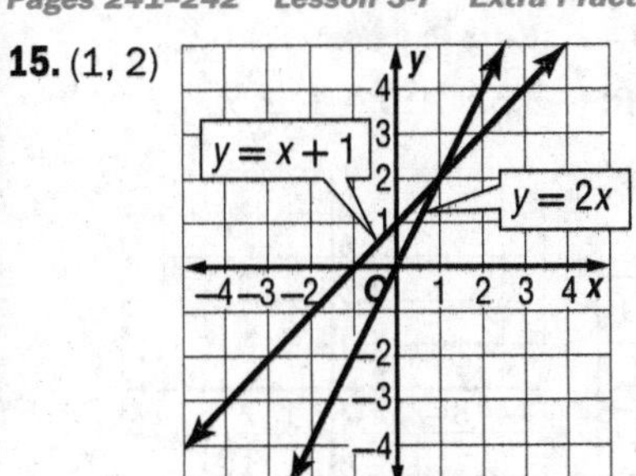

17. no solution

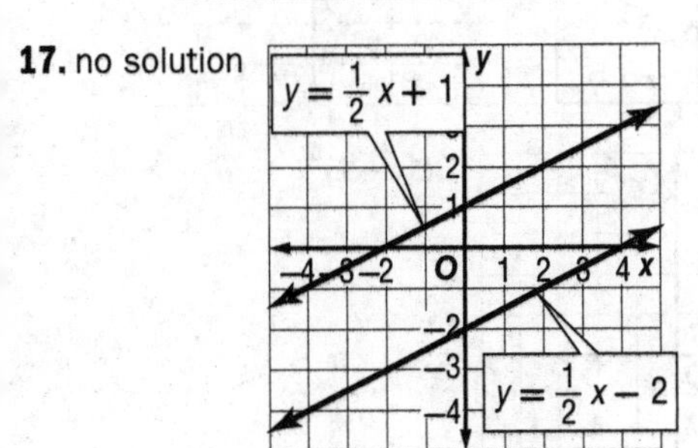

19. (−2, 0); one solution **21.** A **23.** $y = 5$ **25.** $x = -28$ **27.** $y = \frac{1}{4}$

Pages 247–248 Lesson 3-8 Independent Practice

1. (1, 6) **3.** (−2, −12) **5.** (7, 11) **7.** $\left(\frac{1}{2}, 12\frac{1}{2}\right)$ or (0.5, 12.5) **9.** Sample answer: $s + p = 15$; $p + 7 = s$; (4, 11); She bought 11 shirts and 4 pairs of pants. **11.** Sample answer: $8x + 2y = 18$; $3x + y = 7.50$; (1.5, 3); A muffin costs $1.50 and 1 quart of milk costs $3. **13.** Ø; Sample answer: Adding $5x$ to each side of $y = -5x + 8$ results in the equation $5x + y = 8$. Since $5x + y$ cannot equal both 8 and 2, there are no values for x and y that make this system of equations true. **15.** C

Pages 249–250 Lesson 3-8 Extra Practice

17. (−12, −3) **19.** (−3, 0) **21.** (5, −1)
23a. $y = 4x + 95$ and $y = 9x + 75$

23b.

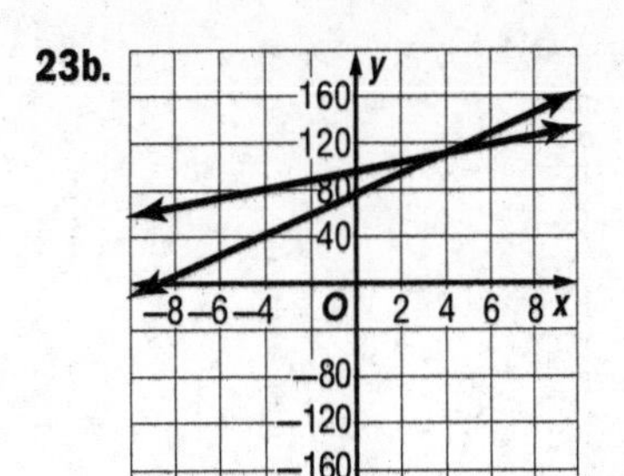

(4, 111); the costs, $111, are the same if 4 students attend either. **23c.** nature center **25.** 12 **27.** 67 **29.** 22.9 **31.** 97 **33.** 78

Page 255 Chapter Review Vocabulary Check

Across

5. direct variation **7.** *x* intercept

Down

1. substitution **3.** *y* intercept

Page 256 Chapter Review Key Concept Check

1. $y = -0.5x + 1$ **3.** $y = 0.5x$ **5.** $x = 5$

Page 257 Chapter Review Problem Solving

1.

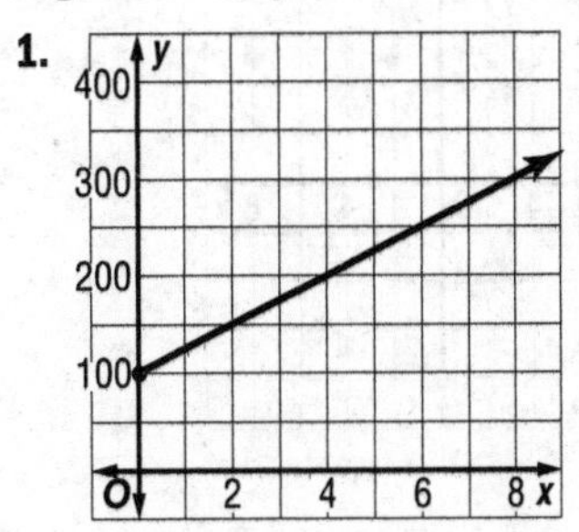

3. Sample answer: Let x = the number that preferred steak and y = the number that preferred pizza; $x + y = 25$, $y = x + 5$; (10, 15); 10 students preferred steak and 15 students preferred pizza.

Chapter 4 Functions

Page 266 Chapter 4 Are You Ready?

1. (1.5, 2.5) **3.** (0, 1.5) **5.** (1, 1) **7.** −18 **9.** −3 **11.** $901

Pages 273–274 Lesson 4-1 Independent Practice

1 a. $b = 45d$; Forty-five baskets are produced every day. **b.** 16,425 baskets **3 a.** $f = 3.5 + 0.15d$

b.

d	$3.5 + 0.15d$	f
10	3.5 + 0.15(10)	5.00
15	3.5 + 0.15(15)	5.75
20	3.5 + 0.15(20)	6.50
25	3.5 + 0.15(25)	7.25

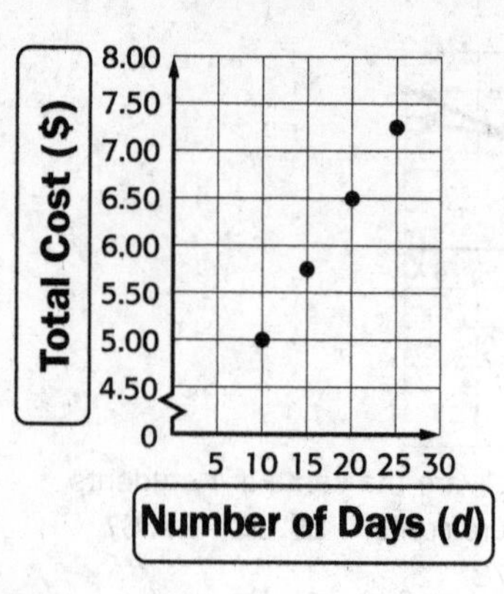

5. Sample answer: $d = 60t$; A car is traveling at a rate of 60 miles per hour. **7.** B

Pages 275–276 Lesson 4-1 Extra Practice

9a. $d = 15w + 5$ **9b.** \$365 **11.** C **13a.** $y = 28.4x$ **13b.** 4,260 g **15.** $8w + 9$

Pages 281–282 Lesson 4-2 Independent Practice

1 D: {−6, 0, 2, 8}; R: {−9, −8, 5}

x	y
8	5
−6	−9
2	5
0	−8

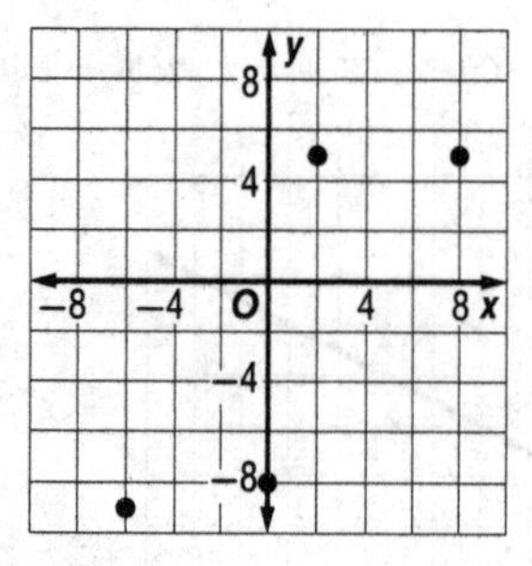

3.

x	825x	y
1	825(1)	825
2	825(2)	1,650
3	825(3)	2,475
4	825(4)	3,300
5	825(5)	4,125

5 **a.** To get the y-value, the x-value was multiplied by itself. **b.** (1, 1), (2, 4), (3, 9), (4, 16), (5, 25)

c.

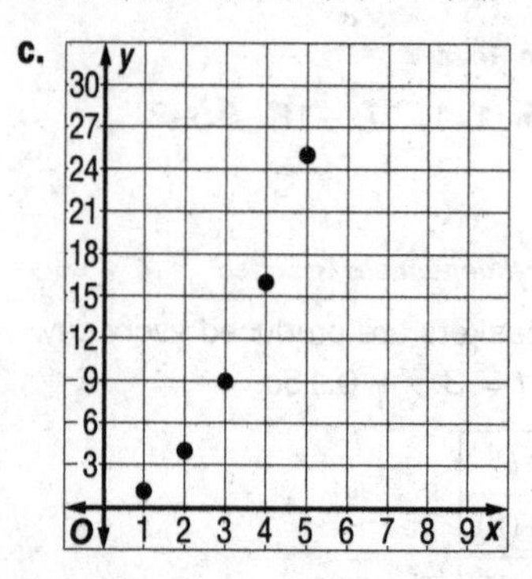

d. Sample answer: This graph curves upward. The points in all of the other graphs in the lesson lie in a straight line.

7a, 7c.

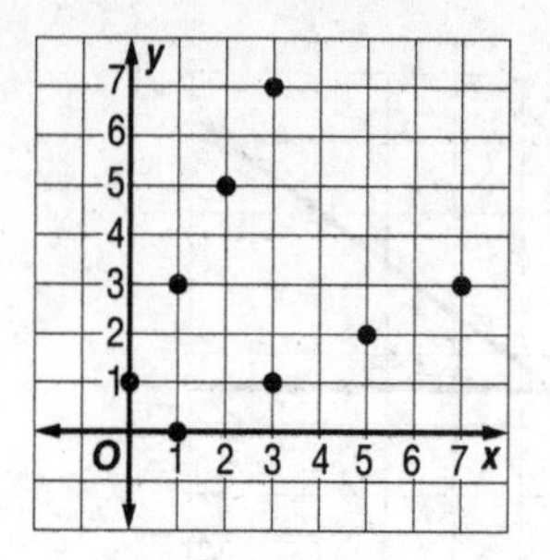

b. (1, 0), (3, 1), (5, 2), (7, 3) **d.** Sample answer: The distance between each point in the original table and the x-axis is the same as the distance between the points with the reversed ordered pairs and the y-axis. **9.** C

Pages 283–284 Lesson 4-2 Extra Practice

11. D: {−1.5, 2.5, 3}; R: {−3.5, −1.5, −1, 3.5}

x	y
−1.5	3.5
2.5	−1.5
3	−1
−1.5	−3.5

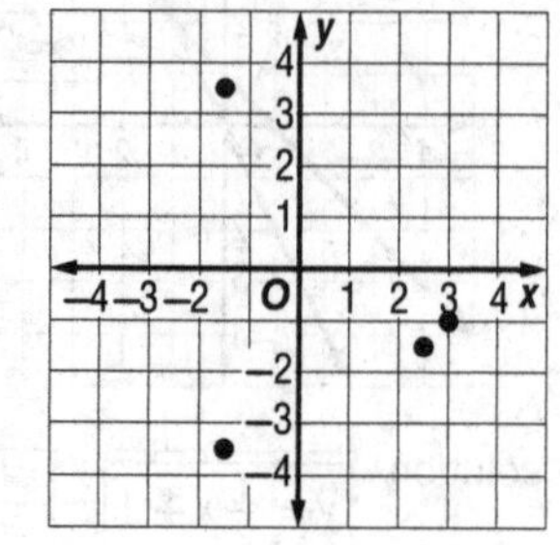

13.

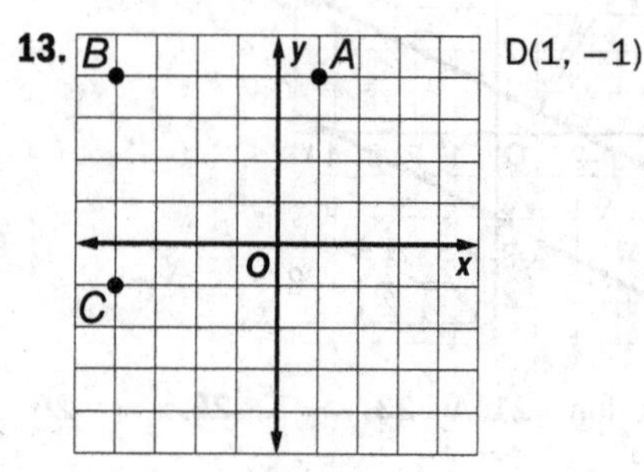

D(1, −1)

15. A **17.** $\left(\frac{3}{4}, \frac{1}{2}\right)$ **19.** $\left(1, -\frac{3}{4}\right)$ **21.** $\left(-\frac{1}{2}, -\frac{1}{2}\right)$ **23.** $\left(-1, \frac{1}{4}\right)$

Pages 291–292 Lesson 4-3 Independent Practice

1. 35 **3** 11

5 Sample answer:

x	5 − 2x	f(x)
−2	5 − 2(−2)	9
0	5 − 2(0)	5
3	5 − 2(3)	−1
5	5 − 2(5)	−5

D: {−2, 0, 3, 5}
R: {9, 5, −1, −5}

7a. The total points $p(g)$ is the dependent variable and the number of games g is the independent variable. **7b.** Only whole numbers between and including 0 and 82 make sense for the domain because you do not want data for a partial game and there are only 82 games in a season. The range will be multiples of 20.7. **7c.** $p(g) = 20.7g$; 186.3 points **9.** 2 **11.** Sample answer: $f(x) = 2x - 2$; $f(0) = -2$, $f(-4) = -10$, $f(3) = 4$ **13.** C ages 293–294 Lesson 4-3 Extra Practice

5. −41

7. Sample answer:

x	$x - 9$	$f(x)$
−2	−2 − 9	−11
−1	−1 − 9	−10
7	7 − 9	−2
12	12 − 9	3

: {−2, −1, 7, 12}
: {−11, −10, −2, 3}

9. Sample answer:

x	$4x + 3$	$f(x)$
−4	4(−4) + 3	−13
−2	4(−2) + 3	−5
3	4(3) + 3	15
5	4(5) + 3	23

: {−4, −2, 3, 5}
: {−13, −5, 15, 23}

1. $m(s) = 5 + 0.50s$; \$20 **23.** H **25.** 3 **27.** 8

ages 301–302 Lesson 4-4 Independent Practice

1

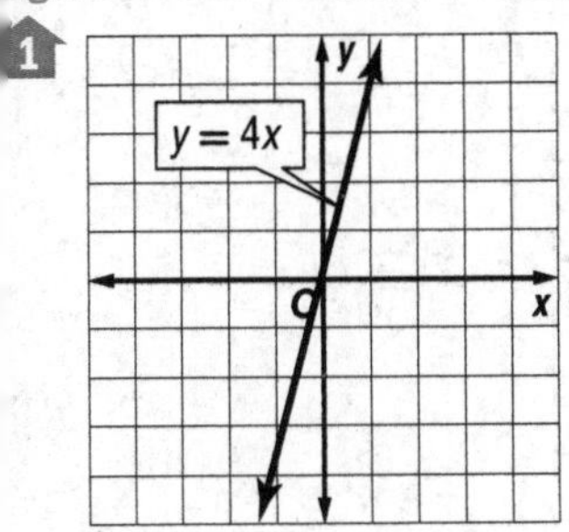

.

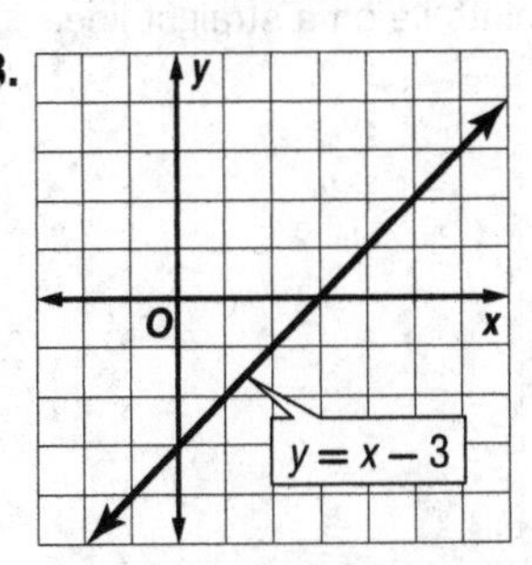

a. bike: $c = 15 + 4.25h$; scooter: $c = 25 + 2.5h$

b.

Mountain Bike Rental		
h	$15 + 4.25h$	c
2	15 + 4.25(2)	23.50
3	15 + 4.25(3)	27.75
4	15 + 4.25(4)	32.00
5	15 + 4.25(5)	36.25

Scooter Rental		
h	$25 + 4.25h$	c
2	25 + 2.5(2)	30.00
3	25 + 2.5(3)	32.50
4	25 + 2.5(4)	35.00
5	25 + 2.5(5)	37.50

5c.

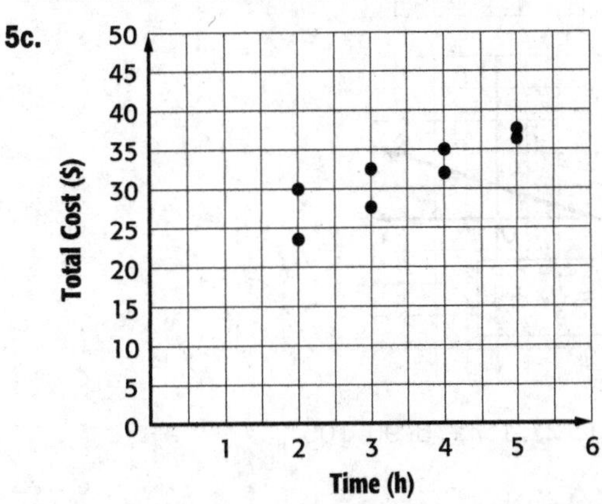

Both situations are discrete because you cannot rent either piece of equipment for a partial hour. **5d.** mountain bike **5e.** \$49

7 25 weeks

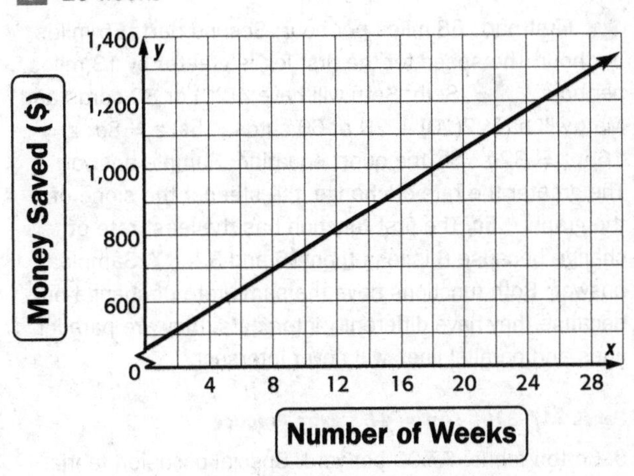

9. Sample answer: (−2, −4), (0, −2), (2, 0), (4, 2); $y = x - 2$

Pages 303–304 Lesson 4-4 Extra Practice

11.

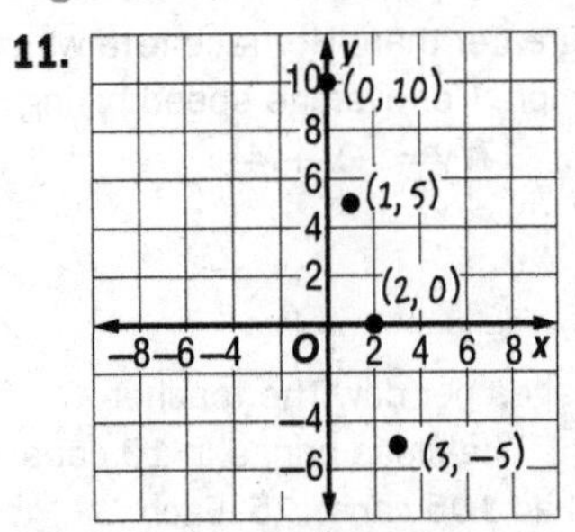

She cannot buy negative amounts. So, she can buy 0 T-shirt packs and 10 shirts individually, 1 T-shirt pack and 5 shirts individually, or 2 T-shirt packs and 0 shirts individually.

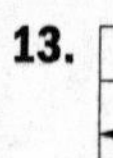

13.

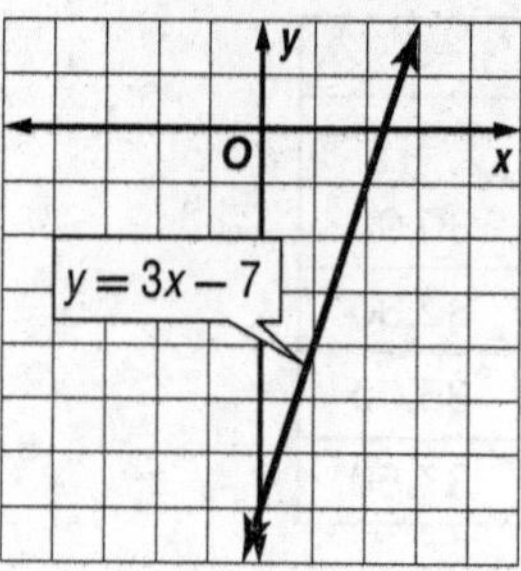

15.

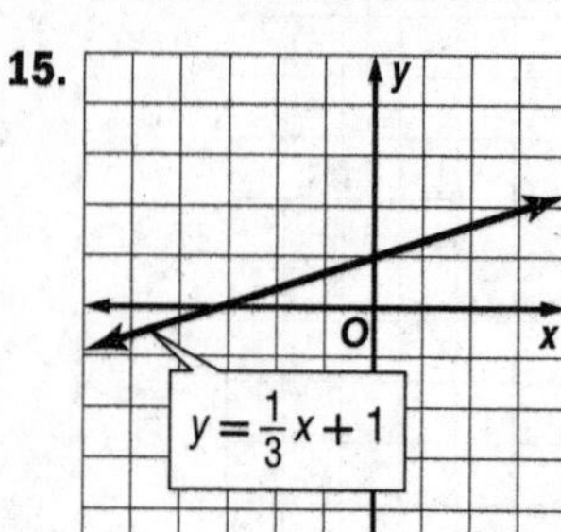

17. D **19.** I **21.** $15n$; 75, 90, 105

Page 307 Problem-Solving Investigation Make a Table

Case 3. 3 ft **Case 5.** 2.3 mi

Pages 315–316 Lesson 4-5 Independent Practice

1 First part: 68 miles per hour; Second part: 55 miles per hour. The speed for the first leg is greater by 13 miles per hour. **3** Seth; Seth will have 4(20) or 80 cards and Matt will have 2(20) + 20 or 60 cards. **5a.** $z = 8c$; $z = 16p$; $z = 32q$ **5b.** the quart equation; Sample answer: The greater the rate of change, the steeper the slope of the graph. **5c.** The first function has the least rate of change because 8 is less than 16 and 32. **7.** Sample answer: Both functions have the same rate of change but because they have different y-intercepts, they are parallel lines and parallel lines will never intersect.

Pages 317–318 Lesson 4-5 Extra Practice

9. Cotton fabric: \$7.00 per yard. Special occasion fabric: $\frac{18-9}{2-1} = \frac{9}{1}$; or \$9.00 per yard. Special occasion fabric has the greater rate of change. **11.** Juan; Sample answer: In 8 weeks Juan will have 5(8) or \$40. Jesse will have saved \$37. **13.** $\frac{3}{11}$ h **15.** greater than; Her race rate was $22.2 \div 2.15$ or about 10.3 mph. Her average speed biking was 13.8 mph; $13.8 > 10.3$. **17.** $y = -x + 4$
19. 0 **21.** 52

Pages 323–324 Lesson 4-6 Independent Practice

1 The teacher read 100 pages per day. The teacher initially read 150 pages. **3** The class brings in 10 cans per day. The teacher initially had 105 cans. **5.** Each month Jonas adds 3 DVDs. He started with 9 DVDs.
7. Sample answer: The rate of change is rerepsented by the ratio $\frac{\text{change in } y}{\text{change in } x}$. For a horizontal line, x can increase or decrease, but y does not change. The numerator is 0 so, the rate of change is 0. **9.** C

Pages 325–326 Lesson 4-6 Extra Practice

11. The family drove 200 miles per day. They drove 280 mile to their grandmother's house. **13.** B **15.** –17 **17.** 14
19–24.

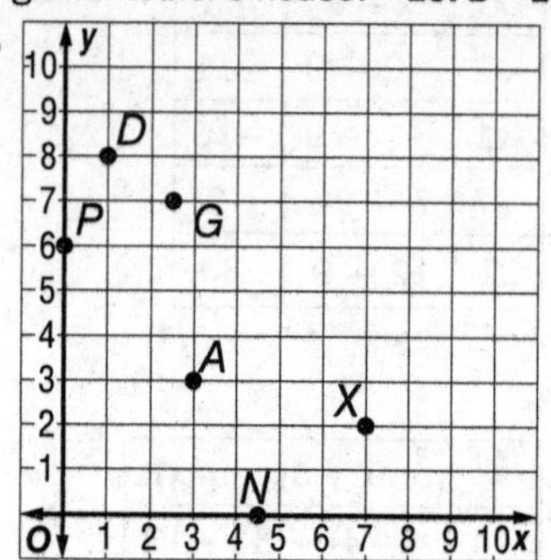

Pages 331–332 Lesson 4-7 Independent Practice

1 Linear; rate of change is constant; as x increases by 2, y increases by 1. **3.** Linear; rate of change is constant as x increases by 5, y increases by 15. **5** Yes; the rate of change is constant; as the time increases by 1 hour, the distance increases by 65 miles. **7.** Linear; sample answer: If you graph the function, the ordered pairs (hours, seconds) lie on a straight line.
9.

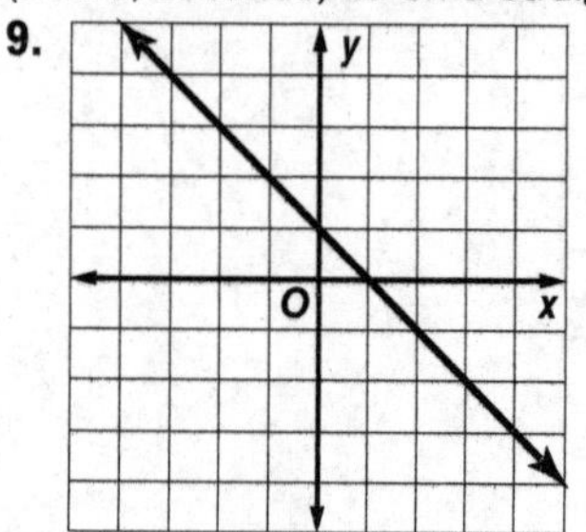

Linear; sample answer: The points lie on a straight line.
11.

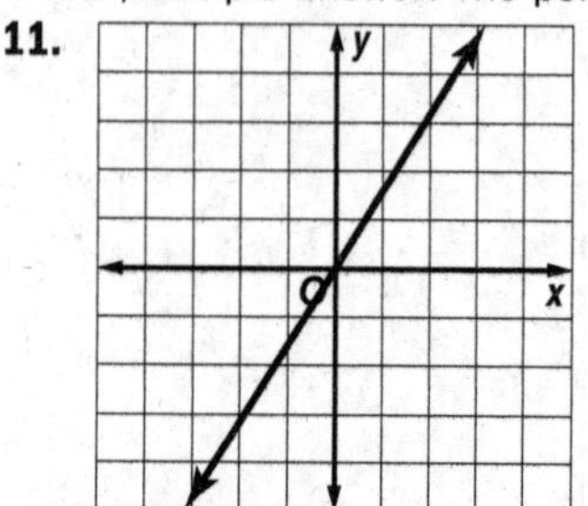

Linear; sample answer: The points lie on a straight line.
13. No; sample answer: the graphs of vertical lines are not functions because there is more than one value of y that corresponds to $x = 2$. **15.** D

Pages 333–334 Lesson 4-7 Extra Practice

17. Linear; rate of change is constant; as x increases by 4, y decreases by 3.
19a.

Radius r	Circumference $2 \cdot \pi \cdot r$	Area πr^2
1	$2 \cdot \pi \cdot 1 \approx 6.28$	$\pi \cdot 1^2 \approx 3.14$
2	$2 \cdot \pi \cdot 2 \approx 12.57$	$\pi \cdot 2^2 \approx 12.57$
3	$2 \cdot \pi \cdot 3 \approx 18.85$	$\pi \cdot 3^2 \approx 28.27$
4	$2 \cdot \pi \cdot 4 \approx 25.13$	$\pi \cdot 4^2 \approx 50.27$
5	$2 \cdot \pi \cdot 5 \approx 31.42$	$\pi \cdot 5^2 \approx 78.54$

19b.

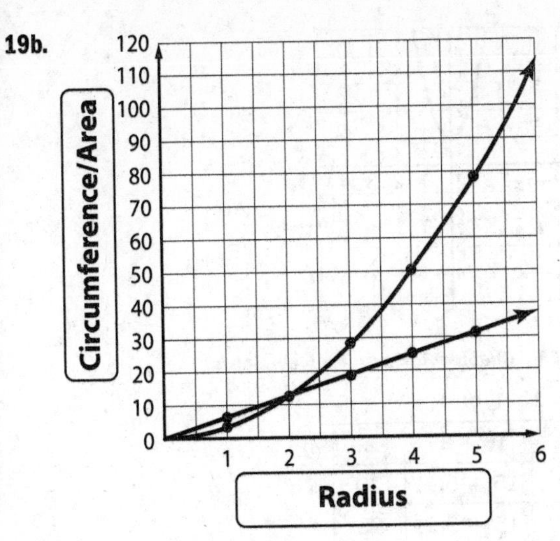

19c. Circumference: linear; sample answer: When the ordered pairs are graphed, the points fall in a line. Area: nonlinear; sample answer: When the ordered pairs are graphed, the points do not fall in a line. **21.** Jung's savings represent a linear function because the rate of change is constant. As the months increase by 1, the savings increase by 10. Miguel's savings represent a nonlinear function because as the month's increase by 1, the total savings increases at a different rate. **23.** −14 **25a.** $c = 5d$; Riley makes an average of 5 phone calls per day. **25b.** 35 phone calls

Pages 339–340 Lesson 4-8 Independent Practice

1

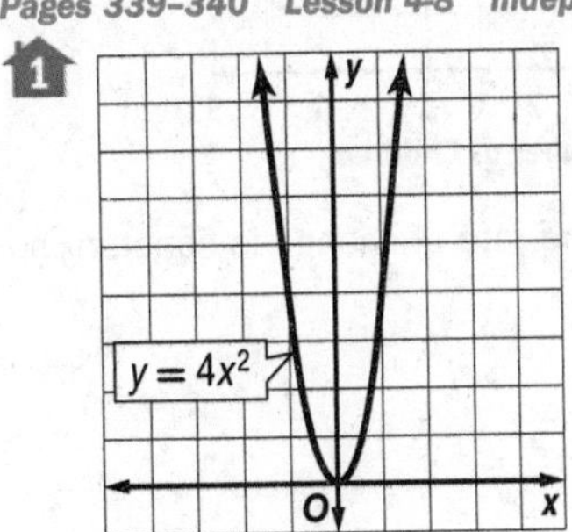

3

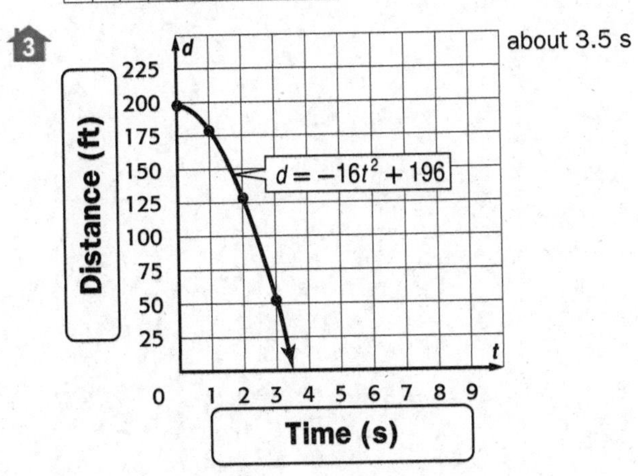

about 3.5 s

5a. $A = 12x - x^2$ **5b.** 6 in. by 6 in. **7.** nonlinear; Sample answer: The function is quadratic. **9.** linear; Sample answer: The equation is written in slope-intercept form so it is a straight line. **11.** nonlinear; Sample answer: The function is quadratic. **13.** Sample answer: $y = x^2 - 3.5$

Pages 341–342 Lesson 4-8 Extra Practice

15. $y = -x^2 + 2$

x	$-x^2 + 2$	y	(x, y)
−2	$-(-2)^2 + 2$	2	(−2, 2)
−1	$-(-1)^2 + 2$	1	(−1, 1)
0	$-(0)^2 + 2$	2	(0, 2)
1	$-(1) + 2$	1	(1, 1)
2	$-(2)^2 + 2$	−2	(2, −2)

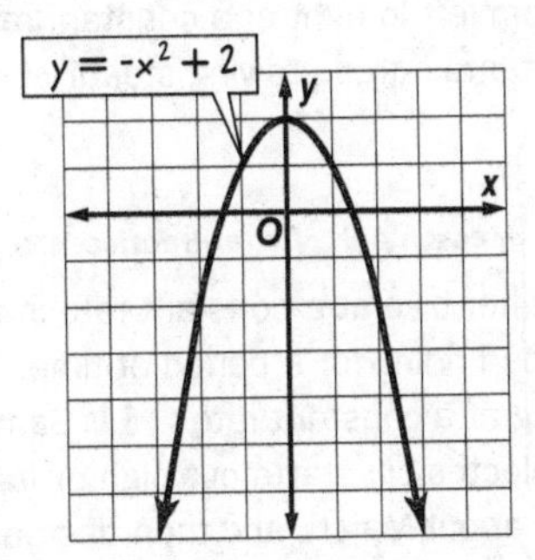

17.

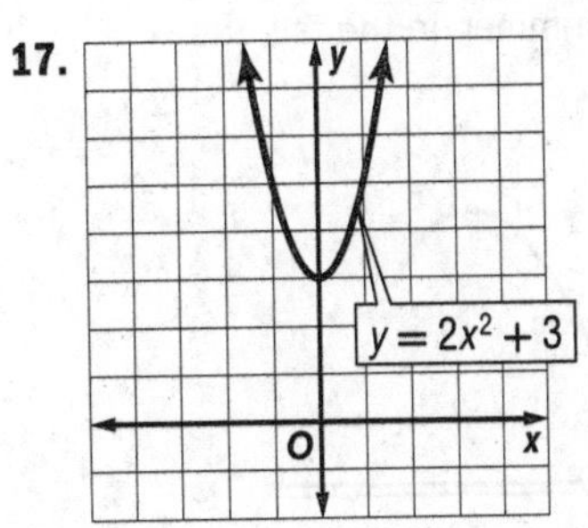

19a. $A = x^2 + 4x$

19b.

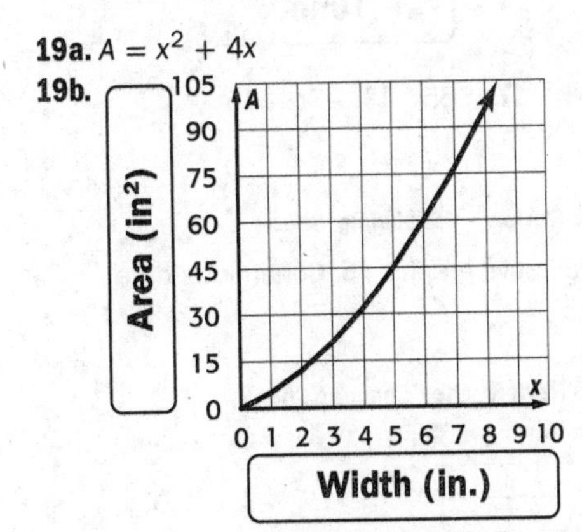

19c. 96 in² **21.** H **23.** 1,296 **25.** 243 **27.** 16 **29.** 36 **31.** 1 **33.** 3 **35.** −2 **37.** −36

Pages 351–352 Lesson 4-9 Independent Practice

1 Sample answer: Luis starts out from his home. He walks away from his home, stops to let the dog run around, and walks further away from home. Then he walks towards home.

3 Sample answer:

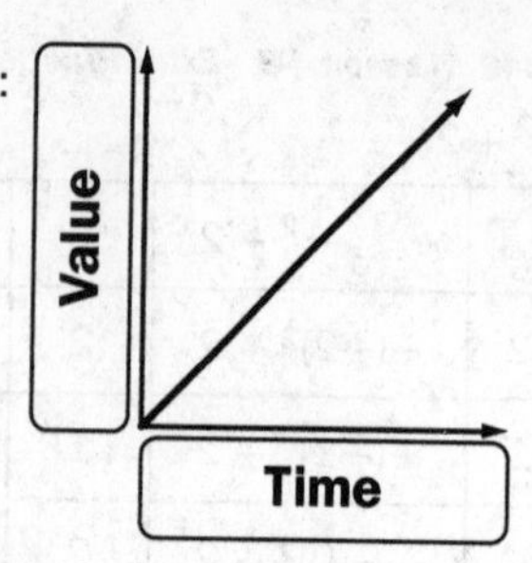

5a. Sample answer: Hector hikes at a steady rate.
5b. Sample answer: Hector suddenly stops hiking.
5c. increase; Sample answer: The graph rises from left to right at the beginning. **7.** Graph A; Sample answer: Graph A increases from left to right at a constant rate then levels off. This represents a tree growing steadily before it stops growing.

Pages 353–354 Lesson 4-9 Extra Practice

9. Justine rode her bike at a constant rate in the beginning. She then stopped riding for a period of time. Then she continued riding at a constant rate. **11.** Sample answer: Mrs. Fraser's electric bill starts out high in January, increases until about March, and then decreases throughout the spring and summer. In the fall, the electric bill increases again.
13. Sample answer:

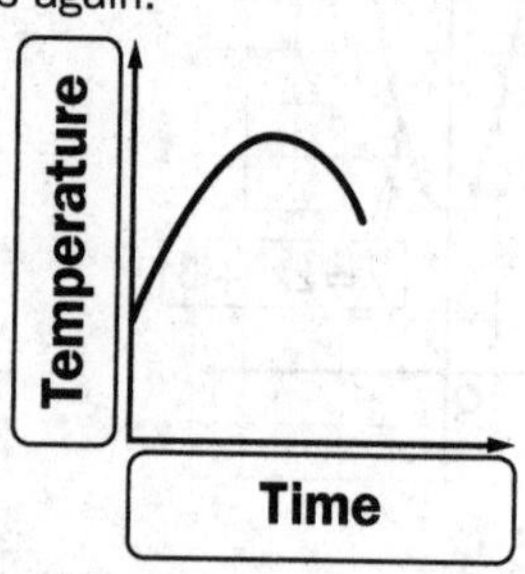

15. A **17a.** $3p + 5$ **17b** \$65 **19.** $-3t - 45$
21 $-5n - 87$

Page 357 Chapter Review Vocabulary Check

1. relation **3.** qualitative graphs **5.** Continuous data
7. function

Page 358 Chapter Review Key Concept Check

1.

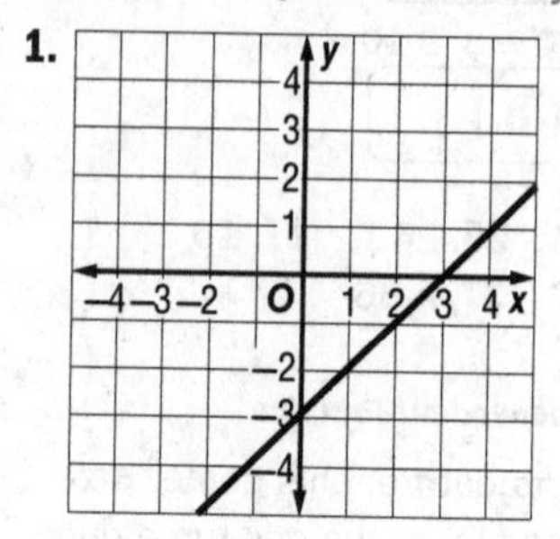

3.

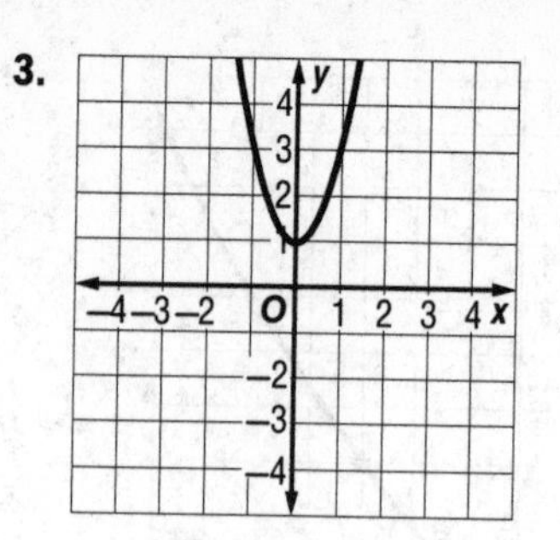

Page 359 Chapter Review Problem Solving

1a. $c = 15 + 5n$
1b.

n	$15 + 5n$	c
1	$15 + 5(1)$	20
2	$15 + 5(2)$	25
3	$15 + 5(3)$	30
4	$15 + 5(4)$	35

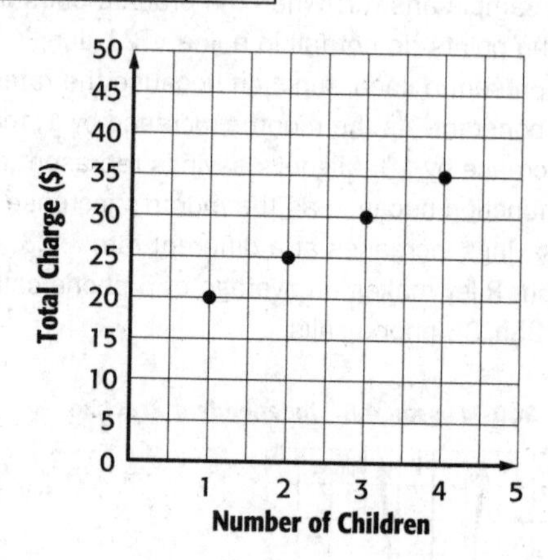

3. linear; Sample answer: the rate of change is constant so the function is linear.

Index

Bb

Cc

Dd

Ee

Ff

Index

Mm

Nn

Qq

Rr

Name

Work Mats

=

Name ______

Name ______________________________

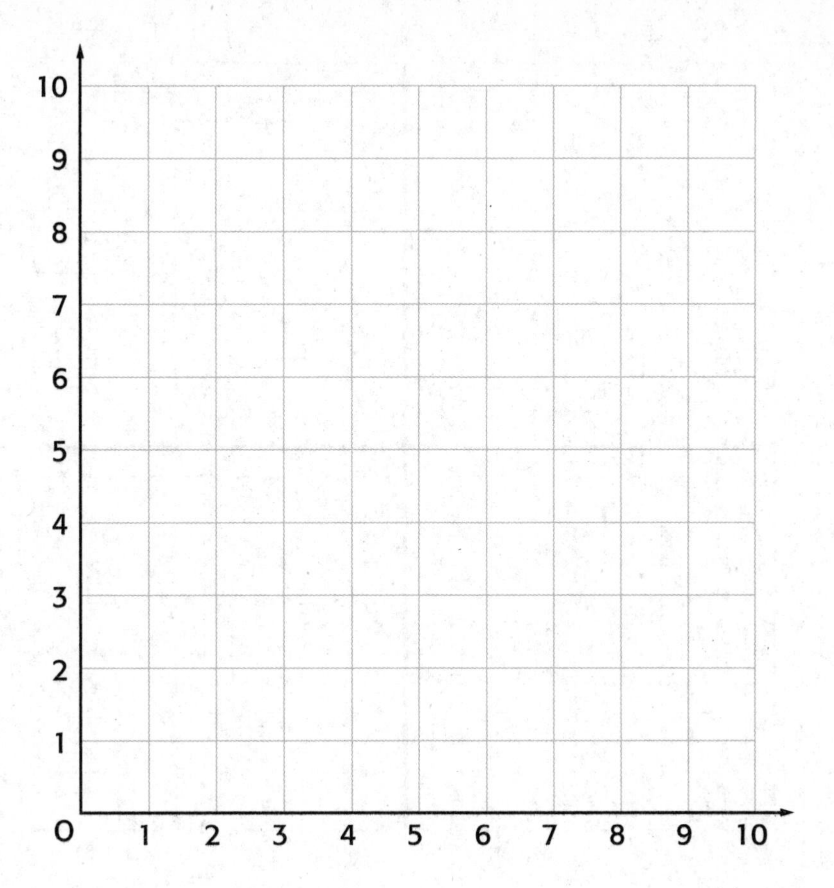

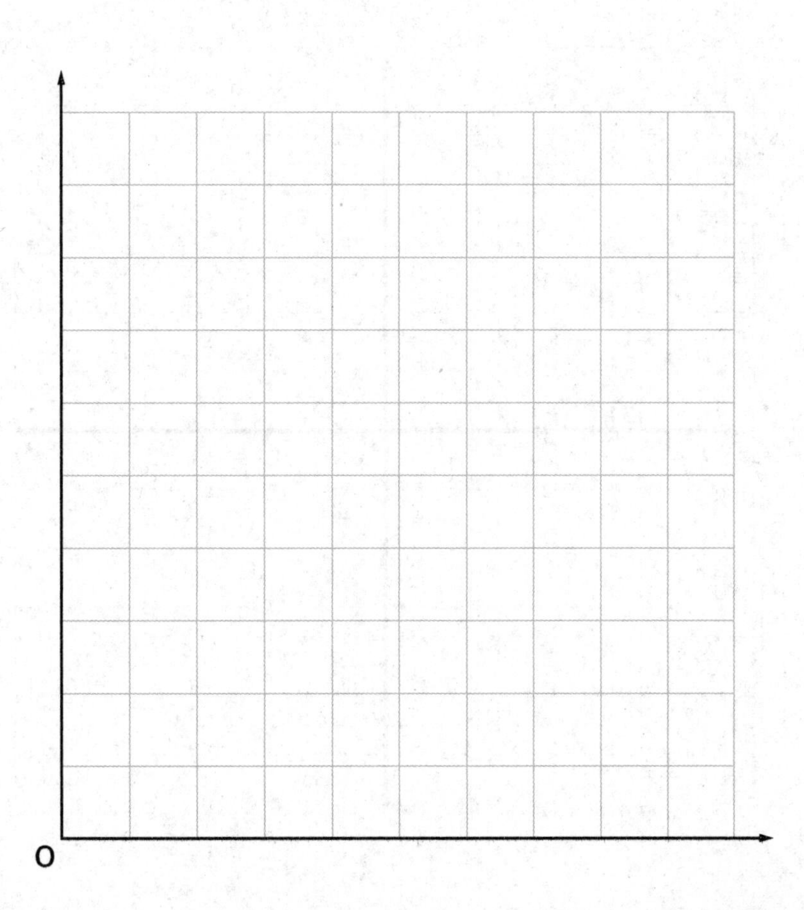

Name ____________________

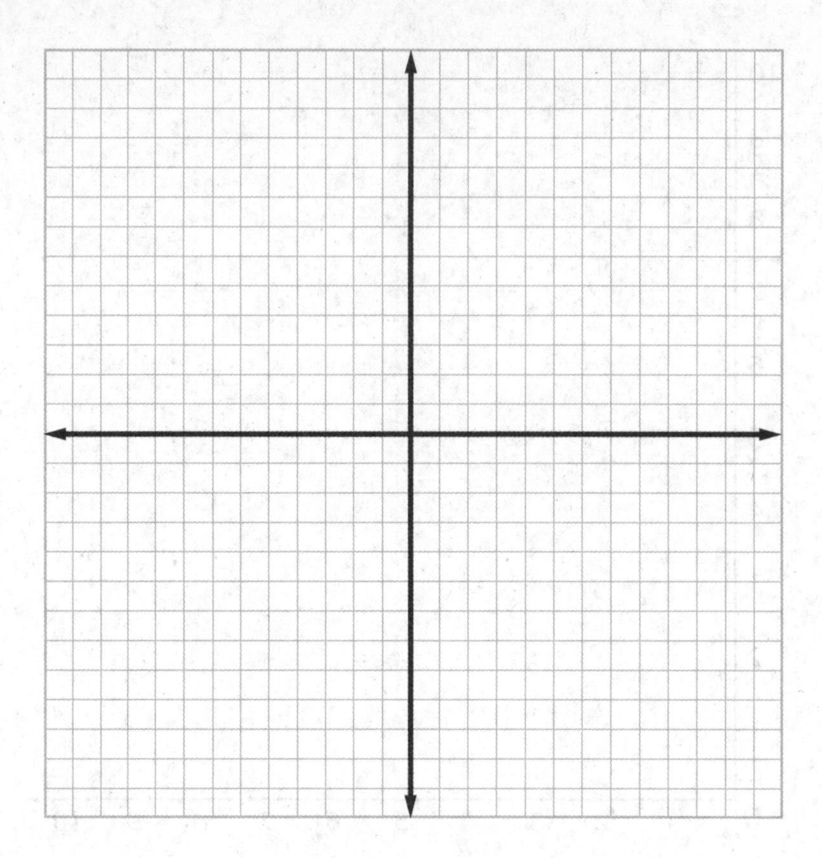

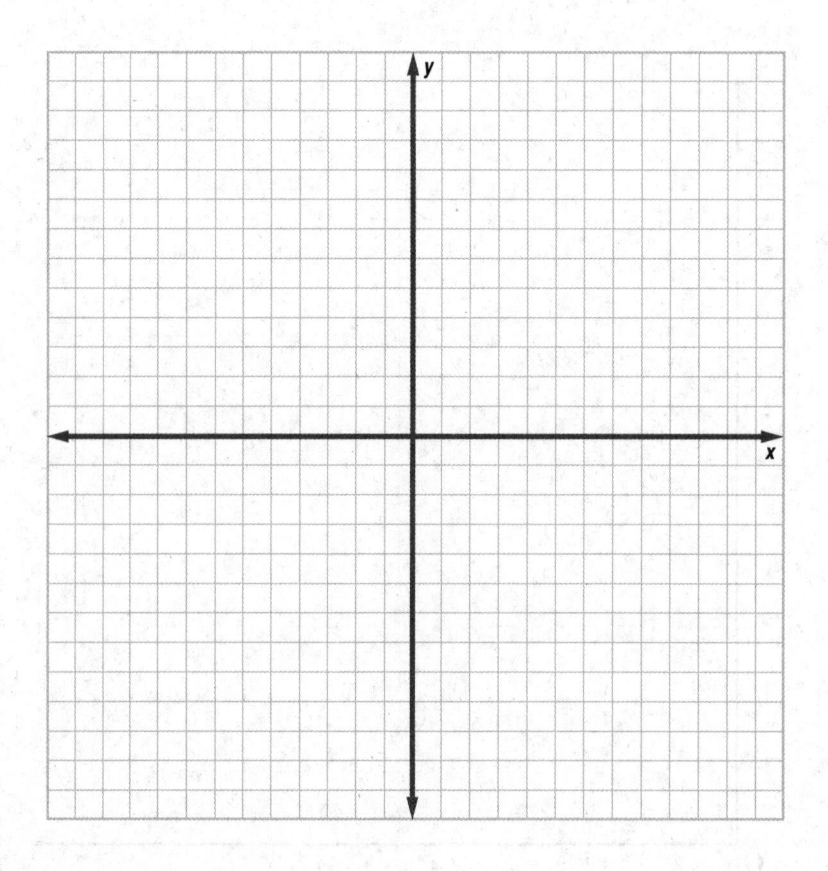

FOLDABLES® Study Organizers

What Are Foldables and How Do I Create Them?

Foldables are three-dimensional graphic organizers that help you create study guides for each chapter in your book.

Step 1 Go to the back of your book to find the Foldable for the chapter you are currently studying. Follow the cutting and assembly instructions at the top of the page.

Step 2 Go to the Key Concept Check at the end of the chapter you are currently studying. Match up the tabs and attach your Foldable to this page. Dotted tabs show where to place your Foldable. Striped tabs indicate where to tape the Foldable.

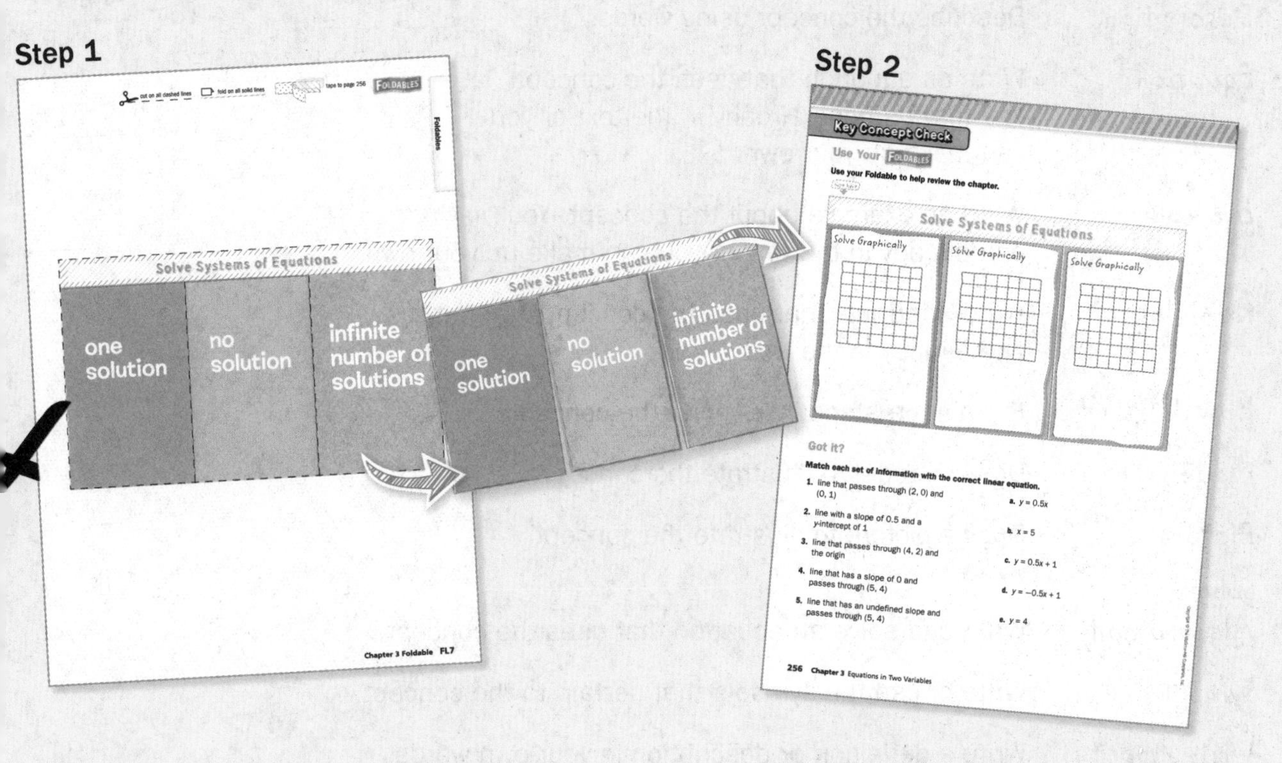

How Will I Know When to Use My Foldable?

When it's time to work on your Foldable, you will see a Foldables logo at the bottom of the **Rate Yourself!** box on the Guided Practice pages. This lets you know that it is time to update it with concepts from that lesson. Once you've completed your Foldable, use it to study for the chapter test.

How Do I Complete My Foldable?

No two Foldables in your book will look alike. However, some will ask you to fill in similar information. Below are some of the instructions you'll see as you complete your Foldable. **HAVE FUN** learning math using Foldables!

Instructions and what they mean

Instruction	Meaning
Best Used to...	Complete the sentence explaining when the concept should be used.
Definition	Write a definition in your own words.
Description	Describe the concept using words.
Equation	Write an equation that uses the concept. You may use one already in the text or you can make up your own.
Example	Write an example about the concept. You may use one already in the text or you can make up your own.
Formulas	Write a formula that uses the concept. You may use one already in the text.
How do I ...?	Explain the steps involved in the concept.
Models	Draw a model to illustrate the concept.
Picture	Draw a picture to illustrate the concept.
Solve Algebraically	Write and solve an equation that uses the concept.
Symbols	Write or use the symbols that pertain to the concept.
Write About It	Write a definition or description in your own words.
Words	Write the words that pertain to the concept.

Meet Foldables Author Dinah Zike

Dinah Zike is known for designing hands-on manipulatives that are used nationally and internationally by teachers and parents. Dinah is an explosion of energy and ideas. Her excitement and joy for learning inspires everyone she touches.

cut on all dashed lines   fold on all solid lines   tape to page 100   FOLDABLES

Laws of Exponents

Product of Powers

Quotient of Powers

Power of Powers

cut on all dashed lines   fold on all solid lines   tape to page 100

FOLDABLES®

Examples

Examples

Examples

page 100

cut on all dashed lines fold on all solid lines tape to page 164 FOLDABLES

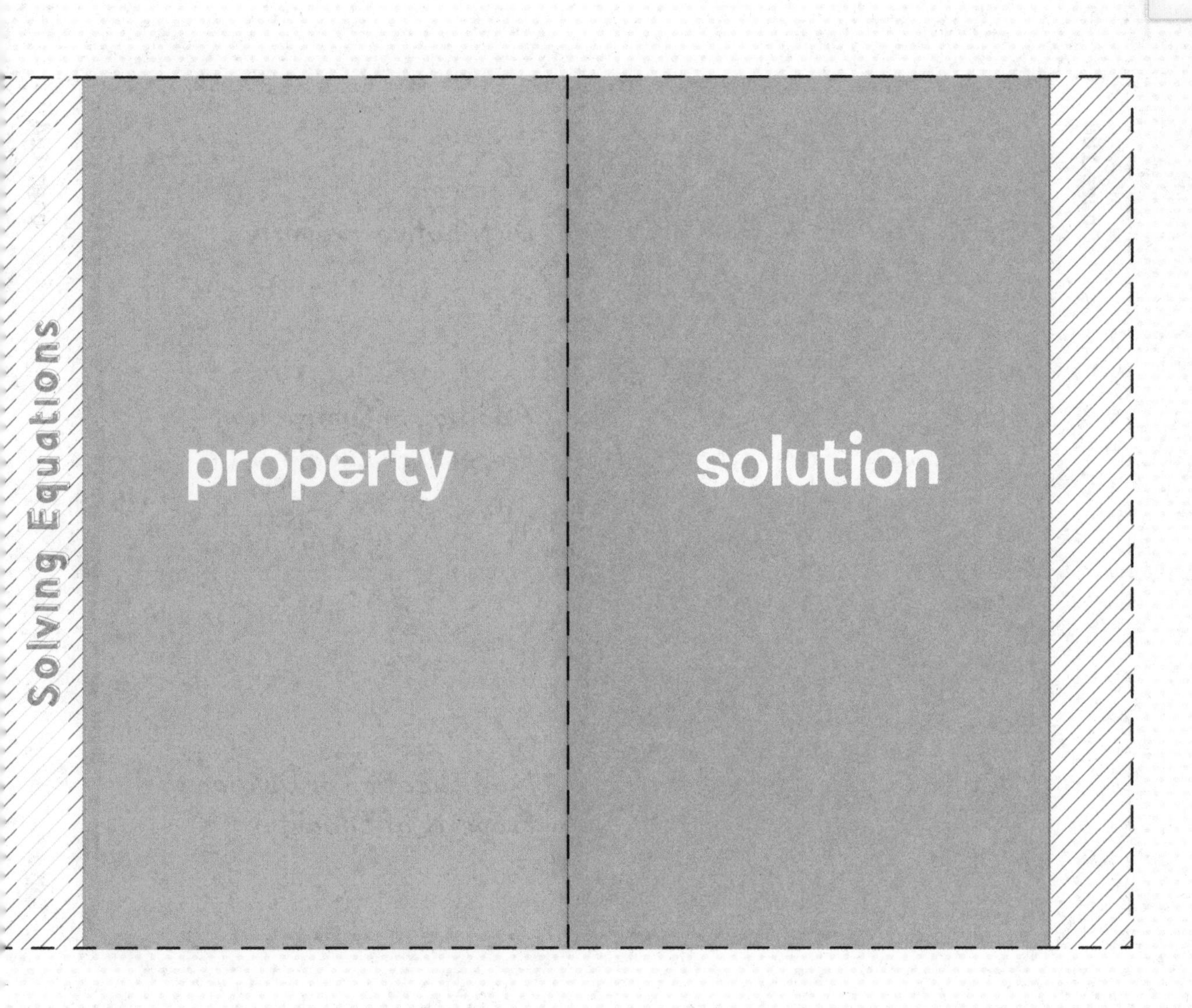

cut on all dashed lines
fold on all solid lines
tape to page 164

page 164

Step 1

Step 2

Step 3

Step 4

Tab 2

Distributive Property

Addition or Subtraction Property of Equality

Multiplication or Division Property of Equality

page 164

Tab 1

cut on all dashed lines | fold on all solid lines | tape to page 256

FOLDABLES

Solve Systems of Equations

one solution	no solution	infinite number of solutions

cut on all dashed lines fold on all solid lines tape to page 256 FOLDABLES

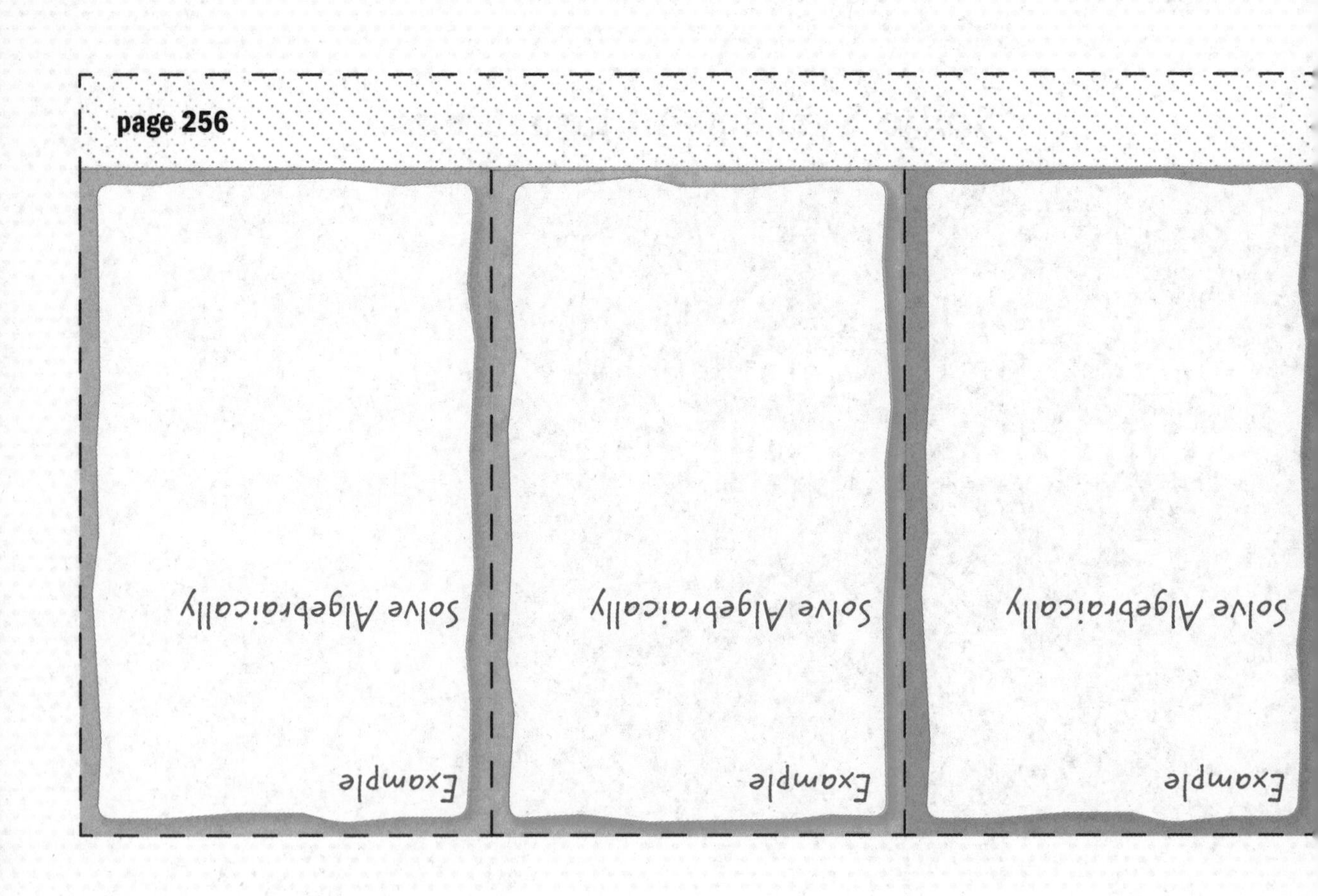

cut on all dashed lines   fold on all solid lines   tape to page 358   FOLDABLES

Foldables

Relations and Functions

relations

functions

linear

nonlinear

cut on all dashed lines | fold on all solid lines | tape to page 358

FOLDABLES

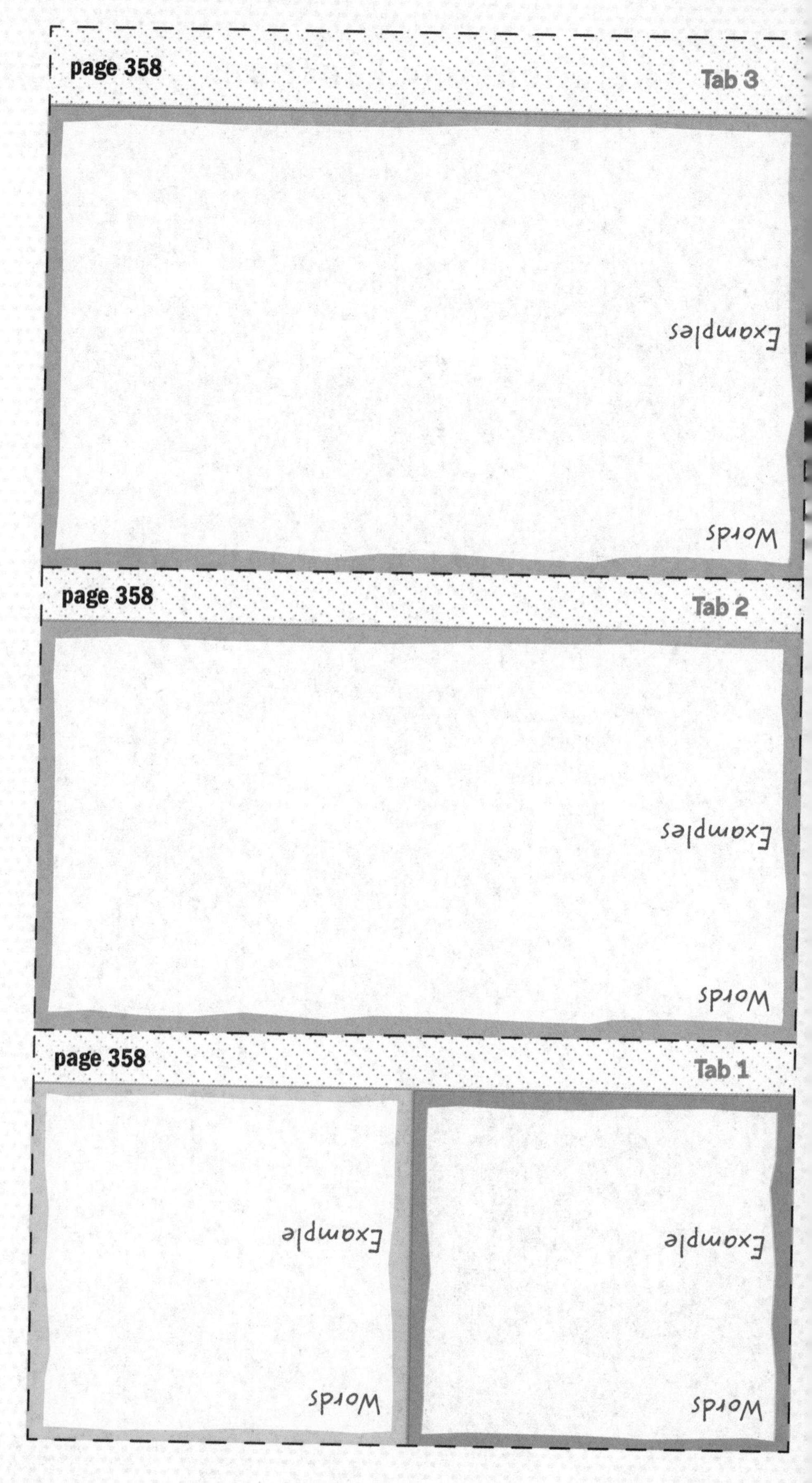